DEVELOPING BASIC SPACE SCIENCE WORLD-WIDE

Developing Basic Space Science World-Wide

A Decade of UN/ESA Workshops

Edited by

Willem Wamsteker
European Space Agency (ESA/VILSPA), Madrid, Spain

Rudolf Albrecht
ST-ECF, ESA/ESO, Garching, Germany

and

Hans J. Haubold
United Nations Office for Outer Space Affairs (UNOOSA), Vienna, Austria

KLUWER ACADEMIC PUBLISHERS
DORDRECHT / BOSTON / LONDON

A C.I.P. Catalogue record for this book is available from the Library of Congress.

ISBN 1-4020-1681-6

Published by Kluwer Academic Publishers,
P.O. Box 17, 3300 AA Dordrecht, The Netherlands.

Sold and distributed in North, Central and South America
by Kluwer Academic Publishers,
101 Philip Drive, Norwell, MA 02061, U.S.A.

In all other countries, sold and distributed
by Kluwer Academic Publishers,
P.O. Box 322, 3300 AH Dordrecht, The Netherlands.

Printed on acid-free paper

Printed in the Netherlands.

Editorial

When the first United Nations/European Space Agency Workshop on Basic Space Science was planned to be held in Bangalore, India, on the invitation of the Indian Space Research Organization (ISRO), few of those involved could have expected that a unique forum was going to be created for the scientific dialogue between scientists from developing nations and from industrialized nations. As the format of the first workshop was on purpose left a bit free with time for presentations, working group sessions and plenary discussions, the workshop was left to find its own dynamics. It deserves to recognize the participants in the first workshop, that they were able to demonstrate the openness to work together, the intellectual discipline to contribute effectively, and the originality to develop a format, which has been transferred to all economic regions on Earth with great success and effectiveness.

This book tries to bring together, the historical activities which have been done in the past, the plans which have been developed over the past decade in the different nations, and the results which have materialized during this time in a large number of developing nations.

We hope that through this compilation, development agencies will be assisted in ways to find more effective tools for the application of development aid, not only helping to mitigate immediate needs but also resolving other necessary, but more complex, long term needs in developing nations. The last section of the book contains a guide for teachers to introduce astrophysics in general physics courses at university level which, we hope, will be of use to teachers in many nations.

Everything that is described in this book is the result of a truly collective effort from all involved in all United Nations/European Space Agency (UN/ESA) workshops. The mutual support from the participants has helped significantly to implement some of the accomplishments described in this book. Rather than organizing this book in a subject driven way, it is essentially organized according to the economic regions on Earth, as considered by the United Nations. This allows better to recognize the importance of a regional and at times global approach to basic space science for the developing nations world-wide. It has also allowed us to highlight very specific scientific investigations which have been completed successfully in many developing nations.

These Workshops and their follow-up actions have been highlighted at the Third United Nations Conference on the Exploration and Peaceful Uses of Outer Space (UNISPACE III) and the Vienna Declaration "The Space Millennium: Space and Human Development", held at the United Nations Office Vienna in 1999.

Special thanks have to be expressed to the Governments of the host countries of the UN/ESA workshops, in chronological order, India, Costa Rica, Colombia, Nigeria, Egypt, Sri Lanka, Germany, Honduras, Jordan, France, Mauritius, Argentina, and P.R. China. Also many space science related organizations have contributed in important ways to the UN/ESA Workshops: Austrian Space Agency (ASA), Committee on Space Research (COSPAR), French Space Agency (CNES), German Space Agency (DLR), International Astronomical Union (IAU), International Centre for Theoretical Physics (ICTP), Institute of Space and Astronautical Science (ISAS) Japan, National Aeronautics and Space Administration (NASA), national Space Developemnt Agency of Japan (NASDA), The Planetary Society (TPS), which each in its own way supported the Workshop series.

We want to mention a special expression of appreciation to Professor Masatoshi Kitamura for his efforts in cooperation with the Government of Japan, resulting in the donation of astronomical telescopes and planetaria to a number of developing countries.

This book would not have been completed without the dedicated support of many people of which we would like to mention here Mrs. Barbara Haubold for her work on the early version of the manuscripts; Mrs. Carmen Rosales for her very professional preparation of the camera-ready version of this book itself. Mr. Juan Carlos Fernández has designed the frontispiece of the book. Ms. Kiani Wamsteker and Mr. Alexander Haubold provided critical support when unusual electronic editing problems arose. Ms. Valerie Hood and Ms. Karen Barbance deserve recognition for their continued enthusiasm in support of the cause of Basic Space Science in developing nations.

The Editors

Contents

PREFACE

J. Andersen[*]
Niels Bohr Institute for Astronomy
Physics and Geophysics Astronomical Observatory
Copenhagen
ja@astro.ku.dk

The development of astronomy worldwide begins at the roots: Already from childhood, humans of all nations and civilizations seem to share an innate fascination with the sky. Yet, people in different regions of the world have vastly different possibilities for pursuing this interest. In wealthy, industrialised societies the way is open to a school or higher education in science, possibly leading to a career in astronomy or basic or applied space science for the benefit of the country as well as the individual. In other regions, neither the financial nor the trained human resources are sufficient to offer that avenue to the future of the young generation, or those intellectual resources to the development of their country. This book addresses ways and means by which these obstacles can be, if not fully overcome, then at least significantly reduced.

The International Astronomical Union (IAU) is the international organization uniting professional astronomers worldwide. As such, the IAU is concerned with the buildup of the knowledge base in the less-privileged parts of the world as well as with the progress of front-line science at the most advanced level. Thus, for more than a quarter of a century, the IAU has conducted training programmes in astronomy in various parts of the world, often with lasting benefits for the host country and community. It is increasingly clear, however, that for such efforts to have optimum positive effects, our resources should be coordinated with those of other organizations promoting education in neighbouring branches of basic or applied science - space science being the most natural partner. Equally importantly, the backing of governmental

[*] General Secretary of the International Astronomical Union 1997-2000

authorities is needed for such initiatives to benefit in the long term not only science itself, but the development of the host country: Bright young scientists need not only brains and training, but also a job and appropriate tools, if they are to be able to use their capabilities effectively in the service of their country.

This simple reasoning may appear self-evident. Yet, it is put into actual practice only too rarely. The IAU therefore welcomes the initiative of organization of the UN/ESA workshops to call the attention of the governments of the world to the great potential of astronomy and space science to fascinate the brightest young minds. Experience shows that many continue to pursue productive careers in science and thereby contribute to the scientific and economic development of their countries. We share the hope that this book, which gives a summary of all workshops, may serve to inspire grassroots and government officials alike to take advantage of the experience and good will of the scientific community in this common enterprise.

SECTION:

1. INTRODUCTION

Chapter 1.1

THE UN/ESA WORKSHOPS ON BASIC SPACE SCIENCE IN THE DEVELOPING COUNTRIES

H.J. Haubold and W. Wamsteker
United Nations Office for Outer Space Affairs
Vienna, Austria
hans@neutrino.aquaphoenix.com

European Space Agency
ESA-VILSPA
Madrid, Spain
willem.wamsteker@esa.int

Our civilization is no more than the sum of all the dreams that earlier ages have brought to fulfillment. And so it must always be, for if men cease to dream, if they turn their backs upon the universe, the story of our race will end.

Arthur C. Clarke

Abstract The origin and evolution of the UN/ESA Workshops on Basic Space Science (BSS) is discussed. We highlight here the motivation and results in global terms. The various new insights in the role of BSS in a sustainable development are indicated.

Introduction

Astronomy and basic space science have advanced rapidly in recent years. This has mainly been driven by the advent of the Space Age over the past five decades. At the same time the needs for a new development process became obvious as countries made efforts to benefit from rapid progress in space science and technology. This process required a complete innovation of the driving forces for development where no previous examples were available. Although new development models have been applied with different levels of success, a large number of countries around the world continue to lack the human, technical, and

W. Wamsteker et al. (eds.), Developing Basic Space Science World-Wide, 3-8.

financial resources to conduct even the most basic activities in this field, such as operating a small astronomical telescope facility in an university environment and making use of such facilities in research and education. The need to make the benefits of basic space science available to all countries is growing more urgent with each passing year.

1. The Workshops

The United Nations Programme on Space Applications, since its inauguration in 1970, has made an effort in furthering knowledge and experience of space applications around the world. Provision of in-country capacity-building, education, research and development support and technical advisory services by the Programme have all helped to create conditions under which the developing countries have started to benefit also from some of the progress made in the BSS by the –at that time- still very small number of developed countries actively involved in space. However, until 1990, this provision focused mainly on applications oriented activities in remote sensing of the Earth, meteorological satellite applications, and satellite communications. Only in the late 1980's, discussions among developing countries and UN/ESA identified the possible need and importance to support the growth of small research groups in universities and research establishments in the developing countries in the field of astronomy and space science (Rao, 1992). This lead to the holding of the UN/ESA workshops on basic space science on an annual basis in the different regions: Asia and the Pacific, Latin America and the Caribbean, Africa, Western Asia, and Europe. Between figure 1.1-1 and figure 1.1-2 the workshops have gone around the world twice. The support of Member States (Governments) and their participation in these workshops, under the auspices of the UN Programme for Space Applications, was and is crucial for the success of the workshops.

The fact that the workshops were held in the different regions allowed a factual focus to be developed, addressing the fundamental questions and their relevance in a regional context. Associated with each workshop were very dynamic working group sessions in which the participating scientists were able to raise the issues perceived to be of relevance and interest in their own environment. Each of the UN/ESA workshops has resulted in a set of unique observations and recommendations identifying the regional needs, strengths and possible outlook for future progress. These observations and recommendations are contained in the UN General Assembly documents for wider distribution among the Member

Figure 1.1-1. Opening ceremony of the First UN/ESA Workshop on Basic Space Science, 1991, Indian Space Research Organization; left to right: K. Kasturirangan, U.R. Rao, M.G.K. Menon, N. Jasentuliyana, M.G. Chandrasekhar (Courtesy of ISRO)

Figure 1.1-2. Closing ceremony of the Eighth UN/ESA Workshop on Basic Space Science, 1999, Mafraq, Hashemite Kingdom of Jordan; left to right: H.M.K. Al-Naimiy, N. Saleh, H.J. Haubold, W. Wamsteker (Courtesy of M.-J. Deutsch)

States representatives (see table 1, where the observations and recommendations are identified with their UN document numbers).

Table **1.1-1. UN/ESA Workshops**

Workshop	Recommendations and observations document	Proceedings
1991 Bangalore, India	A/AC.105/489	Haubold and Khanna,1992
1992 San Juan, Costa Rica & Bogota, Columbia	A/AC.105/530	Fernandez and Haubold, 1993 Haubold and Torres, 1994
1993 Ile-Ife, Nigeria	A/AC.105/560/Add.1	Haubold and Onuora, 1994
1994, Cairo, Egypt	A/AC.105/580	Haubold and Mikhail, 1995a Haubold and Mikhail, 1995b
1995, Colombia, Sri Lanka	A/AC.105/664	
1996, Bonn, Germany	A/AC.105/657	Haubold and Mezger, 1998
1997, Tegucigalpa, Honduras	A/AC.105/682	No Proceedings
1999, Mafraq, Jordan	A/AC.105/723	Haubold and Al-Naimiy, 2000
2000, Toulouse, France	A/AC.105/742	No proceedings
2001, Reduit, Mauritius	A/AC.105/766	Haubold and Rughooputh, 2002
2002, Cordoba, Argentina	A/AC.105/784	Haubold and Rabolli, 2003

Revision of the observations and recommendations in column 2 of table 1 shows that in each of the regions very specific local problems were identified, but the recommended approach to improve the conditions show remarkable similarities. Although the nature of the perceived issues has a strong local cultural and social aspect, with minor adjustments to the local cultural patterns the fundamentals of the underlying problems are common.

To assure that the UN/ESA workshops did not create a duplication of efforts with other activities in the UN framework, an important issue was the definition of the nature of BSS and its involved technologies. These were defined in the first workshop and have been maintained throughout the series (A/AC.105/489). In the context of the needs for the developing countries, basic space science was defined to cover:

- Fundamental physics,
- Astronomy and astrophysics,
- Solar-terrestrial interaction and its influence on terrestrial climate,
- Planetary and atmospheric studies, and
- Origin of life and exo-biology.

With the applicable techniques as tools for the pursuance of basic space science research:

- Ground-based optical, and radio observations,

- Radio and optical telescopes with associated equipment,
- Remote sensing, both from the ground and from space,
- All measurements from ground-inaccessible windows which can only be made through the use of instruments and telescopes in Earth orbit,
- In-situ measurements from rocket, balloon and satellite platforms.

It was only through the commitment of individual hosting nations that the Programme achieved its primary objective: to clarify the importance of BSS for autochthonous involvement in space science to contribute to sustainable economic, social, and cultural development, not just in individual countries, but on a global basis. At all stages of the series of the workshops, the exceptional cooperation among scientists from developing and industrialized countries, the substantive support of the European Space Agency and other national and international space-related organizations, was critical.

2. Implementation Process: Tripod

From the results of the deliberations, an implementation model for the accelerated implementation of BSS associated activities in developing countries has been developed referred to in general as "Tripod". The Tripod concept, identified in the very first workshop in India in 1991, is to assure in the developing countries

1. The availability of research tools of a level where meaningful science can be done, but at a level where the national socio-economical infrastructure can maintain functionality in the university/research laboratory environment, e.g.. a small telescope facility (Kitamura, 2003)
2. Teaching materials allowing basic space science to be introduced at the teaching level of fundamental mathematics, physics and chemistry courses in middle and higher education (Wentzel, 2003).
3. Application materials for original research in basic space science such as e.g. observing programmes for variable stars (Percy and Mattei, 2003).

In a number of chapters and papers published elsewhere, the importance of the Tripodial approach has already been illustrated and although the effects of this will clearly not be immediate, it is quite encouraging that it is already beginning to spread throughout the educational system and also is having its impact on the working conditions for some established researchers with their base in developing countries. The establishment and subsequent operation of small astronomical telescope facilities in Colombia, Egypt, Honduras, Jordan,

Paraguay, the Philippines, and Sri Lanka, is ongoing. Also the introduction of astronomy in the context of basic space science in the university curricula in countries as Honduras as a regional activity is showing the way in this process. The introduction of archives from space missions (Wamsteker et al., 2000) in BSS has already started to introduce a new parameter in this Tripod and surely in the future will contribute a fourth leg to the implementation process. If the communications environment will be able to be brought forward in the developing countries as has been recommended in all workshops, the availability of the Virtual Observatory concepts (Albrecht, 2003) can be an important contribution. And, if the industrialized countries can find agreement on the implementation of the next phase of active observational capabilities, as proposed in the World Space Observatory concept (Wamsteker and Shustov, 2003), it can be expected that the BSS environment in the developing countries will *have made the necessary quantum jump needed for accelerated development.*

The application and results of the application of the Tripodial process in the different regions of the world is the subject of this book where we have collected the information ordered by region. This is expected to present detailed and precise information on past and current achievements in this field worldwide.

References

Albrecht, R., 2003, this volume.
Fernandez,W, and Haubold,H.J., 1993, Earth, Moon and Planets, **63**, No.2.
Haubold,H.J.,and Khanna,R.K., 1992, AIP Conf. Proc., **245.**
Haubold,H.J., and Torres, S., 1994, AP&SS, **214**, No.1-2.
Haubold,H.J., and Onuora, L.I., 1994, AIP Conf. Proc., **320.**
Haubold,H.J., and Mikhail, J.S., 1995a, AP&SS, **228**, No.1-2.
Haubold,H.J., and Mikhail, J.S., 1995b, Earth, Moon and Planets, **70**, No.1-3.
Haubold,H.J., and Mezger, P.G., 1998, AP&SS, **258**, No.1-2.
Haubold,H.J.,and Al-Naimiy, H.M.K., 2000, AP&SS, **273**, No.1-2.
Haubold,H.J.,and Rughooputh,S.D.D.V., 2002,AP&SS,**282**, No.1.
Haubold,H.J.,and Rabolli, M., 2003, AP&SS, *in press.*
Kitamura,M., 2003, this volume.
Percy, J.R., and Mattei, J.A., 2003, this volume.
Rao,U. R., 1992, Basic Space Science, Eds. Haubold, H.J. and Khanna, R.K., AIP Conf. Proc., **245**, 12.
Wamsteker, W., 2000, AP&SS, **273**, 155.
Wamsteker, W. and Shustov B.M., 2003, this volume.
Wentzel, D., 2003, this volume.

Chapter 1.2

THE INTERNATIONAL ASTRONOMICAL UNION

J. Andersen*
Niels Bohr Institute for Astronomy
Physics and Geophysics Astronomical Observatory
Copenhagen
ja@astro.ku.dk

Abstract This chapter presents an overview of the International Astronomical Union (IAU). Particular emphasis is given to the core activities.

Introduction

The International Astronomical Union (IAU) is the international non-governmental organization which unites professional astronomers all over the world. Its mission is to promote and safeguard the science of astronomy in all its aspects through international cooperation. It pursues this goal though a multitude of activities at many levels, the common denominator of which is the effective application of the collective expertise of its members to international scientific issues.

Despite being an organization with over 8,300 individual members and 60 member countries worldwide, the style of operation of the IAU remains family-like and informal, with a minimum of bureaucracy. While based on national membership, the main emphasis of the IAU is on activities for and by its Individual Members. This spirit is in part an historical heritage, as astronomers were among the first to rebel against the restrictions placed on national membership of scientific Unions by the precursor of ICSU when the IAU was formed in 1919. It is reinforced by the inherently global nature of our studies and - for many - the experience of social and scientific exchanges with international colleagues that brighten long cloudy nights on remote observatories.

* General Secretary of the International Astronomical Union 1997-2000

W. Wamsteker et al. (eds.), Developing Basic Space Science World-Wide, 9-15.

The following brief introduction to the IAU is intended to convey the flavour of the many-faceted activities of the Union. All these are subject to the strict observance of the ICSU rules regarding unrestricted access and non-discrimination for all qualified participants. More detailed information can be obtained from the Secretariat at the address above. In addition, the IAU Web page provides ready access to news and information on scientific meetings and other events, policy issues, basic documents and other publications, and links to the IAU Divisions and Commissions which are the direct contact points the many scientific communities within the field of astronomy.

1. Scientific Activities

The programme of Scientific Meetings is the most visible of the scientific activities of the Union. Each year, the IAU sponsors about a dozen high-level scientific Symposia and Colloquia, the main difference between the two categories being the somewhat more specialised nature of the Colloquia. Meetings organised by other Unions on themes including astronomy may also be co-sponsored by the IAU. Every three years, the Symposia are held in conjunction with the Union's General Assemblies, which also contain a very rich spectrum of Invited Discourses and Joint Discussions as well as many scientific and business sessions of the various bodies of the IAU. The 1997 General Assembly was attended by some 2000 participants, and the scientific programme featured a total of ~800 oral and ~1100 poster contributions.

In years between General Assemblies, IAU Regional Meetings are held in the Latin American and Asian-Pacific regions. The main purpose of these is to promote contacts between (especially young) scientists in the region, and a very broad scientific scope for these meetings is therefore accepted and encouraged. For all types of meeting, IAU sponsorship implies recognition of high scientific standards and substantial financial support towards the attendance of qualified participants from all over the world. Hence, competition for such recognition is fierce, and not all quality proposals can be supported, e.g., while regional meetings are also held in other parts of the world than those mentioned above, their need for IAU scientific and financial support is less urgent.

Another important international scientific function of the Union is the rapid dissemination of news about astronomical discoveries by the IAU Central Bureau for Astronomical Telegrams, located at the Smithsonian Astrophysical Observatory in Cambridge, MA, USA. The rapid verification and dissemination of such news ensures that follow-up observations of short-lived phenomena can be made in a timely manner and credit for discoveries be given as appropriate.

Located together with the Telegram Bureau, the IAU Minor Planet Center acts as the world-wide clearing-house for data on the minor bodies of the solar system. Thus, incoming observations are assigned to the proper minor planet, new or improved orbits are computed and predictions made for follow-up observations, and numbers and names are assigned when orbits have been sufficiently securely established. This role is especially important in connection with Near Earth Objects, minor planets (or comets) whose orbits take them close enough to the Earth that a collision sometime in the future is conceivable, and which are therefore of more than purely scientific interest. A special IAU Working Group is responsible for defining optimum search, computation and prediction strategies of the highest scientific standards for this vital task.

In order to promote effective communication with its constituency within well-defined fields, whether focused on specific scientific topics or on key observational techniques, the scientific work of the IAU is organised and coordinated by it 40 Scientific Commissions, themselves grouped into11 Scientific Divisions, each covering a broad scientific discipline. Further, the Divisions and Commissions create Working Groups to organise specific, time-limited scientific or organizational tasks. Before each General Assembly, the Divisions and Commissions report on their work in the preceding triennium; together these reports, collected in the volumes Reports on Astronomy (IAU Transactions, Volumes A), constitute a concise overview of the progress of astronomy since the creation of the IAU in 1919.

Certain specialised Commissions of the IAU are charged with the task of defining international standards in a variety of fields. These may concern definitions of fundamental physical concepts such as time or the inertial reference frame, units of measurement, or practical standards, e.g. for scientific documentation and interchangeable data formats. A special Working Group is responsible for assigning internationally recognised names to newly discovered celestial bodies and surface features (mountains, valleys, craters, etc.) on such bodies. In view of the current fashion of "selling star names", it is worth emphasizing that the IAU is the only internationally recognised authority for assigning such names. And that the Universe is not for sale.

2. Educational Programmes

The IAU is not only concerned with the scientific elite. The development of astronomical education, at all levels and world-wide, is the special charge of a Commission and a Working Group of the

Executive Committee. The motivation for this is not a naive ambition to populate the world with astronomers. There is, in wide segments of the population of all nations, an almost insatiable interest in astronomical phenomena and discoveries and a desire to understand them. For that reason alone, a certain amount of science-based education in public schools is required, also as an antidote against superstition and charlatanry. But in addition, astronomy is valuable both as a means to attract bright young minds to science, and as a way to illustrate many basic physical concepts in interesting and informative ways. Both aspects deserve attention in developed and developing countries alike.

The IAU is active in the educational sphere by promoting international exchange of information on effective curricula, methods, and materials that might be adapted to astronomy educational in other countries. The main effort is, however, focused on programmes in developing countries (regardless of IAU membership) desiring to promote astronomy education for the reasons cited above. The Teaching for Astronomy Development programme is designed for countries with little or no recent scientific activity in astronomy, and aims at helping the formation of adequately trained teachers at the high school level. The form of support is flexible and is adapted to local conditions; it may, e.g., comprise visits by foreign lecturers, travel support to regional courses, and/or provision of teaching material (including books, personal computers, and software).

IAU International Schools for Young Astronomers are held in most years. They feature three-week intensive courses in astronomy for university level students from the host and neighbouring countries, comprising both lectures and practical laboratory and data reduction exercises. Faculty is strongly encouraged to stay for the full course to promote personal contacts and exposure to the often unfamiliar ways of scientific reasoning and argument. After graduation, astronomers of all ages and nationalities are eligible for support for study and research visits abroad from the IAU programme for Exchange of Astronomers, under the published guidelines for that programme.

Currently, the IAU is exploring ways to promote synergy with similar activities in neighbouring fields by other organisations, notably COSPAR and the UN Office of Outer Space Affairs and its partners. The motivation is not only the wish to make optimum use of scarce resources in terms of skilled teachers, quality educational material, and suitable infrastructure, but also a concern for the long-term result of educational programmes such as those mentioned above. Nobody benefits if trainees are just left in a vacuum: For bright young scientists to use their capabilities effectively in the service of their country, not only brains and training, but also a job and appropriate tools are needed. Collaboration between pure and applied sciences, and with governments, is needed to reach this goal.

3. Environmental Issues

Astronomy is among the purest of sciences, with little if any military importance or financial reward. But astronomical research is not immune to events in the outside world. The very faintness of the signals received from the deep reaches of the Universe makes astronomical observations vulnerable to man-made pollution and interference at all wavelengths of the electromagnetic spectrum. Ground-based light pollution from major urban or industrial centres makes deep-sky observations impossible in large parts of the world (and wastes much fossil and nuclear fuel by sending light into the sky rather than where it is needed!). Space debris threatens the survival of scientific as well as commercial satellites, and when illuminated by sunlight ruins astronomical observations with luminous trails. Radio noise, especially from the booming industry of space telecommunications, threatens to drown not only radio astronomy as a science but also all hope of ever detecting intelligent signals from the planets that we have recently begun to detect. Airline communications and safety may be the next victims.

A recent threat is that of bright objects being launched into space for artistic, technological, or advertising purposes. Within reach of powerful corporations and unencumbered by international regulations, such displays would respect no national borders nor the views of countless future generations; the dark night sky would be ruined for all and forever. Astronomers are not so unreasonable as to demand that all useful space activity cease for the sake of a pure, ``useless" science: The IAU simply appeals to the governments of the world that such activities be regulated by international agreements, taking into account the needs of all humankind, including those of international science. An excellent precedent is set by the international treaty concerning the Antarctic continent - which is only apparently closer and more vulnerable than space.

4. Communications

Effective internal and external communication is vital for the IAU to accomplish its mission. Internal communication to Members is maintained through the IAU Information Bulletin, which is published twice per year (normally in January and June) and is sent to all Individual Members and Adhering Organisations as well as to a large number of astronomical institutes and libraries throughout the world. The Bulletin is also available in electronic form from the IAU WWW Page, which provides ready access to a wide range of information about the IAU and its activities, Divisions and Commissions, and an On-Line Membership Directory. Electronic mail is in increasing use, but is still not available or complete for members in all regions of the world.

The external communications of the IAU, to other Scientific Unions as well as to intergovernmental organisations such as those of the United Nations, is primarily maintained through its membership of ICSU: The International Council for Science. Through ICSU, interdisciplinary programmes are initiated and fundamental concerns and issues, such as those related to freedom in the conduct of science or the improvement of science literacy, are addressed and resolved. But the IAU will also, from time to time, address itself directly to the media on significant issues related to the health and progress of astronomy.

5. Membership and Organization

Membership in the IAU exists in two categories, national and individual. Most nations with significant professional research activity in astronomy are National Members, at the Full or Associate level depending on the degree of development of astronomy in the country. The National Members contribute most of the Union's income, in proportions loosely related to the number of Individual Members from each country. Accordingly, the administrative and financial affairs of the Union are decided by vote of the designated national representatives.

Individual Membership in the IAU is open to scientists with a Ph.D. or equivalent in some branch of astronomy, followed by a few years of successful research activity in the field as documented by a number of publications or similar evidence. New Individual Members are admitted by the Executive Committee, normally on the proposal of the Adhering Organisation in their home country. However, individuals may also be nominated by the President of an IAU Division, and all individual members enjoy the same rights and privileges regardless of nationality. The highest body of the IAU is the General Assembly, which is held every three years and decides any modifications of its Statutes and Bye-Laws, elects the Officers, Executive Committee, and Division and Commission Presidents, and approves the budget for the following triennium. The Executive Committee consists of the four Officers and six Vice-Presidents, with the past President and General Secretary as non-voting Advisors. It meets normally once a year and decides on the programme of scientific meetings and other major issues for the coming year. New Individual Members are admitted by the Executive Committee in years of a General Assembly. The Officers of the Union are the President, President-Elect, General Secretary and Assistant

General Secretary (who normally succeeds the General Secretary for the following triennium). They supervise the overall operation of the Union according to the policy set by the General Assembly and the Executive Committee, and normally meet once per year between the meetings of the latter. The General Secretary is responsible for the day-to-day business of the Union, and is its legal representative and responsible for its financial affairs. In these tasks, the General Secretary is assisted by a small Secretariat, located at the Institut d'Astrophysique de Paris, France (see contact information at the beginning of this article).

6. To Learn More

Not all activities of the IAU can be adequately described or even mentioned in this brief presentation. Educational activities, ongoing or planned and with or without direct IAU participation, are developed in further detail in the remainder of this volume (Batten, 2003; McNally, 2003). More information can be obtained from the IAU Secretariat or via the links provided at the IAU WWW page http://www.iau.org.

References

Batten, A.H., 2003, this volume,
McNally, D., 2003, this volume

Chapter 1.3

THE INTERNATIONAL ASTRONOMICAL UNION: HISTORIAL PERSPECTIVE

D. McNally
University of Hertfordshire
College Lane
Hatfield, Herts AL10 9AB, U.K.
dmn@star.herts.ac.uk

Abstract The historical development of the International Astronomical Union (IAU) is summarized. The manner in which the IAU supports astronomical research on an international level is described.

Historical development

The International Astronomical Union (IAU) is one of four founding Scientific Unions in the International Council of Scientific Union (ICSU) family. It was established in 1919, celebrating 75 years of existence at its XXIInd General Assembly in 1994. In the course of its history it has grown from 207 members in 1922 to 8660 at the close of the XXIVrd General Assembly in 2000. Since members must have a demonstrable commitment to astronomy and have made a contribution to the science of astronomy, the membership of the IAU is a measure of the minimum number of people engaged in astronomical or astronomy related activities. A rough guide of a factor of 2 is often used to estimate the total population associated with astronomical activities. In round numbers about 16,000 people would be pursuing astronomical investigations very roughly, one astronomer for every half million people. That high dilution factor for astronomers in the world's population means that astronomers are internationally minded and have been so for many centuries. They have had, by force of circumstances, to keep in contact across national

W. Wamsteker et al. (eds.), Developing Basic Space Science World-Wide, 17-21.

frontiers and jurisdictions. International cooperative ventures were not new to astronomy when G.E. Hale urged the formation of the IAU during the closing stages of the 1st World War. Astronomers had co-operated together on, for example, measuring the size and shape of the Earth, the establishment of a global system of time zones, the study of the Sun, and selected areas of the sky. Despite this long history of international contact, the newly founded IAU did not get off to a truly international start. Certain countries were excluded initially, to the dismay of many astronomers who wanted a wholly international body. That was not achieved until after the establishment of ICSU in 1931. Not every country is as yet a member of the IAU - there are 60 countries with sufficient astronomical activity who deem it worthwhile to adhere to the IAU but IAU members exist in a few other countries which, as yet, do not adhere to the Union. About one in three countries sustain astronomical activity.

1. Tasks and Activities

The purpose of the IAU is to promote and sustain astronomical activity. It approaches this task in a number of very different ways - services to astronomy on an international basis, maintenance of standards and critical data, development of international collaboration, triennial reviews of the state of astronomy and support of international symposia, colloquia, regional meetings and the Union's General Assembly for the discussion and dissemination of new astronomical discoveries. The Union also cooperates with other Scientific Unions of the ICSU family and other nongovernmental and governmental bodies over matters of mutual interest or concern. The Union is governed by Statutes, By-Laws and Working Rules designed to ensure adequate regulation of the Union in its representation of international astronomical activity, in its financial arrangements and in the conduct of its business (*http:www.iau.org*). The operation of the Union is regulated triennially by a General Assembly of its members - both Adhering Countries and individual members. The representatives of the Adhering Countries vote on financial and administrative matters affecting the conduct of the Union, the entire Union membership votes on scientific matters of General Assemblies. The Union is unique in that it has individual membership and that participation of the members is widely diffused within the activities of the Union. Therefore the Union is representative of a very wide consensus of the world's astronomers when it rules on scientific matters. The Union is financed by the contributions of its Adhering Countries. Payment is made on the basis of an annual unit of contribution and Category of Adherence - the annual dues are the product of the unit of contribution and the agreed multiplying factor associated with the Category of Adherence. It is up to each adhering country to choose its Category of Adherence and it is remarkable that proportionality with individual membership is broadly maintained by the Adhering Countries. Between General Assemblies, the affairs of the Union are managed by means of an annual meeting of the Executive Committee (of 10 members and 2 advisers) and in the interim by the General Secretary, assisted by the Assistant General Secretary. The Union is further organized into 40 Commissions divided among 11 Divisions. It is the Commissions and their 70 Working Groups that undertake the main burden of the Union's activities. Most Commissions are devoted to specific scientific tasks, e.g. Stellar Structure, Interstellar Matter, Light of the Night Sky and so on. Such Commissions will have Working Groups to undertake specific tasks which may either be long or short term. Divisions support proposals for the scientific meetings of the Union and organize joint discussion meetings at General Assemblies. Some Commissions exist to support a particular technique, e.g. Radio Astronomy, Photographic Astrometry and explore, refine and develop these techniques. Other Commissions, e.g. Ephemerides, Telegrams, are concerned with particular services for the astronomical community. The Commission on Bibliography is concerned with library matters and the improvement of communication between astronomers, e.g. by agreeing systems which are unambiguous to identify members of a newly discovered class of objects. The Commission on Astronomical Education is concerned with improving and extending knowledge of astronomy at all levels. The Union undertakes the naming of entities within the solar system e.g. planets, their satellites, craters and the like. The reason why it does so, is to avoid ambiguity of nomenclature. The aim is to give a unique name to each solar system feature. A database of suitable names, assembled on an international basis has been formed and is continuously expanded. Success has not always attended the Union's efforts to achieve unique nomenclature - until recently the naming of asteroids rested with the discoverer with the result that some duplication occurred. Procedures are now in place to ensure that such duplication becomes a matter of history.

2. Support of Science Activities

When the first steps in astronomy are being taken in a particular country, the Union can often give support through its TAD programme (Commission 46). This programme can assist astronomers to travel to an institution which is developing its astronomical science to give courses.4 of lectures outside the scope of the parent institutions. Graduate students can be assisted to seek Ph.D. supervision at an institution in another country. The Union's Working Group on the World Wide Development of Astronomy is constantly seeking ways in which the international astronomical community can assist the support and development of astronomy. The support which is needed varies with each circumstance. Some countries may require a regular flow of outside lecturers. Other countries may need library assistance. Still others may need limited funding for overseas travel. The Working Group on World Wide Development of Astronomy explores the possible avenues through which the most effective Union support can be channeled, given its strictly limited resources. On the scientific side, the Union supports a wide range of activities. Its working groups investigate the establishment of standards, the development of sky survey activities, etc. A notable recent achievement was the development of a fundamental Reference Frame for Astronomy. Such matters affect the entire community and can only be achieved on an international basis. Here specialist working groups exist to elucidate a particular problem within the purview of a specific Commission. Standardization of procedures in the observation of a particular class of stars may need to be established and the collaboration of a number of specialists at a range of observatories may be required. The Union also supports a range of scientific meetings for the exchange and discussion of the most recent investigations into astronomical phenomena. The Union supports both a Symposium and Colloquium series of meetings (Colloquia for the more specialized topics) and co-sponsors meetings with other ICSU Unions and Committees. At the triennial General Assemblies there are Symposia and Joint Discussions. The Joint Discussions are organized by the Divisions/Commissions after selection by the Executive Committee. There are Invited Discourses by distinguished astronomers. These Discourses may be about a specific investigation or a life's work. At General Assemblies, Working Groups can meet as can inter-Union groups, such as the Inter-Union Working Group on the Allocation of Frequencies or the Working Group on Adverse Environmental Impacts on Astronomy. The General Assembly also gives Commissions time to review their business, discuss future activities and select their Officers.

3. IAU Scientific Reports

A feature of Union activities is the publication of the triennial review by each Commission of the activities of the triennium in the field of the. Commissions. This is published as the relevant volume of Transactions A (bearing the same number as the General Assembly its publication precedes). This is a useful document to consult when seeking an update on the state of astronomy and a useful guide to reviews and an optimized route into the abstracting services. The Union also publishes an account of its meetings, during a General Assembly in Highlights of Astronomy and Transactions B. Together these publications present a unique record of astronomy as it is understood by its practitioners. They give a series of snapshots of current astronomy differing from other expert reviews or abstracting services in that they give a less individualized and less formalized picture of what is considered important in current astronomy over the entire field of endeavour in astronomical science.

4. International Relations

The value of the Union in promoting international relations must not be overlooked. At each General Assembly astronomers from a wide range of nations of differing cultural traditions gather together in one of the countries adhering to the Union. For two weeks, national, social and cultural differences recede to their proper perspective under the barrage of new, exciting and controversial astronomical results and interpretations. One finds diverse groups in deep debate on equally diverse astronomical topics. Astronomy is not an easy science to pursue from the ground or from space and, as astronomers, we know that we will only succeed if we present a united front to both science and patron alike.

References (*Suggested reading*):

Blaauw, A. 1994, History of the IAU: The Birth and First-Half Century of the International Astronomical Union (Dordrecht: Kluwer Academic Publishers).

IAU Bulletin, published twice yearly in January, June, information for members.

Transactions of the International Astronomical Union; published triennially; A - Reports on astronomy, B - Proceedings of General Assemblies, Transactions are edited by the General Secretary.

Chapter 1.4

ASTRONOMY AROUND THE WORLD

A.H. Batten
Dominion Astrophysical Observatory
Victoria, Canada
alan.batten@nrc.ca

Abstract The needs of astronomers in developing countries, as they appeared to the writer in the late 1990s, are surveyed and summarized. While the situation has changed in some respects since the body of this article was written, and a few revisions have been made to take account of these changes, the isolation and inexperience of many strongly motivated astronomers still prevent them from realizing their full potential. Several possible ways of giving support are discussed. All of them need funding.

Introduction

The 1960s and 1970s saw an unparalleled expansion of astronomy throughout the world. Partly this was stimulated by the successful launch of the first Soviet Sputnik in 1957 and the "space race" that ensued between the then two dominant superpowers of the world. The great technical improvements in the acquisition and reduction of data from observations that began to be introduced into astronomy during those two decades, however, and the amazing discoveries that those developments made possible have surely been at least as important in generating the expansion of our science. The present writer had the good fortune to be in the final year of his graduate studies when Sputnik I was launched and so began his career at a most fortunate time, enjoying what may well have been the best three or four decades for astronomy in the entire history of the world. The expansion has proved too great for even the wealthiest countries to maintain and the 1980s, as well as the present decade, have been a time of cutting back. Observatories have been closed, libraries have reduced their acquisitions, and funds for meetings and travel have become harder to find. Those of us who have enjoyed the

W. Wamsteker et al. (eds.), Developing Basic Space Science World-Wide, 23-30.

good decades are now retiring; there are many highly competent and fully motivated people waiting to succeed us, but it is not clear that all of us will be replaced. We shall be fortunate if we are able to maintain equilibrium between the supply of and demand for astronomers. The situation in other sciences is, of course, similar, but in this document we focus on that in astronomy. A world culture is emerging as we approach the end of the twentieth century in which science seems destined to play a dominant role, yet, precisely in respect of science, the developed part of the world seems in danger of developing the "failure of nerve" that some authorities see as the cause of the decline of classical Greece.

1. Science and the Economy

There is, however, a paradox in all this. While the astronomers of wealthy countries complain about cut-backs and tight funding, they do have access to many telescopes of 3m and 4m aperture, and even several of up to 10m, not to mention instruments in space, including the Hubble telescope. Computing power continues to increase and the development of electronic communication may, to some extent, compensate for the reduced holdings of individual libraries in wealthy countries as a few major libraries make their holdings available on line. The frequent specialist meetings that have been such a feature of the last few decades may seem less necessary, if television conferences can be arranged; it is even no longer always necessary to travel in order to observe with specific instruments. These developments partly compensate astronomers in the developed world for the cuts in funding, but they are changing the way that astronomy is done. Observations are more frequently being obtained by large groups with access to one of the great telescopes, and are less likely to be obtained by lone observers with instruments of the order of 1m aperture despite the acknowledged fact that there is much that small instruments can do more cost-effectively. All these developments are adversely affecting our colleagues working in small groups that are isolated either geographically or electronically. This is not always understood by those in the privileged part of the world and needs to be emphasized. In principle, an astronomer anywhere can join with others of established reputation in submitting observing proposals, and can at least share in the reduction and analysis of the data. In practice it is often hard for astronomers from small developing countries to become well enough known to do so. They do not know whom to approach, and relatively few astronomers in the developed countries are prepared to seek out possible unknown partners. More and more, our modus operandi is making it harder for people not connected with the big centres of astronomical research to make significant contributions of their own. The situation is particularly difficult for those who live in countries from which a scientific tradition is absent, or in which one has been interrupted. The radical political changes of the late 1980s and early 1990s in many formerly socialist countries have added to the numbers of those experiencing difficulties in pursuing research careers (Bochkarev, 1998). In many of these countries, a tradition of scientific research was strong under the old regimes, but the political and economic changes that have taken place have interrupted that tradition {we all hope only in the short term{ and have created much personal hardship for scientists. In fact, scientific research is closely linked to economic progress. A nation must be generating a certain surplus before it can afford to support research, particularly in an area like astronomy, which is often perceived as promising no immediate economic benefit. Developing countries that are making significant progress astronomically are precisely those that are improving their economic status (e.g. South-East Asia) while those that are not making headway in astronomy are not doing so economically either (e.g. much of Africa). It may seem obvious that these are cause and effect, the economic factors being the cause. The relationship may be a two-way one, however: the encouragement of scientific research, even of curiosity-oriented research, can be a stimulus to economic development. Astronomers' demands for large telescopes, for example, can help to stimulate a country's heavy industry, as they have done in India; or the need for instrumentation for those telescopes can create an electronics industry. Moreover, the teaching of astronomy is a basic part of the teaching of all the physical sciences, since the latter grew out of the attempt to understand the motions of the planets. Some competence in astronomy is, therefore, desirable within every nation, regardless of particular economic circumstances. Those who are trying to provide that competence in the smaller and less developed countries are constantly discouraged, however, by the complete lack of books and journals, of access to instruments, and all too often by their inability to keep regularly in touch with colleagues in more fortunate positions.

2. International Collaboration

For many reasons, including their relatively small numbers, astronomers are dependent on international cooperation, and a more effective web of support has grown up than is, perhaps, to be found in many other sciences. The International Astronomical Union (IAU) is one of the oldest and largest international scientific unions, and its tradition of having individual members, as well as adhering countries, enables it to bring isolated workers at least nominally into the international community. Membership of the IAU is a beginning for such people, but the IAU is now a large organization with nearly 9,000 members, and it is easy for a new member to be lost to sight and to exist only as a name in the membership list. Frequently, new members do not know the range of services the IAU offers (McNally, 1998). The Union's Commission 46 on the Teaching of Astronomy, for example, not only provides advice on teaching, but organizes International Schools for Young Astronomers and is now introducing Teaching for Astronomy Development, a programme that will enable new astronomy departments in the universities of developing countries to receive visiting lecturers. This is particularly important for countries in which the teaching generation, for one reason or another, has been cut off from the opportunity to keep abreast of the most recent developments in astronomy. Commission 38 of the IAU, concerned with the exchange of astronomers, provides a limited number of travel grants for astronomers who receive a temporary appointment in a country other than their own. These can be used either to take people from developing countries to a major centre for a period of research, or to bring an established worker to a developing country so that he or she can teach or help to start a research programme. More recently, the IAU has formed its Working Group for the Worldwide Development of Astronomy to try to provide a focus for its activities in this field. The Group can sometimes identify problem areas and alert those organizing the other activities of the Union to them. We are slowly building up a list of the needs felt by astronomers in developing countries and of ideas of what the astronomical community can do to further astronomical science development (Batten, 1994). This is important because there is always a danger that well-intentioned efforts to help will succeed only in imposing a pattern of astronomical research that, however suitable it may be in a developed country, is inappropriate to the needs of a country just beginning. It is far from clear, for example, that the best way for a developing country to create an astronomical tradition is to plunge into modern cosmology. In 2000, the Working Group and Commission 38 were merged into an enlarged Commission 46, in an attempt to make the IAU's efforts to help more effective. The Special Session on "Astronomy for Developing Countries" at the IAU XXIV general Assembly in 2000 resulted in a useful publication devoted to fuller discussion of many of the problems discussed here (Batten, 2001). Besides the IAU, other organizations are working to the same end, particularly the United Nations, through the Committee on Peaceful Uses of Outer Space. This Committee has organized a series of workshops on Basic Space Science for Developing Countries, holding each workshop in a specific region in which there is a need to encourage astronomy and space science. This series of workshops has been arranged in cooperation with the European Space Agency (Haubold, 1998). NASA also has a programme of assistance. Individual initiatives have been taken, such as the network of robotic telescopes proposed by Querci (1995). Perhaps the most valuable way for the astronomical community to proceed, however, is to help to establish and to encourage regional cooperation. The IAU began this in Europe, with regional meetings that have now evolved into the European Astronomical Society. Regional meetings of the IAU continue in the Asia-Pacific and Latin American regions. Sometimes, even more localized cooperation is called for. The six Spanish-speaking countries of the Central American isthmus, for example, are coming together in order to foster astronomical development in that fairly small region. South-East Asia contains many countries at different stages of development, and of differing political systems. The experience of astronomers in Indonesia and the Republic of Korea could, however, be of immense benefit to those who are trying to revive Vietnamese astronomy. The IAU is actively trying to foster this collaboration. Although Africa is a vast continent, there are rather few astronomically active countries on it (Batten, 1995). To this extent it makes sense to try to build up a continent-wide collaboration, particularly because political barriers to full cooperation with the vigorous South African community of astronomers have now been removed. The construction of the Southern African Large Telescope (SALT) is a welcome sign of this cooperation (Martinez, 2001). On the other hand, there is a clear cultural community north of the Sahara that is distinct from that to the south, so that eventually one might try to encourage cooperation in at least two distinct regions on that continent. Another potential region stretches from the eastern end of the Mediterranean to India. There is little doubt that the astronomers scattered through that region are very willing to help each other, but the political barriers to their doing so are still very high. An encouraging sign is the International School for Young Astronomers organized by the IAU in Iran during the summer of 1997.

3. International Exchange Programs

The chief problem faced by small groups of astronomers in developing countries is isolation. In some countries, this isolation has lasted for a generation or more, and no one in the country is fully equipped to give students a proper introduction to modern astronomy. The only way to break this vicious circle is to enable either students or teachers to spend some time in another country: it is rarely feasible to send someone from that other country to spend long enough to have a lasting effect, although the Peoples' Republic of China came close to achieving that goal through its large programme of visiting experts. From the developing country's point of view, the great danger of sending people abroad, especially young students, is that they may be tempted to stay. The attraction of a working environment with more resources and, very often, less bureaucratic control, is very real. If financing could be found for a major scheme to facilitate this sort of exchange, some safeguards would have to be built in to allay the justified fears of a \brain drain". Such a drain probably cannot be entirely prevented, but in most developed countries now, there is so much difficulty finding jobs for qualified nationals that visitors are much less likely to be offered positions than they were a few decades ago. Narlikar's proposal (Narlikar, 2001) for electronic networking would overcome many of the problems of isolation without creating any risk of a brain drain, but this, too, requires funding. On balance, the more programmes there are for exchanging astronomers, the better it will be for those countries wishing to develop their astronomy. Indeed, a similar statement would be true for all the sciences. The twin problems are to devise a suitable scheme, and to find financing for it. Countries that were formerly ruled as colonies by a European power should be aware that that power is often willing to help with such schemes, but astronomy would be only a small (and not necessarily favoured) part.

4. Research Support

Another great need is for journals and modern books. This need is not met, or at best only partly met, by the dispatching of long runs of old journals. The need is for access to what is being published now. The astronomers of many countries simply cannot afford the subscriptions to the major journals {nor can their libraries. The changeover to electronic publishing that has already begun will exacerbate this problem rather than solve it, since the countries concerned are just the ones that are not connected to e-mail networks. Although such networks are being extended to developing countries, many of the astronomers there still have only limited access to them - a situation that too many of us in developed countries fail to understand. Many isolated astronomers place much value on receiving preprints. Developing countries are also usually \instrument-poor": they have no access even to the smallest telescopes with which useful research can be done or students can at least be trained in the techniques of observation. In this context, a Japanese programme of supplying a limited number of 40cm telescopes to developing countries (Kitamura, 2003) is much to be commended, although there is a corresponding need to provide training that will ensure their productive use. Instruments in the range of 40-cm to 1-m are ideal for teaching and useful for research. People who have begun to produce results with such instruments may later graduate to telescopes of aperture 2m or larger. A hurdle here, however, is that the writing of a successful application for time on a large telescope is itself something that needs practice. New observers would do well to ally themselves with some established team. There is a need for a clearing-house to match would-be observers with teams willing to lend a helping hand. It is often pointed out that satellite data are in principle freely available after the original investigators have finished with them (or even failed to use them within a stated period). Astronomers without access to an instrument of their own could, in principle, use these archival data; but such astronomers are often without electronic access to the data or to computers of the sophistication needed for analysis. Once again, there is a need to match such people with teams willing to help. Hearnshaw (1998) has discussed the advantages and disadvantages of these different ways of providing access to observational data.

5. Outlook

There is enough goodwill in the astronomical community at large to enable many of these problems to be solved, or at least eased. Too many people in the developed countries are not aware enough of the needs, and those in the developing countries are not aware of all the avenues through which help might come. In the IAU, and particularly in the Working Group, we are trying to raise consciousness of these matters on both sides. The active involvement of governments in the countries concerned can only help to improve the situation and this United Nations document is an important step toward securing that involvement. Government officials, whether in developing or developed countries, are not always aware of the contributions that astronomy can make to the economy of a country, as mentioned above, or of the potential versatility of a student who has received a thoroughly modern astronomical education. Such a student will have many skills, including familiarity with computers and their interfacing with other systems, knowledge of the technology of detecting low levels of incident light, familiarity with statistical methods and the drawing of inferences from imperfect data, and often a familiarity with other sciences (physics, chemistry, geology and meteorology) depending on the particular area in which the student has specialized. Since most astronomers work in universities, they are also the educators of the next generation. Astronomers can therefore be a real asset to any country, even if their working careers cannot be primarily devoted to purely astronomical problems.

References

Batten, A.H., 1994, Journal of the Royal Astronomical Society of Canada, **88**, 163.

Batten, A.H. (ed.), 2001, Special Session of the 24th IAU GA., , San Francisco, Astronomy for Developing Countries, ASP Conference Series **SPS**.

Batten, A.H., 1995, in Highlights of Astronomy, **10**, I. Appenzeller (Ed.), Dordrecht: Kluwer Academic Publishers, 677.

Bochkarev, N.G., 1998 , in Highlights of Astronomy, **11B**, J. Andersen (Ed.), Dordrecht: Kluwer Academic Publishers, 898.

Hearnshaw, J.B., 1998, in Highlights of Astronomy, **11B** , J. Andersen (Ed.), Dordrecht: Kluwer Academic Publishers, 920.

Haubold, H.J., 1998, in Highlights of Astronomy, **11B** , J. Andersen (Ed.), Dordrecht: Kluwer Academic Publishers, 906.

Kitamura, M., 2003, this volume.

Martinez, P., 2001, in Special Session of the 24th IAU GA, A.H.Batten (Editor), Astronomy for Developing Countries, ASP Conference Series **SPS**, 221.

McNally, D., 1998, in Highlights of Astronomy, **11B**, J. Andersen (Ed.), Dordrecht: Kluwer Academic Publishers, 903.

Narlikar, J.V., 2001, in Special Session of the 24th IAU GA, A.H.Batten (Editor), Astronomy for Developing Countries, ASP Conference Series **SPS**, 324.

Querci, F.R., 1995, in Highlights of Astronomy, **10**, I. Appenzeller (Ed.), Dordrecht: Kluwer Academic Publishers, 677.

Chapter 1.5

RESEARCH AND EDUCATION IN BASIC SPACE SCIENCE

H.M.K. Al-Naimy[1], C.P. Celebre[2], K. Chamcham[3], H.S.P.de Alwis[4], M.C.P. de Carias[5], H.J. Haubold[6], A.E. Troche-Boggino[†7]
[1]Jordan, [2]The Philippines, [3]Morocco, [4]Sri Lanka, [5]Honduras, [6]Austria, [7]Paraguay.
alnaimiy@yahoo.com

Abstract The UN/ESA workshops have been held in the regions of Asia and the Pacific, Latin America and the Caribbean, Africa, Western Asia, and Europe. Additional to the scientific benefits of the workshops and the strengthening of international cooperation, the workshops lead to the establishment of astronomical telescope facilities in Colombia, Egypt, Honduras, Jordan, Paraguay, the Philippines, and Sri Lanka. The annual UN/ESA Workshops continue to pursue an agenda to network these astronomical telescope facilities through similar research and education programmes. Teaching material and hands-on astrophysics material has been developed for the operation of such astronomical telescope facilities in an university environment.

Introduction

Research and education in astronomy and astrophysics are an international enterprise and the astronomical community has long shown leadership in creating international collaborations and cooperation: Because (i) astronomy has deep roots in virtually every human culture, (ii) it helps to understand humanity's place in the vast scale of the universe, and (iii) it teaches humanity about its origins and evolution. Humanity's activity in the quest for the exploration of the universe is reflected in the history of scientific institutions, enterprises, and sensibilities. The institutions that sustain science; the moral, religious, cultural, and philosophical sensibilities of scientists themselves; and the

[†] Shortly after the completion of this paper Professor Alexis E. Troche-Boggino passed away.

W. Wamsteker et al. (eds.), Developing Basic Space Science World-Wide, 31-37.

goal of the scientific enterprise in different regions on Earth are subject of intense study (Pyenson and Sheets-Pyenson, 1999).

The Bahcall report for the last decade of the 20th century (Bahcall, 1991) has been prepared primarily for the North American astronomical community, however, it may have gone unnoticed that this report had also an impact on a broader international scale, as the report can be used, to some extend, as a guide to introduce basic space science, including astronomy and astrophysics, in nations where this field of science is still in its infancy. Attention is drawn to the world-wide-web site at http://www.seas.columbia.edu/~ah297/un-esa/ where the initiative is publicized on how developing nations are making efforts to introduce basic space science into research and education curricula at the university level. This initiative was born in 1990 as a collaborative effort of developing nations, the United Nations (UN), the European Space Agency (ESA), and the Government of Japan, and covers the period of time of the last decade of the 20th century. Through annual workshops and subsequent follow-up projects, particularly the establishment of astronomical telescope facilities, this initiative is gradually bearing results in the regions of Asia and the Pacific, Latin America and the Caribbean, Africa, and Western Asia.

1. Workshops on Basic Space Science

In 1959, the United Nations recognized a new potential for international cooperation and formed a permanent body by establishing the Committee on the Peaceful Uses of Outer Space (COPUOS). In 1970, COPUOS formalized the UN Programme on Space Applications to strengthen cooperation in space science and technology between non-industrialized and industrialized nations. In 1991 the UN, in close cooperation with developing nations, ESA, and the Government of Japan, started under the auspices of COPUOS a series of annual Workshops on Basic Space Science, which were hosted by UN member States in the five economic regions defined by the UN: India, Costa Rica, Colombia, Nigeria, Egypt, Sri Lanka, Germany, Honduras, Jordan, France, Mauritius, Argentina and China (Haubold & Wamsteker, 2003).

Over more than ten years, the workshops established a close interaction between scientists from developing and industrialized nations to discuss research findings at the current front lines in basic space science. The workshops also initiated a direct interaction between scientists from developing nations. In depth discussions in working groups were fostered to allow the identification of the needs - especially common needs, which could be addressed on a larger scale - to enhance the participation of the developing nations in basic space science and to identify the best ways and means in which each nation could accelerate its participation in a meaningful endeavor.

The eighth in the series of UN/ESA Workshops on Basic Space Science, which, among other topics, addressed the feasibility of establishing a World Space Observatory (WSO), was held at Jordan in 1999 (Wamsteker and Shustov, 2003). The UN, ESA, Japan, and international organizations will continue to provide assistance for the establishment and operation of astronomical facilities in Colombia, Egypt, Honduras, Jordan, Paraguay, the Philippines, and Sri Lanka.

2. Astronomical Telescope Facilities

A number of Governments (among them Honduras in 1997 and Jordan in 1999), in cooperation with international partners, have acquired and established astronomical telescope facilities in their countries (Meade 16" Schmidt-Cassegrain models).

In conjunction to the workshops, to support research and education in astronomy, the Government of Japan has donated high-grade equipment to a number of developing nations (among them Sri Lanka 1995, Paraguay 1999, the Philippines 2000) within the scheme of ODA of the Government of Japan (Kitamura 1999). We refer here to 45cm high-grade astronomical telescopes furnished with photoelectric photometer, computer equipment, and spectrograph with CCD detectors. After the installation of the telescope facility by the host country and Japan, in order to operate such high-grade telescopes, young observatory staff members from Sri Lanka, Paraguay and the Philippines have been invited by the Bisei Astronomical Observatory for education and training, sponsored by the Japan International Cooperation Agency [JICA] (Kitamura, 1999; Kogure, 1999).

The research and education programmes at the newly established telescope facilities will focus on time varying phenomena of celestial objects. The 45cm class reflecting telescope with photoelectric photometer attached is able to detect celestial objects up to the 12th magnitude and with a CCD attached up to the 15th magnitude, respectively. Such results have been demonstrated for the light variation of the eclipsing close binary star V505 Sgr, the X-ray binary Cyg X-1, the eclipsing part of the long-period binary ε Aur, the asteroid No.45 Eugenia, and the eclipsing variable RT Cma (Kitamura, 1999). In forthcoming workshops, common observational programmes for variable stars for all the telescope facilities are envisaged.

3. Observing with the Telescopes: Research

In the course of preparing the establishment of the above astronomical telescope facilities, the workshops made intense efforts to identify available material to be used in research and education by utilizing such facilities. It is obvious that variable star observing by photoelectric or CCD photometry can be a prelude to even more advanced astronomical activity. Variable stars are those whose brightness, colour, or some other property varies with time. If measured sufficiently carefully, almost every star turns out to be variable. The variation may be due to geometry, such as the eclipse of one star by a companion star, or the rotation of a spotted star, or it may be due to physical processes such as pulsation, eruption, or explosion. Variable stars provide astronomers with essential information about the internal structure and evolution of the stars. A predominant institution in this specific field of astronomy is the American Association of Variable Star Observers (AAVSO). The AAVSO coordinates variable star observations made by amateur and professional astronomers, compiles, processes, and publishes them, and in turn, makes them available to researchers and educators. The AAVSO receives over 350,000 measurements a year, from more than 550 observers world-wide. The measurements are entered into the AAVSO electronic database, which contains close to 10 million measurements of several thousand stars.

To facilitate the operation of variable star observing programmes and to prepare a common ground for such programmes, AAVSO developed a rather unique package titled "Hands-On Astrophysics" (see also Percy and Mattei, 2003). This includes 45 star charts, 31 slides of five constellations, 14 prints of the Cygnus star field at seven different magnifications, 600,000 measurements of several dozen stars, user-friendly computer programmes to analyze them, and to enter new observations into the database, an instructional video in three segments, and a very comprehensive manual for educators and students (http://www.aavso.org/). Assuming that the telescope is properly operational, variable stars can be observed, measurements can be analyzed and send electronically to the AAVSO.

The flexibility of the "Hands-On Astrophysics" material allows an immediate link to the teaching of astronomy or astrophysics at the university level by using the astronomy, mathematics, and computer elements of this package. It can be used as a basis to involve both the professor and the student to do real science with real observational data. After a careful exploration of "Hands-On Astrophysics" and thanks to the generous cooperation of AAVSO, this was adopted by the above astronomical telescope facilities for their observing programmes (Mattei and Percy, 1999; Percy, 1991).

4. Teaching Astrophysics: Education

Various strategies for introducing the spirit of scientific inquiry to universities, including those in developing nations, have been developed and analyzed (Wentzel, 1999a; AAVSO, 1999). What concerns the spirit of the workshops on basic space science, they have been organized and hosted by Governments and scientific communities which agreed beforehand on the need to introduce or further develop basic space science at the university level and to establish adequate facilities for pursuing such a field of science in practical terms. This implied to operate an astronomical facility for the benefit of the university or research establishment (and to prospectively make the results from the facility available for public educational efforts). Additional to the hosting of the workshops, the Governments agreed to operate such a telescope facility in a sustained manner with the eye on the international community for support and cooperation in devising the research and educational programmes. Gradually, this policy is being implemented for those telescope facilities established through the workshops in cooperation with the UN, ESA, Japan, and other national and international organizations. Organizers of the workshops have acknowledged in the past the desire of the local scientific communities to use educational material adopted and available at the local level (prepared in the local language). However, the workshops have also recommended to explore the possibility to develop educational material (additional to the above mentioned "Hands- On Astrophysics" package) which might be used by university staff in different nations while preserving the specific cultural environment in which astronomy is being taught and the telescope is being used. A first promising step in this direction was made with the project "Astrophysics for University Physics Courses" (Wentzel, 1999b; 2003). This project has been highlighted at the IAU/COSPAR/UN Special Workshop on Education in Astronomy and Basic Space Science, held during the UNISPACE III Conference at the United Nations Office Vienna in 1999 (Isobe, 1999). Additionally, a number of text books and CD-ROM's have been reviewed over the years which, in the view of astronomers from developing nations, are particularly useful in the research and teaching process (for example, just to name three of them: Bennett et al., 2001; for teaching purposes; Lang, 1999, a reference work in the research process; Hamilton, 1996, a CD-ROM for astronomy in the classroom). A World Wide Web Site, offering a wealth of additional material for professors and students, specifically developed for teaching astronomy with the book of Bennett et al. (2001) which is upgraded on a regular basis is also available at: http://www.astrospot.com/. The issue of educational materials and their suitability could be further discussed in the Newsletters, specifically published for the benefit of Africa (Querci and Martinez, 1999), Asia and the Pacific (Isobe, 1999). A new initiative, which has been developed under the auspices of the European Union, and could be very beneficial in this context is the European Association for Astronomy Education (EAAE). More information on this organization which was constituted in 1995 can be found on the website http://www.algonet.se/~sirius/eaae.htm

5. What Next?

To exactly take into account the obstacles encountered and the progress made, as observed in the ten-years long approach pursued in the UN/ESA Workshops as described above. Forthcoming workshops will address the benefits of basic space science to society, particularly developing nations, and the experience with, results from, and the need for networks of astronomical telescopes in terms of common research and education programmes. The astronomical telescopes meant are particularly those mentioned above. The workshops will also address in detail the feasibility to establish a World Space Observatory (WSO), discussed since the workshop in Sri Lanka in 1995, and the participation of developing nations in such an effort both from the point of view of research and education (Wamsteker and Gonzales-Riestra, 1997; United Nations GA Document A/AC.105/723: see Haubold and Wamsteker, 2003).

References

AAVSO, 1999, Setting Up a Variable Star Observing Program (Cambridge, MA: AAVSO).

Bahcall, J.N., 1991, The Decade of Discovery in Astronomy and Astrophysics (Washington, DC: National Academy Press).

Bennett, J., Donahue, M., Schneider, N., and Voit, M., 2001, The Cosmic Perspective 2nd Edition, Menlo Park, CA: Addison Wesley Longman Inc.

Eenens, Ph. and Corral, L., Eds., 2000, Astronomía Latino Americana (ALA), Electronic Bulletin, available at *http://www.astro.ugto.mx/~ala/*.

Hamilton, C.J.,1996, Views of the Solar System, CD-ROM Arlington: National Science Teachers Association.

Haubold, H.J., and Wamsteker,W., 2003, this volume.

Isobe, S., 1999, Bull. Teaching of Astronomy in Asian-Pacific Region, **15**. (available on request from isobesz@cc.nao.ac.jp).

Kitamura, M., 1999, in Proceedings of the Conference on Space Sciences and Technology Application for National Development, Colombo, Sri Lanka, 21-22 January 1999, Ministry of Science and Technology of Sri Lanka, 147.

Kogure, T., 1999, in Proceedings of the Conference on Space Sciences and Technology Application for National Development, Colombo, Sri Lanka, 21-22 January 1999, Colombo: Ministry of Science and Technology.
Lang, K.R., 1999, Astrophysical Formulae, Berlin: Springer-Verlag, Volume **I**: Radiation, Gas Processes and High Energy Astrophysics, Volume **II**: Space, Time, Matter and Cosmology.
Mattei, J. and Percy, J. R., 1997, Hands-On Astrophysics: Variable Stars in Science, Math, and Computer Education (AAVSO, Cambridge,MA). also available at *http://www.aavso.org/* .
Percy, J.R., 1991, Astronomy Education: Current Developments, Future Cooperation, Astronomical Society of the Pacific Conference Series, **89**.
Percy, J.R., and Mattei,J., 2003, this volume.
Pyenson, L. and Sheets-Pyenson, S., 1999, Servants of Nature: A History of Scientific Institutions, Enterprises and Sensibilities (New York: W.W. Norton & Company).
Querci, F.R. and Martinez, P., 2000, Newsletter African Skies/ Cieux Africains, electronic version available at: *http://www.saao.ac.za/~wgssa/*.
Wamsteker, W. and González Riestra, R., 1997, Ultraviolet Astrophysics Beyond the IUE Final Archive, Proceedings of the Conference, Sevilla, Spain, 11-14 November 1997, European Space Agency **SP-413**, 849.
Wamsteker, W. and Shustov B.M., 2003, this volume.
Wentzel, D.G., 1999a, Bull. Teaching of Astronomy in Asian-Pacific Region,, **15**, 4.
Wentzel, D.G., 1999, Astrofísica para Cursos Universitarios de Física (La Paz, Bolivia: Impreso por Librería Popular). English language version of this book publication is available at *http://www.seas.columbia.edu/~ah297/un-esa/astrophysics*.
Wentzel, D.G., 2003, this volume.

SECTION:

2. ASIA AND THE PACIFIC

Chapter 2.1

INDIAN SPACE PROGRAMME

S.C. Chakravarty
ISRO
Space Science data Centre
Antariksh Bhavan, New BEL Road
Bangalore 560 09, India
subhaschandra@hotmail.com

Abstract An overview of the development of the Indian Space Programme is given as implemented by the Indian Space Research Organization ISRO. This programme involves meteorological, communications, and scientific spacecrafts and payloads, as well as a complete Launcher and Rocket development programme.

1. Introduction

The Indian space programme was initiated in 1962 to harness the potentials of space technology and its applications for national development under the leadership of its pioneer, Dr. Vikram A. Sarabhai (figure 2.1-1) who had cherished a dream that India should be second to none in the application of advanced technologies like space, to solve the real problems of man and society. In 1972, the Indian Space Programme was formally organized with the setting up of the Space Commission and the Department of Space in addition to the existing Indian Space Research Organization (ISRO).

2. Satellite Communication

The space communication activities in ISRO were initiated with the establishment of Experimental Satellite Communication Earth Station (ESCES) at Ahmedabad in 1967 which paved the way to conduct Satellite Instructional Television Experiment (SITE) using NASA's ATS-

W. Wamsteker et al. (eds.), Developing Basic Space Science World-Wide, 41-48.

6 satellite made available to India during August, 1975 - July, 1976. Another milestone experiment was STEP (Satellite Tele-communication Experiment Project) during 1977-79 when the Franco-German satellite SYMPHONIE was relocated for a period of one year. The communication services of STEP were continued through the indigenously built geostationary satellite called APPLE (Applications Payload Experiment).

Figure 2.1-1 Vikram A. Sarabhai, whose visionary views on Space gave rise to the indigenous Indian Space Programs as implemented by ISRO.

Figure 2.1-2. One of the INSAT series of satellites. These third generation national satellites (4 in the current series) represent the means to procure a indigenous continued capability in meteorology and communications

The operational phase of satellite communication began with the development of multi-purpose INSAT (Indian National Satellite) system (see e.g. figure 2.1-2). While the first generation INSAT-1 series satellites were procured from Ford Aerospace Corporation, indigenously built and improved INSAT-2 series satellites have been launched using commercial launchers. The current INSAT series, INSAT-3 is aimed toward achieving continuity of services, increased capacity and improved capability. The first 4 in the series, INSAT 3A to 3D weigh ~ 2,500 kg and INSAT-3E is planned to be larger and more capacity with a mass in GEO of about 3,500 kg. INSAT-3D is meant for augmented meteorological observing capabilities, while INSAT-3A to 3C emphasise improved/ enhanced communications capabilities. INSAT-3E will move to much higher power and will have Ku band channels. R&D activities to support the INSAT-3 series are focussed on shaped beam antennas, linearised solid state amplifiers and higher power TWTs, onboard processing, contiguous multiplexing, low noise receivers and improved packaging and harnessing.

3. Earth Resources Survey

Remote sensing activities in ISRO began in the early 70's with the aerial survey of a coconut plantation in Kerala. In 1979 the LANDSAT receiving station was set up at NRSA, Hyderabad. This was followed up by ISRO's programme to develop and launch remote sensing satellites. Bhaskara-I and Bhaskara-II launched in 1979 and 1981 by Soviet launchers were the two experimental earth observation satellites. The encouraging results of Bhaskara projects prompted the indigenous development of the next generation of Indian Remote-sensing Satellites (IRS-1), four of which have been launched from abroad.

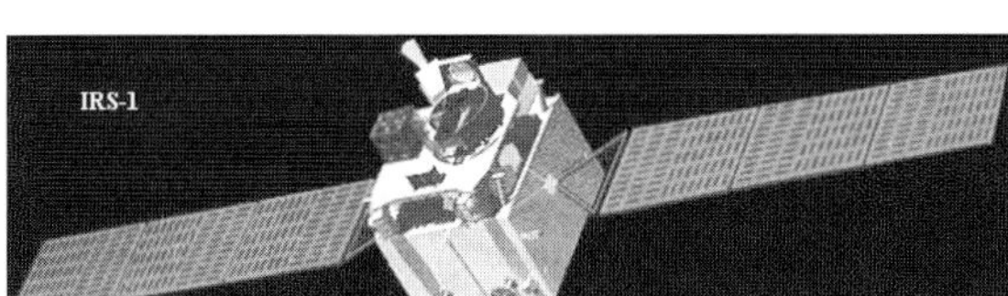

Figure 2.1-3. IRS-1 Satellite. The Indian Remote Sensing satellite system is a critical tool in the integrated sustainable development.

The applications of the satellite remote sensing data to a number of areas of vital national interest have already progressed considerably. Satellite remote sensing along with collateral socio-economic data is the key technique to carry out integrated sustainable development. ISRO has initiated to launch IRS-P series of satellites (see figure 2.1-3) on board its own Polar Satellite Launch Vehicle (PSLV). The series is designed to test and demonstrate new and advanced technologies and applications of space based remote sensing such as those of ocean resources monitoring, climate change, cartography and space sciences. India also provides opportunities for flying payloads/micro satellites from other countries on the IRS-P buss.

IRS-P3 and IRS-P4 satellites have been successfully launched carrying a number of new generation payloads such as Wide Field Sensor (WiFS), Modular Opto-electronic Scanner (MOS), X-ray astronomy telescope, Ocean Colour Monitor (OCM), Multifrequency Scanning Microwave Radiometer (MSMR) etc., and small satellites. As part of the International Geosphere Biosphere Programme (IGBP), ISRO has initiated a programme to study the climate and global change phenomena primarily using satellite remote sensing data of land-ocean-atmosphere along with support from ground based and balloon borne observations and modelling efforts.

4. New Technology Development

ISRO has successfully launched geostationary satellites (GSAT) using indigenously developed GSLV from Sriharikota. The second such mission GSAT-2 was launched on May 8, 2003 carrying payloads for C-band, Ku-band communication and space science related to solar x-rays and imaging of ionospheric total electron content.

Another experiment has been successfully conducted to launch a geostationary satellite using PSLV from Sriharikota, Kalpana-1 geostationary satellite (figure 2.1-4) launched using PSLV carries a complement of advanced meteorological sensors.

The meteorological payload of INSAT series of satellites primarily consists of the Very High Resolution Radiometer (VHRR) in visible (0.55 - 0.75 m) and infrared (10.5-12.5m) channels. The cloud imageries, cloud vector winds, outgoing long-wave radiation (OLR) and sea surface temperature data. The data obtained have been utilised to improve the understanding of monsoon and cyclone predictions. The data have also served as very valuable inputs for the numerical simulation by the national centre for medium range weather forecasting (NCMRWF). INSAT-2E satellite has, in addition to VHRR, a water vapour channel (5.7-7.1m) and a CCD camera in visible (0.63-0.69m), near IR (0.77-0.86m) and shortwave IR (1.55-1.70m) bands with 1 km resolution.

Figure 2.1-4. KALPANA-1 Satellite. These modern satellites represent the meteorological satellite series of ISRO. The earlier versions of ISRO's meteorological satellites were associated with the INSAT series of Communications satellites.

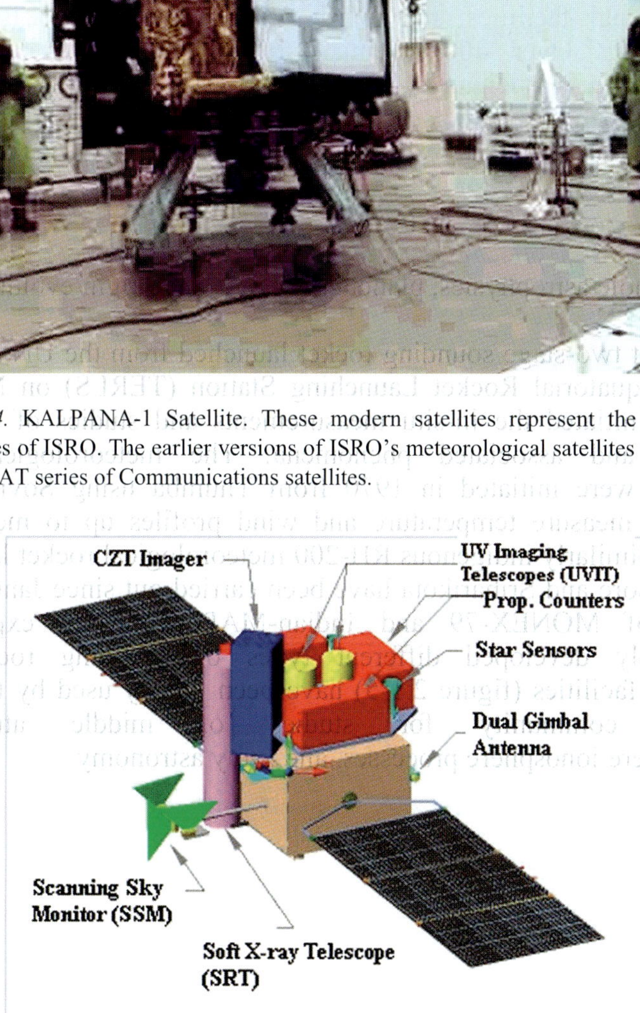

Figure 2.1-6. The future ASTROSAT mission spacecraft (launch 2006). This multi experiment mission with X-Ray and UV/Optical payload is foreseen to make an important contribution to multi-wavelength astrophysics. The combined capabilities of ASTROSAT, together with the ground observatories for the optical, infrared and radio wavelengths of ISRO will supply a unique capability for the study of variable astrophysical phenomena on many time scales.

Space Sciences

The initial thrust for Indian space programme came from the requirement of scientists to carry out investigations in aeronomy as well as in astronomy by conducting space-based experiments. Facilities have been established by developing technologies for conducting sounding rocket and balloon borne scientific experiments. Later scientists have also utilised the payloads of opportunity on board ISRO's SROSS, IRS-P and GSAT satellites. ISRO has also established complementary ground based national facilities, like MST radar, atmospheric LIDARs, IR telescope etc. Over 400 scientists from research institutions and universities are participating in this programme. The major areas of investigation in space sciences have been high energy cosmic ray variability using neutron/meson/Cerenkov monitors, equatorial electrojet and spread-F ionisation irregularities, ozone, aerosol and cloud phenomena, middle atmospheric radiation, dynamics and electrodynamics, solar physics, IR, X-ray, γ ray astronomy, neutron star and black hole astrophysics, planetary science and origin/evolution of life etc.

The first two-stage sounding rocket launched from the UN sponsored Thumba Equatorial Rocket Launching Station (TERLS) on November 21, 1963 initiated the in-situ measurements and studies of equatorial electrojet and associated phenomena. The meteorological rocket soundings were initiated in 1970 from Thumba using Soviet M-100 rockets to measure temperature and wind profiles up to mesospheric altitudes. Similarly indigenous RH-200 meteorological rocket launchings from Balasore and Sriharikota have been carried out since January 1979 as part of MONEX-79 and Indian-MAP campaign experiments. Indigenously developed different types of sounding rockets and associated facilities (figure 2.1-5) have been widely used by the Indian scientific community for studies of middle atmosphere, thermosphere/ionosphere processes, and x-ray astronomy.

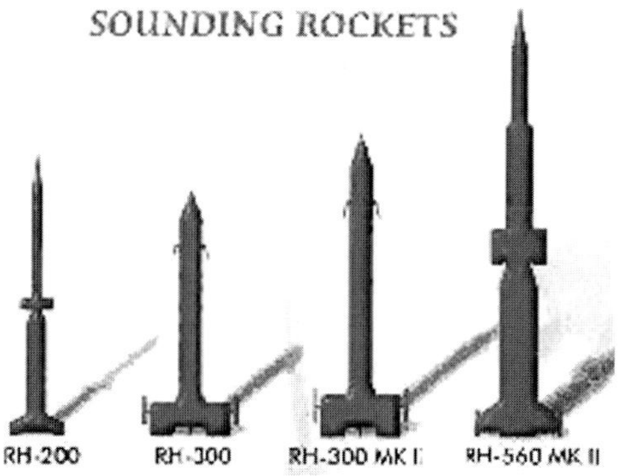

Figure 2.1-5. Sounding Rockets Developed by ISRO

Major campaigns with high altitude RH-300 MK-II and RH-560 MK-II rockets carrying RPA, LP, electric field probe, chemical vapour release (Ba/Sr), ion/neutral mass spectrometer experiments have been conducted to investigate Leonid meteor shower ionisation, F-region ionisation depletions/holes and spread-F irregularities.

Indian scientists had the unique opportunity to carry out space science experiments using dedicated small scientific satellite system called Stretched Rohini Satellite System (SROSS). Two satellites weighing ~100 kg each in the series SROSS-C and SROSS-C2 have been launched during 1992 & 1994 using indigenous ASLV-3 carrying the Retarding Potential Analyses (RPA) payload for in-situ observations of the ionospheric parameters and Gamma Ray Burst (GRB) experiment for detecting celestial gamma ray burst events.

The spectra of more than 25 GRBs observed with SROSS-C2 experiment indicate bimodal distribution in duration with separator duration of about 0.6-6s. In the area of cosmic ray astronomy scientists from TIFR, PRL and ISRO conducted experiments on board SKYLAB mission during 1974 and on space shuttle/SPACELAB-3 during 1985 to study the anomalous cosmic ray fluxes using stacks of lexan polycarbonate plastic detectors.

The IRS-P3 satellite launched during March 1996 carried the Indian X-ray Astronomy Experiment (IXAE) consisting of

(a) 3 pointed mode proportional counters (PPC) in the energy range 2-20 keV and,

(b) an X-ray sky monitor (XSM) operating in the energy range 2-10 keV.

A number of x-ray binary stars including Cyg X-1, GRS 1915+105 and Vela X-1 have been observed by IXAE. Studies have been carried out related to problems in contemporary astrophysics and cosmology including the verification of accretion disk model of the neutron star and black hole. A multiwavelength dedicated satellite (figure 2.1-6) called ASTROSAT for studying high energy stellar astronomy is planned to be launched from Sriharikota by the end of 2006. The satellite is likely to carry the following payloads:

(a) a soft and hard x-ray (0.3-80 keV) detector and imaging systems (LAXPC, CZT detector arrays, conical scan mirrors telescope),

(b) a medium energy (2-10 keV) scanning X-ray sky monitor (SSM), and

(c) a twin UV/optical telescope.

ASTROSAT will be a complex space science mission constituting the IRS buss, 5 science payloads developed by different research institutions, complementary ground based observations in optical, IR and radio regions and science data utilization by a number of university groups.

Chapter 2.2

ASTRONOMY IN THE PHILIPPINES

B.M. Soriano, Jr. and C.P. Celebre
Philippine Atmospherica Geophysical and Astronomical Service Administration
Quezon City
soriano@yahoo.com

Abstract An overview of the historical development of Astronomy in the Philippines is given. The current activities to bring modern Basic Space science into the curricula is being discussed and we indicate the plans for the future.

1. Historical Background

Work in astronomy in the Philippines started in 1897. It was one of the functions of the "Observatorio Meteorologico de Manila" (OMM), which became a government agency on 28 April 1884. Prior to this, it was a private undertaking that began on 1 January 1865. The astronomical dome that housed the telescope of the Observatorio is its most prominent edifice. The observatory performed not only meteorological and astronomical services but also seismological and terrestrial magnetism services. Its astronomical activities were mostly limited to timekeeping and observation of solar and stellar phenomena.

The OMM became the Weather Bureau in 1901 with its observatory in Manila as its central office. During the last world war, the astronomical observatory was destroyed. It was only in 1954 that a new observatory was constructed within the University of the Philippines campus in Quezon City. It has remained there until the present time, now under the Philippine Atmospheric, Geophysical and Astronomical Services Administration (PAGASA), as the only government astronomical observatory. Still performing basically the same functions as its forerunner in Manila, it has, however, updated its equipment. In the past, it disseminated time signals throughout the country, including the meteorological stations, by radio and also operated, beginning in 1970, automatic picture transmission (APT) equipment, the predecessor of the

W. Wamsteker et al. (eds.), Developing Basic Space Science World-Wide, 49-54.

modern telefacsimile or telefax equipment. The APT enabled the reception of satellite and other images over long distances. The PAGASA stopped operating the APT in 1978 because of the interference it creates with other communication network of the government. From 1954, the observatory has been quiescent, but for the construction of a planetarium in the PAGASA Science Garden in Quezon City in September 1977.

2. Resources and Activities in Astronomy

At present, there are two main institutions which undertake activities in astronomy in the Philippines. These are the PAGASA and the Manila Observatory. The former evolved from the OMM while the latter is a private institution under the Ateneo de Manila University in Quezon City. The resources and activities of the PAGASA in astronomy are described in the following paragraphs.

2.1 Resources

The PAGASA, now within the Department of Science and Technology (DOST), is organized in such away that it has nine branches or divisions. Each branch/division is composed of sections. An astronomy section is contained within the Atmospheric, Geophysical and Space Sciences Branch (AGSSB), which is basically the research and training arm of the PAGASA. The Astronomy Research and Development Section (As- RDS) of AGSSB is staffed by 18 professional-level personnel, including employees who hold a college degree, and by 16 sub-professional-level employees.

As mentioned earlier, the PAGASA has an astronomical observatory and a planetarium, both of which are managed by the AsRDS of AGSSB. The observatory has seven (7) telescopes of various sizes, the largest of which is the 30-cm reflector telescope. The other telescopes are of the refractor type, two of which have an aperture of 15 cm and 10 cm respectively, while the rest have a 7.5-cm opening. The largest of the last six telescopes is permanently installed at the observatory.

There are six (6) other telescopes, which are also of the refractor type and have an aperture of 7.5 cm, that are distributed among the meteorological stations in the field of the PAGASA. In addition to these, 38 telescopes are owned by private individuals and institutions. The biggest of these is the 58-cm reflecting Newtonian telescope at the Science Centrum, which is partly being supported by the DOST.

Figure 2.2-1. PAGASA's Astronomical Observatory, The Philippines, 1998; left to right: T. Kogure, M. Kitamura, C.P. Celebre, D.M. Soriano, Jr., D. de la Cruz, M. Raymundo (Courtesy C.P.Celebre).

Not one of the telescopes of the PAGASA, particularly the one installed at the Astronomical Observatory, is equipped with any photographic or recording instrument. A suitable camera is borrowed, when necessary, from the Planetarium of the Philippine Museum, whose curator is a former PAGASA employee. In addition to telescopes, the astronomical observatory is equipped with a quartz clock for determining the time to the nearest tenth of a second. The clock is at least ten years old and needs to be replaced by a modern timekeeping equipment. The transmitter that was used to transmit time to field meteorological stations has been in-operational for almost five years and has not been replaced due to budgetary constraints. A Solar Radiation Center, under the AsRDS, coordinates and supervises 52 solar radiation stations, all of which measure sunshine duration. Some of the stations (number is in parentheses) also record and/or measure global radiation (14), diffuse radiation (3), ultraviolet radiation (1), and infra-red radiation (1).

2.2 Activities

At present, the principal activities of the Astronomical Observatory and the AsRDS of the PAGASA consist of sunspot and lunar occultation observations and solar radiation measurements. It also makes observations of the satellites of Jupiter, the transits of Mercury, comets and the other planets. From 1980 to 1984, the observatory made observations of variable stars. Since then, it has observed astronomical phenomena that are seen in the country such as the three total solar eclipses in this century in which the path of totality crossed the Philippines.

The observations and measurements made by the AsRDS are sent to international data centers and are returned to it in the form of astronomical publications, in addition to other related subscriptions. The astronomy section, in turn, publishes data that are derived from computations based on the publications it receives and subscribes to. Among its important and widely used publications are the Almanac for Geodetic Engineers, Sun and Moon Rise/Set Tables for the Philippines and for Selected Fishing Areas, Daylight Duration Table, and Calendar (Julian) Data.

The other important activity of the astronomy section of the PAGASA is timekeeping. The agency was designated by law to be the official timekeeper of the Philippines. It used to perform this function through the radio transceiver. However, because its radio equipment is in-operational, even the meteorological stations in the field have not been receiving time signals for nearly five years. The only means of disseminating time at present is the telephone and one has to call the Astronomical Observatory for the time and other related information.

Lastly, the PAGASA engages in the promotion of astronomy, including space science, in the Philippines through the public shows in its Planetarium and the publication of posters. Its staff conducts and/or serves as resource speakers in lectures and seminars on astronomy and star-gazing and telescoping sessions in various parts of the country. It coordinates and collaborates with other agencies or institutions in this field, including for example, the organization of astronomical societies in colleges or universities throughout the Philippines.

Currently, there are ten astronomical societies with a total membership of about 300 students. It should be stressed, however, that there is only one educational institution that offers a course related to space science in the country. No college or university gives a full course in astronomy in the Philippines.

All of the preceding activities are performed by personnel who do not have any formal education in astronomy. The knowledge in astronomy that they possess is obtained through in-service training courses conducted by the agency and through books and publications that were procured, usually from overseas sources.

The training courses in astronomy in the Philippines are very infrequent, mainly due to the lack of resource persons or lecturers in this field. Only two courses during the last 15 years have been held. The first of these was conducted by the Manila Observatory for the professional-level "astronomers" of the PAGASA in 1981 while the last one was held by the agency for astronomical observers in 1993. The rarity of training courses can likewise be attributed to the lack of opportunities in astronomy in the Philippines, as there is a very limited number of positions in the AsRDS, the sole unit that deals in astronomy with the PAGASA.

3. Prospects for the Future

Given the preceding information on the past and present resources and activities in astronomy in the Philippines, it is not difficult to make a projection of the status of this science in the near future. In the next decade, the development of astronomy in the country will remain as lethargic as it has been for the past four decades, unless a drastic positive change is implemented.

In the olden past, astronomy was regarded as a pastime, until it became the foundation of modern physics and mathematics. Through observations of the motions of Earth with respect to the sun, stars and other celestial bodies, it was shown that Earth is round and is not the center of the universe. Thus, astronomy became a science.

Today, there are tremendous benefits that are being derived from astronomy and space science. Developed nations have reaped the rewards of these fields, and some developing countries are beginning or have begun to do so as well. Satellites and powerful computers have become common factors in a host of activities such as communications, data and information exchange, remote sensing for environmental monitoring, disaster preparedness and prevention, and resources assessment.

The Philippines is endowed with rich natural resources which include a relatively highly literate population. It would do well to follow the example of the countries that have benefited from the utilization of astronomy and space science. It has to enhance the knowledge and skill of individuals engaged in the two sciences and to improve its facilities, especially its instruments and equipment. It has to provide better opportunities to its present astronomical staff so that they will produce more and perform more efficiently and effectively.

Potential workers in astronomy are many compared to the present staff of the PAGASA, thanks to the organization of astronomical societies in the universities and colleges. These individuals have to be encouraged to pursue a career in the science through attractive salaries, benefits and other opportunities for growth.

The PAGASA has been afforded by the United Nations Office for Outer Space Affairs several opportunities to know and keep abreast of the latest developments in astronomy and space science through its participation in international workshops and congresses. The agency has made and partially implemented plans for modest improvements of its facilities with its scarce financial and human resources. It needs more of these resources to make a big stride in astronomy and space science.

The Philippines has embarked with plans to become a newly industrializing country by the year 2000. The present national leadership is now aware that one of the imperatives to attain this goal is to give a higher priority in fostering the development and application of science and technology in the country. The future, indeed, looks brighter for astronomy and space science in the Philippines.

Chapter 2.3

THE PRESENT STATUS OF ASTRONOMY IN SRI LANKA

I. Medagangoda
Arthur C. Clarke Institute for Modern Technologies
Sri Lanka
indika@accmt.ac.lk

Abstract The experience of the ACCIMT in Astronomy application and the plans for the future are described.

Introduction

As an outcome of the United Nations/European Space Agency workshop held in Colombo, Sri Lanka, in 1995, hosted by the Arthur C Clarke Institute for Modern Technologies (ACCIMT), Sri Lanka was bestowed a 45-cm Cassegrain telescope by the Japanese government under their Cultural Grant Aid programme. Although astronomy used to be taught in schools in Sri Lanka a few decades ago, for reasons that are unclear it has been excluded from the education curriculum until the recent past. For this reason, astronomy was rarely discussed among the general public, and was also not a subject much in demand in universities.

At the time the telescope was installed at the Arthur C Clarke Institute, there were only two scientists working at the Institute. Although they had graduated from renowned universities, neither had a formal education or hands-on experience in astronomy. Having identified the need for the scientists at the Arthur C Clarke Institute to receive some exposure to astronomy abroad, the Besei Astronomical Observatory (BAO) and the Japanese International Co-operation Agency (JICA) came forward to offer training opportunities in Japan. As a result of this, one scientist was trained in photometry. The other was able to receive training at the South African Astronomical Observatory (SAAO) in South Africa. On return to Sri

W. Wamsteker et al. (eds.), Developing Basic Space Science World-Wide, 55-60.

Lanka, these scientists started working with the photometer of the 45-cm reflector. To their disappointment, they found that the photometer could not be brought in focus at the Cassegrain focus of the telescope. They realized that the cause of this problem could be a position mismatch of the secondary mirror of the photometer's telescope. After much correspondence with the telescope manufacturer, the staff members were able to obtain the necessary spacers to rectify the problem.

Having fixed the focusing problem, they found that the telescope site was not suitable for photometric observation because of a high tendency for localized cloud formation in this area (microclimate), and also because of severe light pollution. For these reasons, they abandoned the idea of conducting photometric observations at this site. The fact that the telescope site was inappropriate for photometric observations was highlighted in the report given by the Japanese astronomer, Mr. Osamu Oshima of Besei Astronomical Observatory, who visited ACCIMT last year.

1. CCD Camera Installation

Since the Institute was unable to conduct photometric observations, the only remaining alternative was to try spectroscopic observations. The spectrograph equipped with the telescope used an old photographic method and was bit cumbersome for the novices at ACCIMT. At this juncture, in August 2000, JICA (Japan) and Besei Astronomical Observatory made a donation of materials including two Santa Barbara Instrument Group (SBIG) charge coupled device (CCD) cameras, a laptop computer, astronomical software and books, of about US$8,000 in value, to the Arthur C Clarke Institute to improve the telescope facility. Under the guidance of a Japanese astronomer, Mr. Ohshima, the staff was able to replace the old photographic system of the spectrograph with one of two CCD cameras.

The solar spectrum was the first spectrum to be captured through the ST 7 CCD camera in the vicinity of Hα. We used a spectrograph with a Fe Hollow cathode lamp to calibrate the solar spectrum. We also obtained few spectra of bright stars such as Arcturus, Spica and Vega, in the same wavelength region. We were also able to obtain the spectrum of a star called Delta Scorpii and observed the strong Hα emission line.

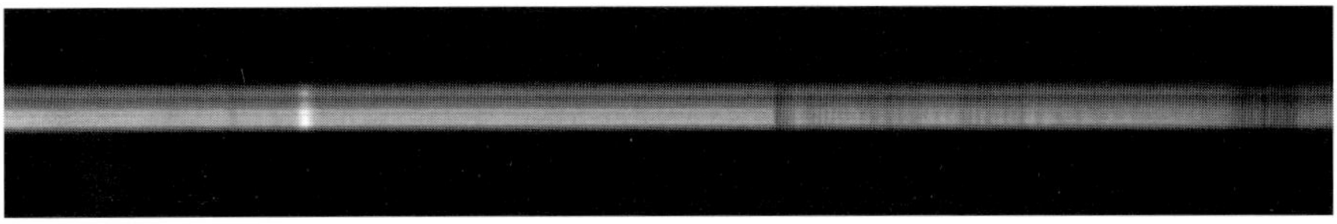

Figure 2.3-1. Spectrum of Delta Scorpii

As none of the staff were familiar with the reduction of CCD images, they had to start from scratch. The image-processing software "IRAF" was successfully installed on a Linux platform and the staff is currently learning how to use it.

An ST 9E CCD camera was first tried out on the telescope in October last year for direct imaging. Magnificent colour images of planets, nebulae and a few galaxies are among our image gallery (see http://www.accimt.ac.uk/space.html).

2. Astronomy Promotional Programmes

With a view to uplifting astronomy education on the island, ACCIMT launched a grass-roots astronomy awareness programme aimed at schoolteachers and children. The programme included activities such as night sky observation camps and dissemination of astronomical information. ACCIMT also decided to set up astronomical societies within schools to streamline activities relating to astronomy conducted by ACCIMT. About 200 school astronomical societies have been registered at ACCIMT so far. Star maps and almanacs prepared by ACCIMT have been distributed among the societies free of charge over the last four years.

About 18 night sky observation camps have been successfully conducted in rural areas of the island, where the most underprivileged people live.

In addition to the above-mentioned activities, ACCIMT regularly conduct lectures and demonstrations for schoolchildren who visit the Institute.

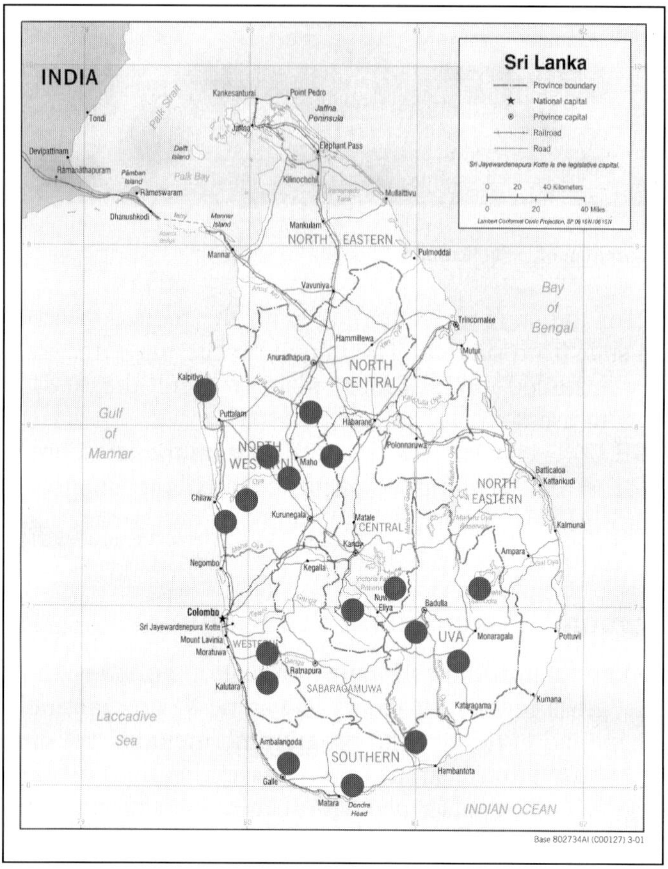

Figure 2.3-2. Locations at which night sky observation camps have been held to date

3. Financial Grant

The Foundation for the International Non-Governmental Development of Space (FINDS) in the United States of America made a grant of US$55,000 for the improvement of space-related activities on the island. ACCIMT purchased a 12" Meade Schmidt Cassegrain telescope with an alt-azimuth mount, with the aim of conducting CCD photometry and CCD imaging at a remote site with a dark night sky. As this telescope has an alt-azimuth mount, an equatorial mount that is suitable for Sri Lanka had to be constructed in order to use this telescope for long-exposure CCD imaging. We were able to design and construct an equatorial wedge for this telescope locally, thereby saving the cost of importing one.

Figure 2.3-3. Images of Jupiter and Saturn taken with the digital camera at ACCIMT. It is clear that such type of images are of considerable quality and could become very useful to maintain a permanent whether record at a scale of 1 arcsec of the very dynamic atmospheres of these planets.

Figure 2.3-4. The library of the ACCIMT in full use by students and visitors. The ACCIMT library is not restricted to the institute only but is open to students from all academic environments, as well as the interested public. Also a remote access to the catalogue is made available on the Internet.

The Institute is also planning to use the FINDS grant to conduct a telescope-building workshop for schoolteachers in the future.

4. Steps to be Taken in the Future

The telescope should be relocated to an appropriate site for astronomy. Staff members at the Institute have identified a few potential sites and are in the process of monitoring their climatic conditions

ACCIMT intends to continue to work collaboratively and closely with foreign astronomers in the future.

Chapter 2.4

A SHORT HISTORY OF ASTRONOMY IN INDONESIA[1]

B. Hidayat
Bosscha Observatory
Lembang, Indonesia
bhidayat@as.itb.ac.id and bhidayat99@hotmail.com

Abstract The history of astronomy in Indonesia is summarized. The historical connections to European astronomy, which were interrupted by the Second World War (WWII), are shown.

Introduction: Transplanting Science

Russel (1934) noticed that the fine and great Dutch work on astronomy has been wholly or partly theoretical, for climatic reasons. Neither in latitude nor in climate are the Netherlands well fitted for great optical telescopes. This national situation leads naturally to the discussion of observations obtained elsewhere. The transfer of astronomy by the Dutch to Indonesia during the first quarter of the twentieth century would conform to the third phase of Basalla's (1967) concept of science activities and its transmission, except during the interruption associated with WW II. This was a disjunction along a single thread of events. According to Hins (1950), three main events blended together to give the impetus to find a powerful observatory in the southern hemisphere.

- The first was the need to open up the southern sky for astronomical research. This had been stimulated by the research of J.C. Kapteyn at Groningen in the Netherlands. During the first two decades of the twentieth century he had convincingly shown the importance of statistical astronomy in order to study the structure of our Universe. His paper (Kapteyn, 1922) on the "*arrangement and motion of the*

[1] Adapted from: Under a Tropical Sky: a History of Astronomy in Indonesia (Hidayat, 2001).

W. Wamsteker et al. (eds.), Developing Basic Space Science World-Wide, 61-76.

sidereal system" reflects the culmination of his work. Here it was shown that the lack of understanding of the structure of our universe was partly a result of the relatively small amount of information that could be gathered from the southern hemisphere. At that time, a large number of observatories were found in the northern hemisphere. It was shown to be necessary to have more southern stations with which modern astronomical observations could be pursued and frontiers could be pushed.

- The second reason was the enthusiasm of Dr. J.G.E.G. Voute, a civil engineer-turned-astronomer. He was assistant in Capetown, South Africa before his appointment to the Meteorological Office in Jakarta in charge of timekeeping at the invitation of van Bemmelen. His interest in double-star astronomy was triggered when he became associated with the IAU Commission of Latitude Determination and was appointed observer at the Leiden Observatory. His 6-year stay (1913-1919) in Capetown not only allowed him to get more insight in astronomical observation, but also was decisive for his outlook of the future. Upon his return to Indonesia he decided to follow on, eventually leading him to establish an astronomical observatory.
- Last but not least, are the accomplishments of Mr. K.A.R. Bosscha, who was at that time, the administrator of a flourishing tea estate in Java. His support was the most influential non-astronomical factor that helped to shape and materialize the idea of an astronomical observatory in the tropics. Bosscha has become a legendary figure in the scientific community in Indonesia through his generous support when Voute approached him for help. A complete account of Mr. K.A.R. Bosscha has been written by van der Hucht and Kerkhoven (1982).

These were the three factors, that finally paved the way for the foundation of the "Dutch East Indies Astronomical Association" in 1920. Its primary task was to build an observatory and, in a widest sense of the words, to promote the progress of astronomical science in Indonesia. Initially, through the use of a small 7-inch telescope donated by Professor de Sitter of Leiden Observatory, parallax observations started at Lembang, now the home of the Bosscha Observatory, in the early 1920s.

When the observatory was still in statu nascendi Professor van de Sande Bakhuysen, retired Director of Leiden Observatory, donated his extensive collection of books on astronomy, at the request of Bosscha. These were soon to become the core of the Bosscha Observatory library which, thanks to the Leids-Kerkhoven Bosscha Foundation, which maintained its support up until now, is still the prime astronomical library in Indonesia. In the 70's the Warner and Swasey Observatory, in

Cleveland, Ohio, USA helped the Bosscha Observatory to update its library material. This contribution extended and expanded far into the 80's with its generous supply of the photographic plates needed for galactic structure research.

In 1928, after 5 years of hard work, the main instrument was finished. This was the 60-cm double-refractor made by Carl Zeiss, Jena. The technical details concerning the instrumentation will not be covered here, as it has already been described elsewhere (Voute, 1933). It is however necessary to note that the completion of the 60-cm double-refractor marked the beginning of "modern" astronomical research in Indonesia. Pannekoek (1929) anticipated (in his account of Bosscha Observatory) *"... the legacy has been left to the (Bosscha) Observatory destined to extend hospitality as "research associates" to astronomers from Europe and America ....* ". He also saw that *" ... the (Bosscha) Observatory bids fair to develop into an important center of scientific research ...* " In later years we have seen that some of these goals were achieved, however much still has to be done to uplift its status.

1. Controversies

Looking back at the founding process one should realize that the road to the establishment of a southern observatory was not as smooth as it appears at first sight. The Director of the Royal Magnetic and Meteorological Observatory, Professor van Bemmelen, continually complained to de Sitter about the poor conditions for observing, even though they were far better than those at Leiden. Professor van Bemmelen appeared rather reluctant to see another scientific institution, besides his own, faring well. He also believed that it would be better if the new observatory would be part of his institute (Pyenson, 1987).

Another problem that arose at that time was caused by Voute's anticipation of an independent director in the "Indies," even though he saw no other alternative but to follow closely the research programs defined in Leiden. De Sitter did not actually object to this, as he was not interested in a power-sharing arrangement with Groningen. In practice, he anticipated that the director in Java would make all the local decisions and so ... "*it would be better to allow and recognize his power*" (Pyenson, loc.cit.).

Another problem relates to the fact that the Leiden-South Africa connection had become very strong. By that time many Dutch astronomers had worked in South Africa and served in the observatories there. As a matter of fact in 1923, de Sitter arrived at a bilateral exchange agreement with Dr. Innes of South Africa to make facilities at Johannesburg and Leiden available to astronomers from both institutions.

These agreements facilitated the northern astronomers to observe the southern sky, but not necessarily from Java. Although the Leiden astronomers Professor de Sitter and Professor Hertzsprung, corresponded correctly with Voute and with Bosscha, both during pre-and post-construction of the observatory in Java, the exchange was formalistic only. Even at the opening ceremony of the Observatory in Lembang, no astronomer from Leiden was present. Besides the photographic observations of visual double stars by the first director Voute (1934), several other interesting investigations were conducted at the new observatory, which was baptized "Bosscha Observatory". Russel (1934) in his "Tribulations of the Tropical Astronomer" reiterated the work of Voute on Proxima Centauri. Based on 40 plates being taken with the Bosscha Zeiss refractor and measured at the Observatory, Voute established without doubt that the faint star, discovered earlier by Dr. Innes, which shares the fast proper motion of Alpha Centauri, is indeed related to Alpha Centauri despite its large distance of about 10° away from Alpha Centauri. The parallax determined by Voute to Proxima Centauri was 0".746 ± 0.006. Voute has put much of his efforts for parallax measurements. He discovered in 1930's 2 faint dwarf stars nearer than 16 light years.

In 1928 Prof. Pannekoek of Amsterdam spent several months at the observatory to extend his work on the drawing of brightness isophotes of Milky Way to the southern sky. Many years later this monumental work was employed as a basis of the well known "Skalnate Pleso" Atlas, which is indispensable for the student of galactic structure. The observatory honored the visit of this prominent scientist by naming one of its "observer houses" as "Rumah Pannekoek" (Pannekoek House). Many of the visiting astronomers still stay there when they are making observations at Lembang.

Other visitors were Dr. Paul Ten Bruggencate of Göttingen, Germany and Dr. E.A. Wallenquist of Uppsala, Sweden. The first studied globular clusters and carried out a photometric and spectroscopic studies of variable stars. Dr. Wallenquist introduced theoretical studies at the observatory. He investigated some southern galactic clusters by means of their colors and studied the distribution of stars, in particular those which are found in the Sagittarius and Ophiuchus regions. Another noted visitor of pre-war time was Dr. E.A. Kreiken from the Amsterdam School of Astronomy whose interest and contribution ranged from theoretical studies of physical double stars to the general distribution of stars in the Milky Way, in particular in Scutum. Later he determined the Oort Constant and derived from that the potential energy of the galactic system (Kreiken, 1950). The value he found of $1.5x10^{59}$ ergs, has been adopted in Allen's Astrophysical Quantities (third edition, 1973).

Some notes on personal accounts may be relevant in this context. Wallenquist after a successful undertaking of 10 years at Bosscha returned to Sweden to become an astronomy professor at Uppsala. In his account (1987) he described the good weather condition prevailing at the Observatory, averaging about 50% suitable for observing. This percentage is of course higher in the dry season during April to October and reaches about 80%. Kreiken did not stay long at the Observatory and became a teacher in the HBS (Dutch Secondary High School) in Central Java. After WWII he became head of the Department of Higher Education in Jakarta, before moving to Ankara, Turkey. There he established a school of astronomy. According to van Albada-van Dien (1994) he was instrumental in UNESCO's donation of a new telescope to the Bosscha Observatory in 1959. Volume IV (1940) of the Annals that were published until the middle 1940's, contained the results of double star observations conducted by many astronomers. Besides Voute and the three scientists already mentioned above, G. Simonov participated in observational programmes from 1936.

In 1939, A. de Sitter and later Chr. Martin, both from the Leiden Observatory joined the staff of the observatory until the WWII broke out. Unfortunately both of them perished at young age during the war. Voute reached retirement age in the same year and was replaced by de Sitter.

In 1943-1945 Prof. Masasi Miyadi, a young Japanese army captain who later (in the 80's) became Director of the Tokyo Astronomical Observatory, was in charge of the Bosscha Observatory under the Japanese military government. His presence saved the observatory from improper treatment that might have otherwise taken place. According to his account (Miyadi, 1979), he also put Voute back to work in Lembang because of his administrative ability, rather than leaving him in a prisoner-of-war camp. Miyadi's act was not a single event as there were many other Dutch nationals in the Japanese occupied territory who were asked to continue administrative duties under the supervision of the military. This persisted at least until 1943 when the war situation became intolerable and Dutch nationals could no longer be employed. Voute, however, worked at the observatory until the end of World War II, and so could receive the "transfer" of authority back from Miyadi. The account of this unofficial transfer was very sentimental. The Second World War and subsequent conflicts were disastrous to the observatory.

Not only did three leading members of the staff perish during this time, but also many instruments were found to have deteriorated. According to Hins (1950) who was sent to Indonesia from the

Netherlands to assist in restoring the observatory, he found a jungle-like condition on the observatory's ground. A discussion soon took place between him and Ir. Poldervaart, of the Army Triangulation Brigade and Professor Dr. H. P. Berlage, the Director of Meteorology and Geodesy Office in Jakarta, to decide the necessary steps to put the observatory into effective use again. It took almost three years to rehabilitate the observatory and by August 1949 photographic observations were resumed. Elsa van Dien (later van Albada-van Dien), and G.B. van Albada (from Amsterdam) took charge of the observatory. After World War II the last had a post-doctoral position with J.J.Nassau at the Warner and Swasey Observatory where he learned the science of near infrared spectral classification of faint stars.

Nassau and Van Albada (1947) established luminosity criteria for low-dispersion spectra for stars from K0 to K5. Within the constraints of time and situation van Albada decided to continue the double star observation and laid down important plans for the 1950's. In this period Iwan Nikoloff of Bulgaria joined the staff and conducted some double-star observations. Santosa Nitisastro served as an assistant astronomer.

2. Education of Astronomy and the New Telescope

The importance of teaching astronomy at the university level was realized as far back as 1948. The idea received the attention of the Dean of the Faculty of Mathematics and Natural Sciences, of the University of Indonesia, Prof. M. Th. Leeman. As Dean he could make the necessary arrangements to transfer the observatory from a private institution, namely the "Dutch East Indies Astronomical Association", into the university confine. In October 1951 the observatory was officially transferred to the Faculty of Mathematics and Natural Sciences (later to become the Faculty of Mathematics and Natural Sciences of the Bandung Institute of Technology), and G.B. van Albada was nominated the first Professor of the Department of Astronomy.

This marked the beginning of an association with the University, which, not only would ensure the supply of astronomers, but also allowed astronomy to be introduced in the physics curriculum at the University. It was also the first time that Indonesia incorporated a curriculum of astronomy in its tertiary level of education. Before that time astronomy was taught only at secondary level schools, under the name of "Cosmography". The model of astronomical education was derived from the Dutch university system of education commonly practiced in Leiden, Utrecht, Amsterdam and Groningen, in which the first three years heavily emphasized courses in physics and mathematics, with a relatively small amount of general astronomy. Only after the student passed successfully the first three years could he embark on specialized astronomy courses to obtain a "doctorandus" degree (Masters equivalent). This part of the study required about 2 to 2.5 years. This system of education persisted until about 1965, when tertiary education in Indonesia became more structured in the Anglo-Saxon model, where "bachelor" and "master" degrees were introduced.

Realizing the isolated position of Lembang as far as astronomy is concerned, young graduates are still stimulated to obtain a higher degree in astronomy abroad. This practice allows younger colleagues to experience contact with a wider circle of activities and thus avoid a certain amount of intellectual "inbreeding", often associated with such isolation. With this attitude in mind the astronomers purposefully seek to establish cooperation with astronomers from other countries.

A new direction of research started in 1959 when the new Schmidt-type telescope (f/3.5, 51 cm) became available to the observatory. The telescope was donated through a UNESCO project grant that was started in 1951, envisioned by van Albada who saw the suitable geographic and climatic situation for galactic research work. It involved the cooperation of astronomers, opticians and engineers from Indonesia, the Netherlands, especially Prof. Oort (Katgert-Merkelijn and Damen, 2000), and the United States. The optical parts for the telescope where donated by UNESCO and were prepared at the Yerkes Observatory of University of Chicago under supervision of G. P. Kuiper (originally from Leiden). The mechanical parts contracted by the Indonesian government with the engineering firm Rademakers in Rotterdam under the supervision of B.G. Hooghoudt. The assembly, adjustment and testing were completed successfully by V.M. Blanco (Blanco, 1961) of the Case Institute of Technology's Department of Astronomy (Cleveland, USA), and by Pik Sin The (The, 1961), now at Amsterdam. Officially inaugurated on May 28, 1960, the telescope was used immediately to survey Hα emission-line stars in the direction of the galactic center and the third and fourth quadrants of the Milky Way. Equipped with a 6° objective prism it yields the dispersion of 312Å/mm at Hγ. The large field of view (5° × 5°) which makes it eminently suitable for survey work and for galactic structure research. This led to the decision to concentrate research on two main topics:

1. Survey of Hα emission-line stars along the southern Milky way belt. Particular emphasis was given to bright and dark nebulae in which T Tauri type stars were expected to be found. In subsequent years surveys of other type of emission-line stars, in particular of planetary nebulae in the quadrant which includes the galactic center, was done by The (1963). He describes the result of the surveys in the field of

NGC 6334, NGC 6357, the dark complexes of Aquila and Scutum, and the dark line of Lupus and Scorpio. The result of the survey in the areas of southern Coalsack by Hidayat (1962) and in the nebulosities surrounding the star M Cen by The (1963) revealed 78 new emission-line stars. The aim and the result of the planetary nebulae surveys have been discussed by The (1967). Some ten new planetaries have been reported by him. In the 1990's surveys for emission-line stars at higher galactic latitude were initiated.

2. Study of large scale structure of our Milky Way though the study of stellar space distribution. Two main approaches have been pursued:
   - To determine the space distribution of groups of stars which have approximately the same absolute magnitude. Emphasis have been given to study the distribution of dA and gK stars (age between 1 to 5×10^8 years). These objects are thought to be the "fossil" of the galactic spiral arms. As part of a larger project, Hidayat and Radiman (1973) presented the results of their study in the Palomar-Groningen Variable Stars Field no.2. A similar study in the Field no.3 is still under way.
   - Making use of the high luminosity of giant M stars which can be observed al large distances, a study of galactic structure from the distribution of this type of stars was made by Vleeming and The (1975). Their studies were concentrated in the Warner and Swasey Observatory's Southern Luminosity Function Fields nos. 13, 14 and 15, in which they determined the spatial distribution of giant M2 to M9 stars up to the distance of approximately 6 kpc. The extinction in the direction of these fields was also established. In later years further areas were surveyed for M stars (Raharto et al.,1984; Raharto, 1996). It is worth to mention that the Schmidt telescope was also used to probe the South Galactic Pole by The and Staller. They discovered numerous M dwarf stars.

Non-survey type work that has been done with the Schmidt-type telescope included photometry of galactic clusters and stars groupings. Occasionally studies of individual objects, whose spectra were of special interest, were also carried out. These were made possible because of the fine spectra that the Schmidt telescope produces at 312 Å/mm at Hγ. On nights when the "seeing" is best (slightly more than 1") the spectra obtained offer great detail. One can easily recognize strong lines that are usually only detected in slit spectra. Examples of these are the spectra of R Vol and HD 114586 described by The (1968). Spectroscopic studies of Novae at various stages of decline have also been undertaken (e.g.

Hidayat and Wiramihardja, 1978). The large collection of Bosscha plates has been summarized by Hidayat et al (2000).

Systematic studies of old nova have been started to determine accurately the position of the dwarf novae, and to search for dwarf novae among the catalogued Mira stars. Astrometric work on the galactic cluster Blanco 1, located near the Galactic Pole, has been published. The material and moral support from the staff of the Warner and Swasey Observatory, during the difficult periods of the late 1960's must be mentioned here. This important cooperation is manifested by numerous papers (see Mc Cuskey, 1976). In the 80's photographic materials were received through the Leids Kerkhoven Bosscha Foundation. Its support enabled the Observatory to continue to undertake the double stars (Jasinta, 1997) as well as galactic structure studies.

3. International Cooperation and the Future

The interest of Indonesia in international cooperation may be illustrated by its readiness to host international meetings in the country. In 1963 an international symposium on "*Stellar Photometry and Spectral Classification*" was organized by Pik-Sin The and hosted by the Institute of Technology, Bandung, the parent institute of Bosscha Observatory. In 1973 the IAU entrusted Indonesia to organize an IAU School for Young Astronomers (Andersen, 2003). Twelve participants from 5 countries participated. Similar activities were organized again in 1983, this time held in conjunction with the 60th Anniversary of the Bosscha Observatory and the long total solar eclipse (6 min.) that could be seen from many parts of the Indonesian Archipelago. The 1983 School for Young Astronomers was attended by 20 astronomers from 6 countries. The school had 14 teaching staff from 7 countries. Prof. Cees de Jager of Utrecht, in his capacity as the general secretary of the IAU has been very supportive of the cause of astronomy in Indonesia.

Only in 1979 became Indonesia a member of the IAU. Two years after that the Second Regional Asia-Pacific Meeting on Astronomy was organized in Bandung. This meeting, which was attended by over 115 astronomers, discussed the problems of galactic structure, extra-galactic astronomy, binary and variable stars and education on astronomy. These were all areas of great relevance for the Bosscha Observatory. Since that time the issue of astronomy education has become an essential feature in regional meetings.

The year 1983 commemorated the 60th Anniversary of Bosscha Observatory. At that time, the IAU sponsored a Colloquium (No. 80) on Double Stars, with emphasis on physical properties and generic relations (Hidayat, Kopal and Rahe, 1984). For both occasions as well as many other opportunities the Leids-Kerkhoven Bosscha Foundation has strongly supported astronomical endeavors. The ICTP of Trieste and UNESCO also supported this meeting. Such cooperative gestures are very important for the community of astronomy in Indonesia and it is more then appropriate to reiterate here our appreciation.

Another aspect of international cooperation is the link between astronomy in Indonesia and Japan that started in 1978. These cooperative activities were structured around a programme to study galactic structure. This collaboration was later extended into other branches of astronomy. With the help of Japan Society for the Promotion of Science (JSPS), the University of Kyoto and the University of Tokyo, some Indonesian astronomers obtained their Ph.D. in astronomy. The subject areas include: Solar Astronomy, Binary and Variable Stars, Galactic Structure, Celestial Mechanics and Dynamics of Star Systems (see Kogure and Hidayat, 1986). A noted example that should be cited is the Indonesian-French treaty on education which in the years of 1970-1990 has produced astronomers with higher academic degree in many branches of astronomy, from cosmology to planetary science.

On every Mars opposition since 1986 until 1994, Dr. Iwasaki and some Indonesian astronomers observed the behavior of Martian Polar caps (for example, Iwasaki et al,,1988 and 1993). The observations were made with the Zeiss double-refractor (D=60 cm; f/17). A special planetary camera was designed to extend the focal ratio of the telescope up to f/150. Observing Mars at the Observatory, due to its favorable geographic position is not new to the Observatory. Before the space-exploration era, the Observatory engaged in Mars observations during the opposition in 1954 and 1956 and in the 70's the associates of Dr. G.P Kuiper (Lunar and Planetary Laboratory in Arizona) observed Mars and Jupiter (Larson, 1971) using the Zeiss Double Refractor of the Bosscha Observatory.

A revival of Dutch-Indonesian cooperation in Astronomy has been established formally in the framework of the "Indonesian-Netherlands Astrophysics" (INA) agreement, which has operated under the cultural treaty between the Indonesian and Dutch governments. Four main areas of research areas have been targeted for cooperative work: studies of visual binaries; studies of spatial distribution of the hottest and coolest stars; high- and low-energy astrophysical studies of evolved massive binaries; investigation of shock waves and particles in curved space time corresponding to plasma and condensed media

Under this cooperation, Indonesian students have had the opportunity to study in the Netherlands to obtain higher degrees in astronomy. The Indonesian community of astronomy is naturally looking forward to the realization of other projects, which would enable them to expand their activities and research. There existed already a close link between the Astronomy Departments in Indonesia with Amsterdam, lead by E.P.J. van den Heuvel (see Sutantyo, 2001, Dewi, 2002) as well as with Utrecht and with Groningen. They have been working respectively on the binary millisecond pulsars with relatively high surface dipole magnetic field and on neutron stars companion in close binary systems. These astronomers are employed by Institut Teknologi Bandung. It is pleasing to notc that there are quite a number of astronomers who have attained various degrees in astronomy are working with the National Space Institute (LAPAN). Ratag, for example, who obtained his degree at the University in Groningen, is now working on long-term climatic predictions at LAPAN while maintaining his research on planetary nebulae. Some other former students of the Department of Astronomy are pursuing their careers at Aerospace Industry. They are in charge of orbital calculations and celestial mechanics.

Since the Schmidt telescope in the 1960s, a new 45 cm telescope was commissioned in 1989. Thanks to the efforts of Professors Kitamura, Kogure and Ishida this telescope was obtained through the Indo-Japan project (see also Kitamura, 2003). It is dedicated to photometry of close binary stars. A versatile instrument, this telescope is also used for spectroscopy. A special compact spectrograph is designed by Malasan and the group at the Gunma Observatory in Japan (Malasan et al, 2001), and the telescope continues to provide significant astronomical data.

Another anticipated project is the acquisition of a large optical telescope. The original idea was launched by van der Hucht in 1984, who suggested that a 2.5 meter class telescope would be suitable for Indonesia. The idea was received with great and sincere enthusiasm in Indonesia. This concept needs considerable further development to implement and should be followed up.

Astronomy education for would-be astronomers as mentioned above is offered at the Department of Astronomy, Institut Teknologi Bandung. The curriculum at the department is closely linked with the curriculum at the Faculty of Physics and Mathematics. Enrollment to the Department of Astronomy (which belongs to the Faculty of Natural Sciences and Mathematics) is 10-15 students every year. The enrollment takes place after the students have attended the so called common first-year at the institute. At the second year, astronomy students take 4 credits (which is equivalent to 4 class hours per week) of astronomy classes, out of 18 credits per semester. The fourteen other credits include 2 credits civics (maintained throughout the studies) and the rest are equally shared by Mathematics and Physics. The proportion of astronomy courses increases in later years. The credits from Mathematics and Physics are reduced in such a way that students take 3 credits for Mathematics and Physics at the last stage of their study (after 4 years at the University). Courses and works on theoretical astronomy are now part of academic activities.

Astronomy students are encouraged to attend (and take credits) on modern physics courses given by the physics faculty. Astronomy is considered a part of the physical sciences and therefore is also on the curriculum at other universities and the Institute of Teacher's Colleges across the country. The instructors in the latter institutions usually acquired their knowledge of astronomy through postgraduate courses that are provided by their respective institutions. For example, at the Education University of Indonesia (UPI), astronomy is provided for those who attend postgraduate training for the Doctorate of Education in Physical Sciences. The programme not only employs astronomers but also uses the facilities of the Observatory.

It is worth mentioning here the efforts of the Ministry of Education and Culture. Over the course of years, high school teachers from all over the country in the field of physical geography and sciences are required to attend refresher courses for up to 2 months. The institution that is charged with conducting these courses has always included astronomy in its curriculum. The course material given here is multi-disciplinary, with an emphasis on new findings in the fields of Earth and Space Sciences (i.e Basic Space Science).

Astronomy is also taught at secondary and primary schools as part of physics or geography. At the last grade of primary schools and in all grades of secondary education, astronomy is taught for 2 to 4 hours a week respectively, usually in the context of Earth and Basic Space Science. Of course, there is a varying degree of the depth in the ways teachers teach their students. In rural areas, the teaching of natural sciences is generally less intensive than in big cities, where exposure to new findings are also brought to the attention of the people through media coverage. On the other hand in rural areas there exists often a significant awareness of the importance of the matters addressed in the Basic Space Sciences, due to the fact that life in rural areas is much more directly connected to many different sidereal phenomena.

In 1980 a new regulation on teaching at tertiary education has been adopted. This includes a requirement to offer courses on basic natural sciences for non-sciences departments at all universities, during the first 2 semesters of university education. The philosophy underlying this principle is to provide a sense and appreciation of science as culture. In the designed course, Earth science and basic space sciences constitute up to 5% of the curriculum content. The rest will include modern biology, chemistry and ecology. The paramount question is how to achieve an equally high standard across the nation. This initiative has been tried at some selected universities, the result of which will be scrutinized carefully.

4. In Retrospect

Ever since the explosive growth of science in the Western Europe in the 17th Century, it has become a permanent engine of progress and a determining force in the pursuit of progress. The hegemony and exploration of the west in general, and the Dutch in particular, has brought Earth, meteorological, biological, and astronomical sciences into the confines of Indonesia. In principle, the people of the east have accepted the message of modern science and technology, but it can not be denied that the implementation and the effective application of modern science, at times conflicts with traditions and cultural attitudes. As a consequence of this, the existing social structure in Indonesia may affect the growth of basic space science in the country. 50 years ago, when Indonesia proclaimed its independence, a set of good functioning scientific institutions was present. However, the time to recover the needed manpower -absent in these years- to steer the high and noble enterprise of basic science, has negatively influenced the continued development in science and technology. In the meantime, during the same period, new avenues of activities and new priorities relating to social institutions and welfare have grown into our working domain.

It was mentioned earlier that from the cradle of Western Europe, science has been following a wave of western domination. In this evolution, science as many other cultural propagations, has not been exempted from the general rule. Filtering processes must have been operating at many cultural borders, whether it be called adaptation or adoption, reluctance or enthusiasm, they manifest themselves in the ways science progresses in science-receiving nations.

Indonesia has now opted for science as a way to elevate national status, because of the conviction that it is important for mankind. Therefore Indonesia built universities and institutions of higher learning. From a number of about 5 before independence, to more than 100 state universities and many more private higher-learning institutions now. The needs of the population of Indonesia are increasingly being accommodated. With these, the path is open for Indonesia to participate more active and aggressively to the development of knowledge.

Of the basic space sciences, astronomy has found its place in Indonesia, thanks to the efforts of many individuals and, in particular, the cooperation of some Dutch and American astronomers.

Like any other field of basic space science it can grow and develop only, if it finds good soil in which to prosper. The soil is the people of Indonesia, who will become interested and support the cause of astronomy. I have therefore included in this contribution the background in which "modern" astronomy evolved in Indonesia. Some new innovations, as well as the future of the Schmidt telescope, have been discussed at length (Chapman et al., 1995). The work at Bosscha Observatory and the problem of education were reported by Raharto (2001). Education of astronomy in Indonesia reached its golden jubileum in 2001. Reflection about it has been presented by Taufiq Hidayat et al (2001). It is a sin of omission if the present report does not mention the involvement of many astronomers who are involved in public teaching by providing the newest results on astronomy research through public media and public lectures. This is to raise public interest in science.

It is important to maintain the best of the past, but the sign of progress is development of thought, leading eventually to action. If astronomy and basic space science is to survive for the next decade and beyond, we should take account the dynamic process and look to the challenges of the future, about which some thoughts are presented here. Cooperation with our Japanese Colleagues in the past 15 years has greatly assisted in the growth of the study of modern astronomy in Indonesia. Theoretical works (Sutantyo, 2001, Premadi et al., 2001, Ardi Erliani et al., 1999) as well as solar observations (Herdiwijaya, 2003), to name a few, have found their place in our astronomy community. Ardi Erliani together with her colleagues are working on, among other things, the heating of galaxy disks and on reformulation of scattering in an expanding universe. Premadi dealt with problem of light propagation in homogeneous universe. There are naturally many things still to be done in order to bring our efforts at the cutting-edge of modern basic space science. A Project as the World Space Observatory (Wamsteker and Shustov, 2003) for astronomy would generate many benefits, also for developing countries, extending well beyond the scientific aspects alone. Other important developments are the implementation of data centers for basic space science (see also Albrecht, 2003). Also the earlier suggestion of the formation of an IUE data center (Wamsteker et al., 2000) requiring not much more than a modernization of the computer communications infrastructure for the University and the Observatory, would have impact extending well beyond the basic space sciences alone. Thanks to the Leids Kerkhoven Bosscha Foundation and the Institute of Technology, Bandung, the realization of the first steps is within reach by 2003. All such proposals are appealing because of their scientific merit and have, at the same time, important implications for the general population of the region and the nation. Like many other observatories located not far from a populated regions (10 km from Bandung, a city of 2 million people) the Bosscha Observatory has to look for an innovative routes to dedicate its time for astronomy as well as to reach an independent competitive level in the future (see Senja et. al., 2000). Therefore, suggestions such as the World Space Observatory and the others are matters require serious attention of the funding agencies, also at the National Indonesian level.

References

Albrecht,R., 2003, this volume.

Andersen, J., 2003, this volume.

Ardi, Erliani, T. Tsuchiya, S. Inagaki, 1999, Publ. Astron. Soc. Japan, **51**, 211.

Basalla, G., 1967, in " The spread of Western science", Science**, 156**, 611.

Blanco, V.M., 1961, "The Bima Sakti", Courier, Unesco.

Chapman, J., Russel Canon, Sandra Harisson, B. Hidayat, 1995, Eds. "The Future Utilization of Schmidt Telescopes", PASP Conf. Series, Vol. **84**.

Dewi, J.M.D., O.R. Pols, G.J. Savonije and E.P.J. van den Heuvel, 2002, MNRAS, **331**, 1027.

Herdiwidjaya, D., 2003, in "The 8th IAU Asian-Pasific Meeting", Vol. **II**, 427.

Hidayat, B., 1994, in "400-year Dutch-Indonesian Relation. Erasmus Huis, Jakarta. Unpublished.

Hidayat, B., 2000", in "Encyclop. for Astronomy in non-Western Culture", Editor: H. Selin, Dordrecht: Kluwer Academic Press, 2000.

Hidayat, B., Z. Kopal, J. Rahe, 1984, AP&SS, **99**, no.1 and 2.

Hidayat, B., M. et al., 2000, in " Treasure – Hunting in Astronomical Plate Archieves; Eds.: Peter Kroll; C. la Dous, Hans-Jurgen Braner, Acta Historica Astronomie, **6**, 116.

Hidayat, T., et al., 2001, "50-year Astronomy Education in Indonesia", U. Press, ITB, Bandung, Indonesia

Hins, C., 1950, in Een Eeuw Natuurwetenschap in Indonesie. Konink Natuurk Vereeniging, Batavia (in Dutch), Eds. P.J.W. MacDonald, P. Braber, and H.G. Derx.

Iwasaki, K., Y., et al., 1988, JGR, **95**, 14, 751.

Iwasaki, K., Y., et al., 1993, Proceedings of the 26th ISAS, on Lunar and Planetary Sysmposium, Tokyo.

Jasinta, D.M.D, 1997, in "A Data Base for Visual Double Stars Observed at Bosscha Observatory", ASP Conf. Series **130**, 135.

Katgert-Merkelijn, J. and Damen, J, 2000, "Jan Oort, astronomer", Leiden University Library.

Kitamura, M.,2003, this volume.

Kogure, T., and Hidayat, B., 1985, in "Galactic Structure and Variable Stars", University of Kyoto, Dept. of Astronomy, Kyoto,

Larson, S., 1971, "The Bosscha Observatory in Indonesia:, Sky and Tel., **42**, no.2.

Malasan, H.L., M.A. Senja, B. Hidayat, M. Raharto, 2001, in "Preserving the Astronomical Sky" Editors: R.J. Cohen and W.T. Sullivan III, IAU. Symp., **196**, 147.

Malasan, H.L., et al., 2001, in Proceedings Indonesian-German Conf., 159, Bandung.
McCuskey, S.W., 1964, A.J., **69**, 104.
Miyadi, Masasi, 1975, The transfer of the Bosscha Observatory, Astronomical Herald, **68**:319 (In Japanese; kindly translated by Prof. N. Sato).
Nassau, J.J. and van Albada, G.B., 1947, Ap.J., **106**,200.
Premadi, P., H. Martel, R. Matzner, T. Futamase, 2001, Ap. J. Suppl., **135**, 7.
Pyenson, L., 1989, Empire of Reason, Brill, Leiden.
Pyenson, L., 1990, Why science may serve political ends: cultural imperialism and the mission to civilize, . Wissenschafts-Geschichte, **13**, 65.
Raharto, M., 1996, " Study of Galactic Structure Based on M-type Stars", Dissertation, University of Tokyo.
Raharto, M., 2001, in Special Session of the 24th IAU GA., Ed. A.H. Batten, San Francisco, Astronomy for Developing Countries, ASP Conference Series **SPS**.
Raharto, M, K. Hamajima, T. Ichikawa; K. Ishida; and B. Hidayat, 1984, Annals of the Tokyo Astronomical Observatory, 2nd series, .**XIX**, No.4.
Russel, H.N., 1934, Scientific American **150**, 180.
Sutantyo, W., X.D.Li, 2000, Astron. Astroph., **360**, 633.
The, Pik-Sin, 1961, Bosscha New Schmidt Telescope, Contributions from the Bosscha Observatory, **17**.
The, Pik-Sin, 1967, in "Planetary Nebulae", Eds: P. Osterb-rock and C.R. O'Dell, Reidel Publ., Dordrecht, pg 36.
van Albada-van Dien, E., 1994, in The Future Utilization of Schmidt Telescopes, Eds. J. Chapman, R. Cannon, S. Harrison, and B. Hidayat, San Francisco, Astronomical Society of the Pacific, ASP Conference Series, **84**, 15.
van der Hucht, K.A. and Kerkhoven, C.L.M., 1982, Zenit, July-Aug., **292** (in Dutch).
van der Hucht, Karel A., 1984, in Double Stars, Eds. B. Hidayat, J. Rahe and Z. Kopal, AP&SS, **99**, 409
Voute, J., 1933, Ann. v.d. Bosscha Sterrenwacht, Gebrs. Kleijne & Co, NV, vol. **I**.
Wamsteker et al., 2000, AP&SS, **273**,155.
Wamsteker, W. and Shustov B.M., 2003, this volume.
Wallenquist, E., 1987, "Over het leven op Bosscha Sterrenwacht bij Lembang een halve eeuw geleden", Orion, Nov., pp.6-7 (in Dutch).

Chapter 2.5

SPACE ASTRONOMY IN CHINA

Ai Guoxiang and Yan Yihua
National Astronomical Observatories
Chinese Academy of Sciences, Beijing 100012
aigx@bao.ac.cn and yyh@bao.ac.cn

Abstract We review the development of space astronomy in China. We mainly introduce the researches relevant to the developments of instruments for space astronomy in China. In particular, we describe in details the Chinese Space Solar Telescope. It should be emphasized that the data analysis for the observations taken from spacecrafts is an important part in the space astronomy and Chinese astronomers have made significant progresses in analyzing data taken from foreign spacecrafts in different branches in astronomy, which are not included in this review, however. The contents of this paper are as follows.

1. The un-completed project "Astronomy-1" satellite
2. Hard X-ray, Soft γ-ray and Infrared Astronomy by Balloon Flights
3. Cosmic Gamma-Ray Burst Spectrometers onboard Spacecrafts
4. Hard X-ray Modulation Telescope (HXMT)
5. Lunar Exploration Project of China
6. The Space Solar Telescope
7. The Long-term Plans
8. Ground-based Supporting Systems for Space Astronomy

1. The un-completed project "Astronomy-1" satellite

The space astronomy in China began in later 1970s when "Practice-1" scientific experimental satellite was launched and space technology in China was in developing. On board "Practice-1" there was a Geiger counter and Solar X-ray detector when it was launched in 1971. In 1976 Purple Mountain Observatory (PMO began to develop Chinese "Astronomy-1" satellite for Solar Maximum then (NSFC, 1997). Later, the non-solar tasks were included. The detectors included Solar Grazing

W. Wamsteker et al. (eds.), Developing Basic Space Science World-Wide, 77-93.

Incidence X-ray Telescope, Solar Hydrogen Lα Ionization Chamber, Solar Soft X-ray Gas Proportional Counter, Solar X-ray Proportion Counter, Cosmic Gamma-Ray Burst Scintillation Detector, Cosmic Gamma-Ray Burst Soft X-ray Detector, and Cosmic X-ray Source Proportion Counter. This project was, however, interrupted in 1984 due to the Governmental Policy then. The project was shifted to industrial departments and the research for space astronomy in China was actually quitted then. Nevertheless, some prototypes that passed environment experiments were able to fly on balloons. The space astronomy in China began in later 1970s when "Practice-1" scientific experimental satellite was launched and space technology in China was in developing. On board "Practice-1" there was a Geiger counter and Solar X-ray detector when it was launched in 1971. In 1976 Purple Mountain Observatory (PMO began to develop Chinese "Astronomy-1" satellite for Solar Maximum then (NSFC, 1997). Later, the non-solar tasks were included. The detectors included Solar Grazing Incidence X-ray Telescope, Solar Hydrogen Lα Ionization Chamber, Solar Soft X-ray Gas Proportional Counter, Solar X-ray Proportion Counter, Cosmic Gamma-Ray Burst Scintillation Detector, Cosmic Gamma-Ray Burst Soft X-ray Detector, and Cosmic X-ray Source Proportion Counter. This project was, however, interrupted in 1984 due to the Governmental Policy then. The project was shifted to industrial departments and the research for space astronomy in China was actually quitted then. Nevertheless, some prototypes that passed environment experiments were able to fly on balloons.

2. Hard X-ray, Soft γ-ray and Infrared Astronomy by Balloon Flights

In 1982 and 1983, the Institute of High Energy Physics (IHEP) and PMO, both of Chinese Academy of Sciences, used CsI/NaI Composite Crystal Detector and Proportion Counter respectively to measure the hard X-ray background and cosmic ray background in North China, heralded the era of balloon-borne hard X-ray and gamma-ray observations in China.

In 1984, IHEP using a CsI/NaI detector with 140 cm^2 detection area (HAPI-2) observed hard X-ray emission from Crab Nebula Pulsar. This was the first time that China had made such observations successfully. In 1985 IHEP used HAPI-2 to observe Cyg X-1 and obtained the energy spectra. In 1986 IHEP developed a CsI/NaI detector with 300 cm^2 detection area (HAPI-3) and observed Hercules X-1 in 1987.

During 1984 to 1986, PMO and IHEP developed proportional counters with detection area of about 500 cm^2 for Hard X-ray Astronomy. In 1988 during a Sino-Japanese Cooperated Balloon Flight, PMO using this detector observed Cyg X-1, Cyg 2 and the energy spectra of a possible transient source. In 1987, 1988 and 1989, IHEP used the same system to observe Cyg 1 and Cyg 3 with success. In 1990 IHEP also used the system for solar flare observations but without success due to short flight time (7 hours).

In 1993, a new balloon borne hard X-ray detection system named HAPI-4, constructed by IHEP, was put into practical operation of observation (Gu et al 1994). HAPI- 4 consists of two collimators, one high pressure Xenon MWPC (1950cmx 15cm) and eight NaI(TI) (144 x 144 x 6mm) CsI(Na) (144 x 144 x 30mm) phoswich detectors. Its FWHM of FOV is 3x3 degree. In Sept. 25, 1993, at the altitude of about 35km, two hours scanning observation was performed by HAPI-4 to Cygnus region (only MWPC was used in the flight). The scanning region was 10° by 10° with right ascension from 294° to 304° and declination 30° to 40°. The step intervals were 2.5^{o} and 2.0^{o} for right ascension and declination respectively. Then there were totally 20 scanning points, with 6 minutes for each point. The total observation time was about two hours. During the observation period, the arriving moment, energy of each detected event had been recorded. The housekeeping data were also recorded, including azimuth, elevation of the telescope orientation, the geographical position and altitude of the balloon, etc. Processing of the observation data according to the principle of object reconstruction by direct demodulation method (Li and Wu, 1994, Lu et al., 1996, Chen et al., 1998). Imaging results for Cygnus region was obtained. It is the first time to realize high accuracy space imaging observation by using non-imaging telescope, and this experiment has confirmed the possibility of new concept scanning imaging method to be used in the space observation on the orbit (Gu et al 1994).

In later 1970s Shanghai Astronomical Observatory (SHAO) of Chinese Academy of Sciences started to develop 15 cm Newtonian Infrared Telescope with a FOV of 9' in 4.6μm and 18μm wavelengths. After a test flight in 1981, it was successfully used to observe the Sun in 1982 and 1983 for two times. A brightness temperature of 4519±109K at 18μm IR wavelength was obtained. This was the first successful balloon-borne IR observation in China (NSFC, 1997). Later with an improved system the scanning observation at the Galactic center was carried out in 1985. Another two observations were carried out during the Sino-Japanese Cooperated Balloon Flights (aperture modified to 10 cm) in 1987 and 1988 respectively.

PMO and Astronomical Instrument Institute of Chinese Academy of Sciences started to develop 30 cm balloon-borne multi-channel IR and XIR telescope system (BIT). The wavelengths were 20, 40, 60 100 and 170μm ($\lambda/\Delta\lambda=2$). The gondola stable pointing control was improved to 1' with the help of star trackers and other advanced technologies. During its first flight in 1990 the IR bright object α Sco was observed but it failed to observe the planned target Comet Austin.

3. Cosmic Gamma-Ray Burst Spectrometers onboard Spacecrafts

As an important part of whole space astronomy, the cosmic Gamma-Ray Burst Spectrometer (GRBS) project was put forward several years ago. GRBS is a high energy radiation detection system which consists of three kinds of detectors that are soft X-ray gas proportional counter (thin film window)-SXRD, hard X-ray scintillation detector (HXRD) with 1 ms high time resolution, and soft gamma-ray detector (GRD) covering a wide energy range (Zhang, 1996, Gan et al., 2000).

Since 1995 PMO and IHEP have begun to design and make this spectrometer which is used to research for cosmic gamma-ray burst.

The scientific objectives of GRBS are as follows:

1. Whole energy spectrum of cosmic gamma-ray bursts.
2. Evolution of high energy radiation with high temporal resolution.
3. Spatial distribution of cosmic gamma-ray burst and its relationship with soft X-ray radiation.

The GRBS has been used to observe solar x-ray and gamma-ray bursts during the 23 cycle solar maximum. It was mounted on the *Shenzhou (Magic Vessel)* 2 spacecraft and launched into orbit on 10 January 2001 successfully. During the fly period of *Shenzhou* 2 spacecraft from 10 January to 25 June 2001, six identified cosmic gamma-ray bursts and 13 gamma-ray solar flares have been recorded. Figure 2.5-1 shows one example for solar gamma-ray burst event on 6 April 2001 (Zhang et al., 2002).

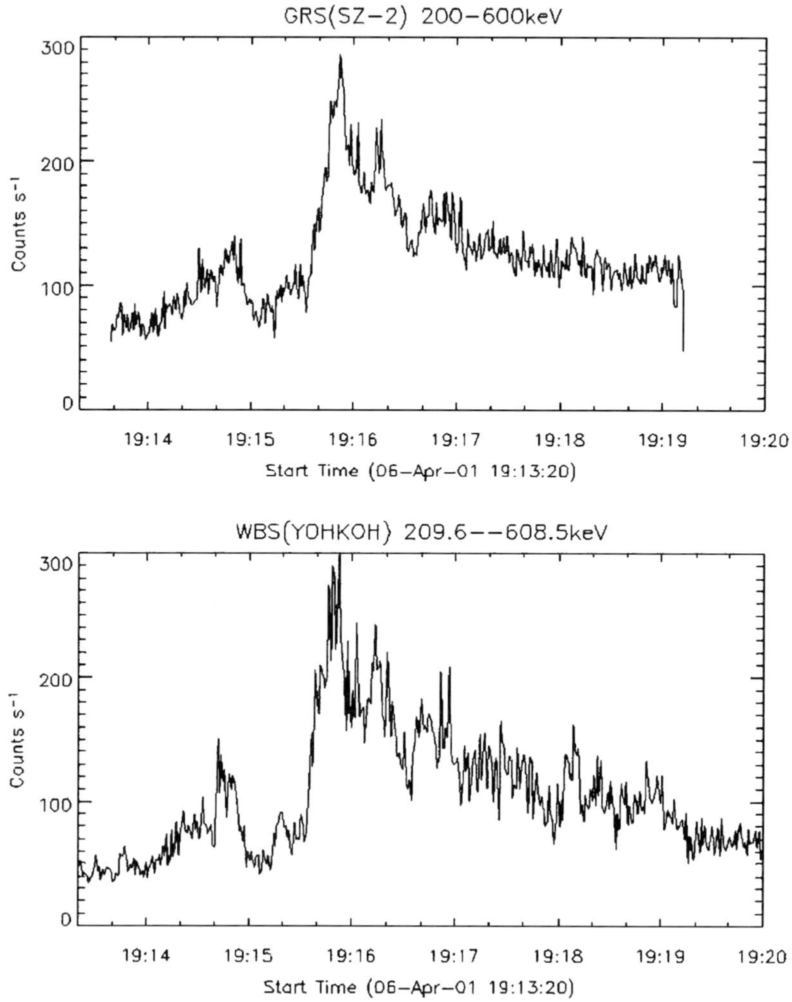

Figure 2.5-1. Comparison of the time profiles of the solar gamma-ray burst event on 6 April 2001, detected by the experiments on the Shenzhou 2(top) and Yohkoh (bottom) satellites (Courtesy of Dr. W.Q. Gan).

Ye et al. (1992) reported the observations for the solar proton events (Sept. 4, 1990, Feb. I and Feb. 2, 1991) by the cosmic ray composition detector on board the Chinese "Fengyun-1" meteorological satellite. It was the first time in China to observe solar protons, α, carbon, oxygen, nitrogen, and iron particles from the space.

It should be mentioned that two astronomical instruments: Solar Ultra-Violet Radiation Sensor and Solar X-ray Sensor were on board satellite "Practice-2" which was launched in 1981. Some observational data obtained during the flight and it helped the development of Space Astronomy in China (Lin et al., 2000).

***Table** 2.5-1.* **Summarizes the main parameters of these instruments (Gan et al 2000)**

Instrument	SXRD	HXRD	GRD
Type	Gas proportion counter	NaI	BGO
Energy range	0.2-2 keV	XD1: 10-200 keV XD2: 40-800 keV	0.3-lO MeV
Effective area	20.4 cm^2	120 $cm^2 \times 2$	45 cm^2
Energy channel	6	64	64
Time resolution	40 ms	40 ms	80 ms
Energy resolution	50% at 1.49 keV	<16% at 661 keV <25% at 662 keV	<25% at 662 keV

4. 4. Hard X-ray Modulation Telescope (HXMT)

Based on the development of methodology and instrumentation of space hard X-ray observation in China, a Chinese High-energy astrophysics mission Hard X-ray Modulation Telescope (HXMT) has been proposed. HXMT can make a hard X-ray all-sky survey with both high sensitivity and high spatial resolution for imaging, and oriented spectroscopy and time observation to scientific hot spot sources and special celestial regions as well (Li, 1998).

The detector of the HXMT consists of 18 same hexagonal prism NaI(Tl)/CsI(Na) phoswiches, the area of single module is 286cm^2, then the total detecting area is 5148cm^2. The primary detector of each module is a NaI(TI) crystal with the thickness of 3mm, and a 50m thick Be slice is used as the incident window. A Csl(Na) crystal of 3cm thick is placed at the back of the primary crystal to act as an anticoincidence shield, it stops the hard X-ray and gamma-ray background of the lower 2π solid angle (assuming the telescope is pointing upwards) and reduces the effect of Compton scattering in the primary counter. The two kinds of crystals are optically coupled. Fluorescence photons are received by a 5" photomultiplier (PMT) for each phoswich module. A pulse shape discrimination circuit (PSD) is used to distinguish the two kinds of pulses with different fluorescent decay times. In front of the primary detector a plastic scintillator with thickness 2mm is used to distinguish charged particles.

In high sensitivity observation, the angular resolution is very important to avoid the confusion of point sources. The principal objective of the HXMT is to obtain the map of hard X-ray sky by means of the direct demodulation technique (Li and Wu, 1994; Lu et al., 1996, Chen and Wu, 1998). The designed 2' angular resolution and 0.2' source location precision of the HXMT can help to accomplish the identification of sources in radio, optical and soft X-ray bands, on the other hand, images of the extended sources like the Galactic plane, the Galactic halo, clusters and SNRs can also be obtained. Imaging for a wide field and increasing exposure time during scanning observation require a wide field of view (FOV). However, a too wide FOV will damage the resolution of imaging, precision of source location and sensitivity of the instrument. After a comprehensive consideration of the requirement of imaging and flexibility of technology, the view field of the HXMT is chosen to be $5^{o}x5^{o}$ (FWHM), which consists of collimators with non-symmetric FOV of $5^{o}x0.5^{o}$ (FWHM) placed with a cross angle of 10° one another.

The background was calculated according to Monte-Carlo simulations and experimental results. The solid angle of collimator of HXMT is 7 times smaller than that of IMAGER/INTEGRAL, which is much beneficial to reducing background of hard X-ray photons. The estimated total background rate in the energy range 10-200keV is $0.02cm^{-2}s^{-2}$. The expected sensitivity of HXMT at 100keV (3σ, in 10^{5}s) is $3x10^{-7}cm^{-2}s^{-1}$ for continuum, which is three times better than that of IMAGER/INTEGRAL.

***Table** 2.5-2.* **Key Performance Parameters of HXMT Mission (Li 1998)**

Energy range	10-200 keV
Energy resolution	~18 % @ 60 keV
Angular resolution	2'
Source location(20σ source)	0.2'
Sensitivity(3σ, in10^{5}s@100keV)	$3\times10^{-7}cm^{-2}s^{-1}keV^{-1}$(continuum)
	$1\times10^{-5}cm^{-2}s^{-1}$ (narrow line)
Orbit	Altitude ~550km circular
	Inclination ~43°
Altitude	Three-axis stabilization
	Control precision±0.25°
	Measurement accuracy±0.01°
Data rate	~30 kbps
Mass	Science instrument~600kg
	Total pay load~1400kg
Nominal mission lifetime	2 years

Table 2 gives a summary of the HXMT mission. The HXMT project has been supported by the Chinese Academy of Sciences and by the Ministry of Science and Technology of China. The prototype model of

HXMT was performed balloon flights several times and obtained hard X-ray images with high resolution successfully.

5. Lunar Exploration Project of China

China is going to send a space probe to the Moon around 2005, which is named Lunar Resource Explorer (LUREX; Li et al., 1999). The LUREX will be a three-axis stabilized lunar satellite, with a mass of about 500 kg and a payload of four instruments. The robotic spacecraft will fly a polar orbit at a working altitude of 100 km above the surface of the Moon for at least half a year. Its main objective is to survey potential resources on the Moon.

5.1 Scientific Objectives

The LUREX will be a targeted lunar-only mission and put science as its highest priority. The main scientific priorities are:

1. To acquire three- dimensional atlas of the Moon’s surface, and spectrally image the (entire) Moon. The spectral imaging and 3D topographic surveys will reveal some of the Moon's enduring mysteries, and provide rough estimates of crustal composition, potential resources and topography.
2. To investigate abundances and distributions of usable elements, such as thorium, potassium, uranium, iron, oxygen, silicon, aluminum, calcium, magnesium, titanium, chromium, manganese, sodium and rare earth elements (REEs).
3. To survey distribution and thickness of dusty, loose lunar regolith, in order to estimate potential gas resource, especially helium, which will probably be a good energy resource for not only the future Moon base, but also the Earth.
4. To learn more about space environment near the Moon, including solar wind, cosmic rays, and interplanetary magnetic field. The scientific return from the LUREX mission will also make major contributions toward understanding the origin, evolution and current state of the Moon itself, even of Earth and the entire Solar System (Lin and Ouyang, 1995: Ouyang, 1997; 2000; Zou et al., 2000).

5.2 Scientific Instruments

The LUREX will survey the Moon using proven technologies. Four series of instruments will be installed in the spacecraft:

1. Optic Imaging System: including an SMHSI spectral imager and two PAN CCD cameras. The former will spectrally image the (entire)

Moon using visible spectra; the latter will be specially allocated to acquire three-dimensional pictures of the Moon's surface.

2. Gamma Ray Spectrometer: On the Moon's surface, solar wind and cosmic rays may induce such usable elements as thorium, potassium, uranium, iron, oxygen, silicon, aluminum, calcium, magnesium, and titanium to produce characteristic gamma rays. The instrument will measure the gamma ray counts and estimate elemental abundances. The data will help scientists understand geology and origin of the Moon, and may also tell future explorers where to find useful resources.
3. Microwave Radiometer and Radar Altimeter: The two instruments will be combined to measure the thickness and distribution of dusty, loose regolith on the lunar surface. The data will also probably reveal some feature of sub-ground of the lunar surface, and possible information about water ice on permanent shadow of lunar crater. The data from the radar altimeter will also be used to acquire rough topography of the Moon's surface.
4. Space Environment detection System: The system will be composed of three instruments: a cosmic ray inspector, a solar-wind plasma detector, and a Magnetometer.

***Table** 2.5-3.* **Main parameters of the LUREX(Li et al., 1999)**

MASS	ABOUT 500 kg (WITH PROPELANT)
Characteristics:	Three-axis stabilized Fixed high-gain antenna Fixed Sensors Combined active and passive thermal design
Orbit	Polar
Working orbit altitude	100 km
Life time	0.5 ~ 1 year
Data Rate:	Downlink selectable about 100 kb/s

Table 2.5-4. **Payload Descriptions (Li et al. 1999)**

SCIENTIFIC INSTRUMENTS	MASS (kg)	POWER (W)	DATA RATE (bps)
Visible Cameras (2)	20	20	87,800
Spectral imager			
Gamma Ray Spectrometer	8.6	3	690
Microwave Radiometer	15	20	200
Radar Altimeter	9	15	1000
Cosmic ray inspector	1.2	3	256
Solar-wind plasma detector	3.5	3.5	300
Magnetometer	2.5	2.4	720
TOTALS	59.8	66.9	88,196

6. The Space Solar Telescope

The development of solar space-based observations went through two steps in the spatial resolution: point resolution (sun as a star) in the 1960s-1970s, and middle resolution (1″-10″) in 1980s-1990s. The next step will be high spatial resolution (~0.1″) in the beginning of 21st century. In the research of solar magnetic field, there are two very important aspects, one is the high accurate measurement of vector magnetic field (Wang 1994,1999); the other is the character of the solar magnetic element with high spatial resolution of 0.1″~0.2″ (Ai and Deng, 1998, Wang, 1995). In the space solar telescope, the two aspects will be the most important contents that are probably achieved a break through advance. The main scientific objectives are to achieve a break through advance in solar physics through coordinated, high resolution observations of transient and steady state solar hydrodynamic and magneto-hydrodynamic processes over 2-D real time polarizing spectrum, UV, hard X-ray, soft-ray and H_{α} image, and continuous time evolution. The space solar telescope has been proposed since 1992 (Ai, 1993; Ai and Zhang, 1997; Ai 1996), the phase A (assessment study) completed in 1995-1998, the design (phase B) basically completed in the passed two years. Now the manufactory is under development.

6.1 Basic parameters of the satellite

Total weight: ~2.0 T, power: 1200W, $16M^2$ solar cell panel; orbit: altitude is: 730km, sun synchronous polar circular, 6am/6 pm nodal crossing; attitude: 3 axis stabilized, pointing to solar disc, pointing accuracy is ±5″ (solar disc), and ±40″ (ecliptic pole), attitude control stability: ±3″/s (solar disc), and no constraint (ecliptic pole), data recorder: 6 GB/day, (after data compression 5-10), data storage: 4GB; telemetry down link rate: 30Mb/s, X waveband, 8200MHz; up link rate: 4Kb/s, S waveband, 1700MHz; rocket: LM-4B; size: 5×2×2 M^3 ; mission life: 3 years; launch date: 7-9, 2004. The Chinese Academy of Space Technology (CAST) is responsible for the design and manufactory of the satellite.

6.2 Payloads

6.2.1 Main Optical Telescope (MOT)

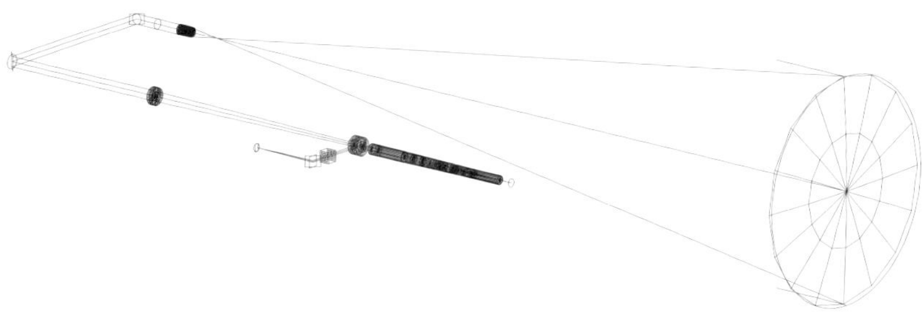

Figure 2.5-2. Optical system of SST main telescope

Optical design of SST main telescope has been principally completed. As shown in the figure 2.5-2, the present defined system consists of a F/3.5 parabolic primary with 1m in diameter, a collimator of 5 lens and a F/39 imaging objective of 2 lens. As designed result, that includes three parts of content:

- Parameters and specifications: telescope clear aperture: ϕ985mm with stop at the primary, center obstruction: 17% in diameter, system focal length: 38500mm at 633nm, filed of view: 2.8' × 1.5', detector CCD size: 2048×1024 with 0.014mm×0.014mm/pix, operating wavelength: any wavelength within 393nm-656nm spectral ranges with a band width of 2nm;

- Optical performance: The calculated PSF results for the main telescope are shown in the table and figure 2.5-3. The results demonstrate that the designed system image quality, on any working wavelength and over a filed of view of 1.6' in diameter, is very close to the limited diffraction imaging (S.R.>0.85). With such optical image quality, it will be permitted to realize its required spatial resolution of 0.15" at wavelength 600nm.

- Alignment tolerances studies: The alignment tolerances given in the table correspond to the displacement and tilt that introduce in the system a spatial resolution degradation of 0.08"

***Table** 2.5-5.* **Strehl Ratio obtained by calculation at F/39 focal plane for the telescope**

Wavelength	393.3nm	422.6nm	517.8nm	524.9nm	532.4nm	587.6nm	656.3nm	632.8nm
S.R. (0.0)	0.948	0.998	0.989	0.988	0.988	0.986	0.989	0.989
S.R.(0.707)	0.913	0.985	0.952	0.950	0.948	0.952	0.965	0.962
S.R (1.0)	0.850	0.950	0.931	0.935	0.938	0.952	0.966	0.962

***Table** 2.5-6.* **Alignment tolerance for SST main telescope**

	Title	De-centre	De-focus
Primary	4"	0.04mm	
Collimator	1'	0.04mm	0.04mm
Imaging objective	2'	0.2mm	0.1mm

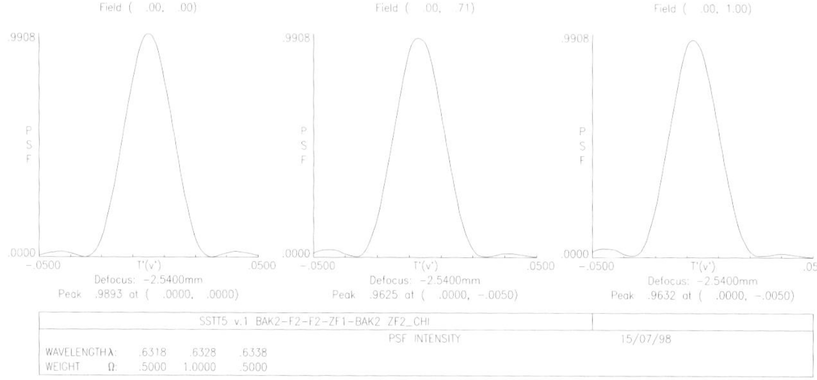

Figure 2.5-3. Calculated PSF at 633nm and with a field of view 3.24' in diameter for SST main telescope.

Assembly, integration and optical test configuration had been defined and will be carried out in National Astronomical Observatories. It concerts a vibration isolation stand onto which the main telescope and all necessary test equipment may be installed. Auto-collimation method is chosen for testing telescope output wave-front errors introduced by either disalignement, between the primary and lens collimator, or the primary surface distortion. In order to characterize wavefront of the telescope operating in full 1m apertures, a plane mirror with the same diameter as the primary of surface quality λ/60(RMS) will be used as re-reflector in the auto-collimation test optics. The gravity load effect and solar thermo effect simulation for primary at their integration on ground, has been also studied. The 1 m paraboloid primary and 1m flat reflector has been manufactured. Zerodur is chosen as mirror glass and both two mirrors demand to be polished to a surface accuracy of λ/60(RMS).

There is 2-D real time polarization spectrograph for 2-D simultaneously Stokes parameter profile, tunable wavelength range: 3900-7000Å, spectral resolution: $\Delta\lambda\sim0.075$Å (λ5300Å), 8-channels distributed in a spectral line or several spectral lines, spectral distance between two close channels can be selected for 1Δλor 3Δλor 5Δλ, correspond wavelength wide: 0.80Å or 2.40Å or 4.00Å, 8 sets of CCD with 2048×2048 and 0.075"/pix, accuracy of polarization analyzer: 2×10^{-4}. The wide band filter-graph: (3800-5500Å, $\Delta\lambda\sim30$Å) CCD: 1024pixels×1024pixels, 0.05"/pix, exposure time: 10^{-4}S, accuracy of correlation tracker: ~ ±0.01", 80Hz. National Astronomical Observatories of the Chinese Academy of Sciences (NAOC, Beijing) is responsible for this system.

6.2.2 EUV Imager for the Solar Telescope (EUT)

The Extreme Ultraviolet Telescope (EUT) is a main payload next to the Main Optical Telescope (MOT). It consists of a bundle of four normal incidence astronomical telescopes with multi-layer coatings. The original scientific objective of EUT is to observe the fine structure of the active high temperature corona. However, since the design of EUT in 1995, some missions like SOHO and TRACE have got new scientific results. One achievement made by TRACE telescope is that it revealed a wealth of fine structure in the corona, which is outside the expectation of theoretical works. Based on this scientific progress and for the reason that SST will be on aboard during the solar minimum (June 2004) and also for being more closely response to our main objective of MOT whose resolution is 0.1"/pixel, we decided to adjust the scientific objective of EUT from studying the active high temperature corona to studying the base of the corona, that is, the relatively low temperature corona with temperature around 1MK. At the time, in response to this adjustment of scientific objective, we decided to remove full disk observation in EIV band. The four EUV telescopes will all pose high-resolution observations (0.25"/pixel). Lines used will change to 171(1.0MK), 195(1.3MK), 211(2.0MK) and 128(10MK).

6.2.3 Wide band spectrometer (WIS)

The overall detector characteristics are as follows: •Soft-X-ray Spectrometer (SXS), detector: gas proportional counter Xe+CO_2, energy range: 2-30 keV, number of energy channels: 64, time resolution: 1 s, •Hard X-ray Spectrometer (HXS), detector: NaI, Ø7.6cm×2.0cm, energy range: 15-450 keV, number of energy channels: 32, time resolution: 1 s. •Gamma-ray Spectrometer (GRS), detector: NaI, Ø7.6cm×7.6cm, energy range: 0.3-14 MeV, number of energy channels: 256, time resolution: 4 s,

6.2.4 H-alpha and white light Telescope (HAT)

12cm, full disk by wedge prisms (FoV~1°), 0.5Å (λ6563Å), white-light: ~5500 Å, 2 CCD: 2048/pixels ×2048/pixels, 1"/pix,

6.2.5 Solar and interplanetary radio-spectrometry(SIRA)

The SIRA-instrument shall determine the flux density and the degree of circular polarization of the solar radio emission in two orthogonal components. The spectrometer will be designed with the following characteristics: Frequency range: 2MHz to 50 MHz, integration time: 100ms, number of channels: 2×120.

7. The Long-term Plans

In 1999, some long-term programs for next 5 to 15 years were drawn up by various departments of the Chinese Academy of Sciences or other governmental ministries and commissions. Among the programs of space sciences, Space Astronomy occupies an outstanding position. So now, we can hope that the new system of Space Astronomy will soon be set up; within 5 to 10 years, the first Chinese astronomical satellite will be launched; and advanced studies for future missions will be encouraged. NAOC have supported several proposals for concept researches.

Radio Image Interferometry at Low Frequency The earth-directed CMEs may have more impact influences on the space weather near the Earth. In order to monitor the CMEs travel towards the earth, it is desirable to use Radio Interferometry to image the CMEs at low frequencies. However, due to the influence of ionosphere, such instruments can only be realized in space. As a first step, BAO has initiated a project on the Radio Interferometry at frequencies between 1-2MHz to 30-50 MHz with minimum elements to form an Interferometer in space for monitoring CMEs at their early stages. The aim is a participation in future spatial missions for imaging low frequency radio sources (f<30MHz) detected in the terrestrial magnetosphere, the solar corona and the whole universe. The objective of this study is to review the radio-interferometry techniques, to investigate the possibility of the concept, to select a basic scientific payload for each satellite, and to develop the radio-interferometry techniques (Yan & Fu 2000, Yan et al. 2002).

Solar High Energy Micro Satellite Project This project aims at the next Solar Maximum with high spectra resolution, high temporal resolution and high spatial resolution to observe the Solar Hard X-ray and Gamma-ray radiations. It may also be used for non-solar researches to detect cosmic X-ray radiation and Gamma-ray burst observations. It is planned to apply the principle of Fourier Transform for imaging observation and use CZT detector at the focus plane.

Micro-satellite as Space-based Multi-band Variable Objects Monitor A micro-satellite carrying a 30cm UV/optical telescope, a wide-field X-ray telescope and a wide-field optical monitor is proposed as a space-based multi-band variable objects monitor. Its scientific goals are asteroseismology, searching for exo-planets ,studying the variation of young solar-like stars ,and monitoring the variation of X-ray sources and transient phenomena in the X-ray band.

Mini-ASTROD: mission concept Advances in laser physics and its applications triggered the proposition and development of Laser Astrodynamics. Mini-ASTROD is a down-scaled version of ASTROD (Astrodynamical Space Test of Relativity using Optical Devices). This mission concept has one spacecraft carrying a payload of a telescope, six lasers, and a clock together with ground stations (ODSN: Optical Deep Space Network) to test the optical scheme and yet give important scientific results. These scientific results include a better measurement of the relativistic parameters (gamma to 1 ppm, beta to a few ppm and others with improvement), a better sensitivity in using optical Doppler tracking method for detecting gravitational waves, a potential of measuring the solar angular momentum via Lense-Thirring effect and measurement of many solar system parameters more precisely. These enable us to build a more precise ephemeris and astrodynamics. The weight of this spacecraft is estimated to be about 300 - 350 kg with a payload of about 100 - 120 kg. The spacecraft goes into an inner solar orbit with several options (Ni et al., 2002).

World Space Observatory/UV Space Telescope is another topic that has been approved by the NAOC for cooperative studies (Cheng & Wang, 2001).

Acknowledgements

The work is supported by the Chinese Academy of Sciences, the Natural Science Foundation of China, Ministry of Science and Technology and the Office of 863-2. The authors would like to thank Profs. W. Gan, Y. Han, J. Hu, S. Jin, C. Li, B. Lu, J. Wang, S. Wang, L. Wu, F. Yang, S. Yang, R. Xu, B. Ye, S. Zhang and X. Zhang for their assistances. The paper has been revised based on an invited talk at The 164th Xiangshan Science Conference, May 9-11, 2001, Beijing, China.

References

Ai, G., 1993, in H. Zirin et al. (eds), The magnetic and velocity fields of solar active regions, IAU Colloquium No. **141**, 149.
Ai, G., Zhang, Heqi and Zhang Borong, 1993, in Hongqi Zhang (ed), Solar magnetic and velocity fields.
Ai, G., 1996, Adv. Space Res., **17**, No 4/5, 343.
Ai, G and Deng. Y., 1998, in Xixian Li (ed.), Puzzles of the 21st Century, Jinlin Publication Press, pg. 126.
Chen, Y., Li, T.-P., Wu, M.,1998, BAAS, **128**, 1.
Cheng, F.-Z, Wang, T.-G., 2001, Progress in UV Astronomy in China, The 164th Xiangshan Science Conference, 53.
Gan, W.-Q., Zhang, N., Ghang, J., 2000, Chin. J. Space Sci. Suppl., **20**, 16.
Gu, Y.-D., et al., 1994, Chin. J. Space Sci. Suppl., **14**, 65.
Li, C.-L.,et al., 1999, Scientific Development Reports Chinese Academy of Sciences, Beijing: Science Press, 146, (in Chinese).
Li,T.-P., 1998, Chin. J. Space Sci. Suppl., **18**, 12.
Li,T.-p.,Wu, M., 1994, Ap&SS, **215**, 213.
Lin W.-Z. and Ouyang, Z.-Y., 1995, Geology & Geochemistry **4**, 38, (in Chinese),.
Lin, Y.-L., Zhu, G.-W., Cao, J.-B., Liu, Z.-X., 2000, Chin. J. Space Sci. Suppl., **20**, 123.
Lu, F.-J., et al., 1996, Ap&SS, **115**, 395.
NSFC, 1997, National Natural Science Foundation of China, Astronomy, Science Press, Beijing, 251.
Ni, W.-T, et al. 2002, Publ. Yunnan Obs., **91**, No. 3, 123.
Ouyang, Z.-Y., 1997, Golden Triangle **2**, 20, (in Chinese).
Ouyang, Z.-Y., 2000, in "Resources, Environment and continuable Development", edited by Zhao Dapeng, Wuhan Chinese Geological University Publishing House, 132, (in Chinese).
Wang D., Ouyang, Z.-Y., 1980, Discussion on chemical constitutes of Apollo-17(70017-291) high titanium mare basalt and its origin, Geochemistry 365, (in Chinese).
Wang, J., 1994, Sol. Phys., **155**, 285.
Wang, J., et al., 1995, Sol. Phys., **160**, 227.
Wang, J., 1999, Fund. Cos. phys., **20**, 251.
Xu, Y.-L, et al., 1991, Science in China, **34A**, 209.
Xu, Y.-L., et al., 1992, Chin. J. Space Sci., **12**, 57.
Yan, Y. and Fu Q., 2000, Astrophysics Reports-Publ. Beijing Astron. Obs., No. **36**, 81
Yan Y., Fu Q., Huang G., Qin Z., & Zhang J., 2002, On the Solar Radio Spectro-interferometry at Low Frequency, COSPAR Col. Ser., **14**, 349.
Ye, Z. H. et al., 1992, Chin. J. Space Sci., **12**, 157.
Zhang H Q., 1996, Chin. J. Space Sci. Suppl., **16**, 5.

Zhang N., Tang H., Chang J., 2002, Chin. Astron. Astrop., **26**, 141.
Zou Yongliao, Ouyang Ziyuan and Li Chunlai, 2000, Chin. J. Space Sci. Suppl. II,**20**, 93.

Chapter 2.6

ASTRONOMY IN KOREA

I.-S. Nha
Yonsei Univ Observ
Seoul 120 449, Korea RP
slisnha@chollian.net

Abstract A short history (1900 – present) of the development of Astronomy and Space Sciences in Korea is given.

A Short History of Astronomy in Korea

Interaction between Korean and occidental astronomy goes back probably to the late 17th century. Astronomers of the Royal Observatory used to have regular official meetings with the Chinese in Beijing every year. While they were in Beijing, some Korean astronomers tried to make personal contact with the Jesuit astronomers who were actively engaged in astronomical work in Beijing at that time, but success was limited. The Chinese government of the time did not want the Korean astronomers to get together with the Jesuits, and guarded the latter's residences while the Koreans were there. Nevertheless, Korean envoys were able to bring telescopes, books and the occidental calendar back to their home country, even until the end of the nineteenth century. When the Japanese colonized Korea in 1910, the Royal Observatory was closed and all astronomical activities by Koreans were stopped. Koreans were not allowed to be astronomers until the end of World War II. The first modern astronomer in Korea, Dr W. Carl Rufus, came as a Christian missionary from the University of Michigan and worked in Korea for ten years (1907- 1917). He taught introductory astronomy at the Choson Christian College (CCC, now Yonsei University) for two years, 1915-17, but he returned to Michigan in protest against the Japanese policies of eliminating Bible study and reducing science courses. He did, however, invite one of the best graduates of CCC, D.W. Lee, to Michigan. In 1926,

W. Wamsteker et al. (eds.), Developing Basic Space Science World-Wide, 95-97.

Lee was able to complete a Ph.D. in Michigan, under the supervision of Rufus, and he returned to Korea to join the faculty of CCC. His career as an astronomer did not last long: he was briefly imprisoned by the Japanese and then classified as an unqualified educator. As a result Dr. Lee had to stay at home until 1945.

After 1945, for nearly three decades, astronomy was almost non-existent in Korea. During the last two decades, however, Koreans have re-established that science in their country. This was made possible by external help, particularly from the U.S.A. Our historical heritage, once comparable to that of China (particularly in the fifteenth century) was also useful. The first bachelor's programme offered in a university of the Republic of Korea was established in 1958 at Seoul National University, followed ten years later by Yonsei University. Both these institutions now offer full master's and doctoral programmes. Since 1980, at least another five universities have begun graduate programmes in astronomy. There are two professional societies: the Korean Astronomical Society, founded in 1965, and the Korean Space Science Society, founded in 1984. The first of these publishes the Journal of the Korean Astronomical Society twice yearly and Publications and a Bulletin at irregular intervals. The second publishes the Journal of Astronomy and Space Science and the Bulletin of the Korean Space Science Society, each twice yearly. The growth of Korean astronomy during the twentieth century is illustrated by the attached tables of manpower and observatories. The Republic of Korea joined the IAU in 1973[1] and now over thirty astronomers in that country are members of the Union.

***Table** 2.6-1.* **Numbers of Korean Astronomers**

	Number of degree holders			Notes
Period	Ph.D.	M.S.	B.S.	
1900-1910	0	1	0	W.C. Rufus
1911-1920	1	0	1	WCR and D.W. Lee
1921-1930	1	0	0	DWL
1931-1940	1	0	0	DWL
1941-1950	1	0	1	DWL and Tokyo U. graduate
1951-1960	1	0	0	Tokyo graduate to N. Korea

[1] The Democratic People's Republic adhered separately to the IAU from 1961 and currently has about twenty members. –The Editors

	Number of degree holders			Notes
1961-1970	0	3?	10	DWL died in 1963
1971-1980	~4	~20	~100	
1981-1990	~30	~80	~300	
1991-	~45	~100	~400	

***Table** 2.6-2.* **Optical Telescopes in Republic of Korea**

Aperture (m)	Institute	Present Status
1.8	National Observatory	Operation just begun
1.0	National Observatory	Under testing
0.75	Kyonghi Observatory	Mainly for teaching
0.75	Sejong University	Mainly for teaching
0.6	National Observatory	CCD photometry
0.6	Seoul National Univ.	Mainly for teaching
0.6	Yonsei University	PE photometry
0.4	Busan National Univ.	Mainly for teaching
0.4	Chonbuk National U.	Mainly for teaching
0.4	Ewa Women's Univ.	Mainly for teaching
0.4	I.-S. Nha Observatory	PE photometry
0.4	Kongju National U.	Mainly for teaching
0.4	National U. of Education	Mainly for teaching
0.4	Seoul National Univ.	Mainly for teaching
0.4	Yonsei University	Mainly for teaching

There is one 16m radio dish at the National Observatory

SECTION:

3. LATIN AMERICA AND THE CARIBBEAN

Chapter 3.1

THE ASTRONOMICAL OBSERVATORY OF HONDURAS: A PROJECT OF INTERNATIONAL COOPERATION

M.C. Pineda de Carias
Observatorio Astronómico Centroamericano de Suyapa
Universidad Nacional Autónoma de Honduras
Apartado Postal 4432, Tegucigalpa, Honduras
Email: mcarias@hondutel.hn

Abstract The history, current situation and achievements in Astronomy and Astrophysics as academicals fields of the National Autonomous University of Honduras is described. The first activity was the Project “An Astronomical Observatory for Central America: a realistic way of strengthening basic space science in developing countries”, because it is the frame incorporating all the initial work for the development of the different academic activities, on the basis of which the Central American Suyapa Astronomical Observatory was brought forward in a project of permanent international cooperation. The associated education, research and outreach activities are described to illustrate the well-organized model of this academic unit with international recognition. Finally we comment on the regional and international scope of this Central American Astronomy Project in Honduras.

Introduction

In Central America, the initiative to create the first astronomical observatory of modern times was raised in Honduras at the beginning of the last decade of the twentieth century. The project, "*An Astronomical Observatory for Central America: A realistic way of strengthening basic space science in developing countries*" (Pineda de Carias, 1995) contains the basis for the establishment of a regional astronomical centre. For the development of this regional project, a strategy based on permanent international cooperation has been organized around the following main activities:

- Central American Assembly of Astronomers;

W. Wamsteker et al. (eds.), Developing Basic Space Science World-Wide, 101-111.

- Central American Courses on Astronomy and Astrophysics;
- Central American Master Program in Astronomy and Astrophysics;
- Central American observational facilities; and,
- International Agreements of Cooperation.

In July 1994, the authorities of the National Autonomous University of Honduras (UNAH) approved the establishment of the University's Astronomical Observatory (OA/UNAH) as an academic unit responsible for:

In July 1994, the authorities of the National Autonomous University of Honduras (UNAH) approved the establishment of the University's Astronomical Observatory (OA/UNAH) as an academic unit responsible for:

1. Carrying out observations and scientific research projects in the field of astronomy and other related areas;
2. Managing and developing instrumentation facilities for observations, reduction and analysis, and for receiving astronomical images;
3. Providing basic astronomical services to students of different levels of the National Educational System and to general public;
4. Organizing and coordinating outreach activities and Education in Astronomy Programs for the dissemination of astronomical knowledge.

In June 1997, within the frame of the 7th UN/ESA Workshop on Basic Space Science held in Tegucigalpa, Honduras, (Report on the Seventh UN/ESA Workshop, 1997) and the participation of 80 scientists from about 30 Agencies and Organizations, the Observatory OA/UNAH was renamed to the broader and international "*Central American Suyapa Astronomical Observatory (OACS)*", opening a new Era of development and establishment of Astronomy and Astrophysics in Central America. The first act was the inauguration of the first regional observational facility, the "Rene Sagastume Castillo" Telescope, a 42 centimeters Schmidt Cassegrain optical telescope, immediately followed by the opening of the Master in Astronomy and Astrophysics program, a post-graduate program for engineers, physicists and mathematicians that, after having completed a 5 years university career, want to continue with Astronomy as a professional career in Central America.

At present, OACS/UNAH is the only astronomical observation and research centre at a Central American university. The level of organization it has achieved now it can be considered as a model for academic development for the establishment of Space Science, trough education, research, outreach and administrative affairs. The work done at OACS/UNAH is very important and highly recognized. It has contributed:

- To give presence to Honduras and Central American countries in the development of space science and technologies, including Astronomy and Astrophysics.
- To raise the Central America human resources indices for training to the highest level;
- To increase research and technical production in Central America;
- To open new fields and to create new training opportunities for new generations;
- To strengthen, in general, international cooperation links between universities, and in particular among astronomical observatories and research centers.

It still remains in agenda to insert Central America into research and space exploration projects. However, it should be recognized that some steps have already be done at the university level (OACS/UNAH), in order to continue with national science and technology policies looking to worldwide projects such as the World Space Observatory (Wamsteker and Shustov, 2003).

1. The Astronomical Observatory of Honduras

The Central America Suyapa Astronomical Observatory of Honduras (see figure 3.1-1) is located at the main campus of the National Autonomous University of Honduras (Lat: 14° 05' N; 87° 09' W; Alt: 1076 meter over sea level). The site of the observatory was chosen within the university campus in order to provide students with facilities to make their own observations and also to provide public access to the observatory so that local citizens could learn about astronomy.

The main telescope of OACS/UNAH is a Schmidt-Cassegrain LX200 with an equatorial mount (Latitude: 14 North). The size of the primary mirror is 42 centimeters and the focal ratio is f/10 or f/6. The telescope is equipped with a CCD camera and a filter wheel. Scientific journals, textbooks, astronomical catalogs and specialized software (i.e. IRAF over Linux platform) are available for data reduction and analysis of astronomical images obtained at the OACS/UNAH or in any other observational center, supported by permanent Internet connections (24 hours per day, 7 days per week).

Figure 3.1-1 Central America Suyapa Astronomical Observatory, inaugurated on 18 June 1997 at the 7th UN/ESA Workshop on Basic Space Science in Tegucigalpa, Honduras. In this picture we can see the dome housing the 42 centimeters optical telescope, the offices and lectures buildings.

Figure 3.4-1 Before (left, 1998) and after(right , 2000) views of the Location of the Astronomical Observatory at National University of Asuncion, Campus at San Lorenzo, Paraguay.

The main working activities of the OACS/UNAH are: education, research and outreach.

a) Education. In the field of education in Astronomy, OACS/UNAH offers and actively supports the following:

- *Master in Astronomy and Astrophysics.* A postgraduate regional program intended to train professionals in charge of the establishment and development of Astronomy and Astrophysics within the region, through the permanent international cooperation. Since 1998, when the program was opened, collaboration of astronomers coming from universities and astronomical observatories of Argentina, Brazil, France, Spain, United States and Mexico had been available; with some institutions having signed International Agreements of Cooperation. In visits of one or two months, these astronomers come to the OACS/UNAH as staff to teach different courses in their owns field of interest, and also as thesis projects advisors. So far, three cohorts of students, MAA-98, MAA-99 and MAA-2001, every 2 years have being taken the 20 courses that comprises the whole Master Program Syllabus, and have initiated their research project, with some of them having the opportunity to travel to Argentina, from one up to six months, to the foreign institution where their advisor belongs. In 2002 the first postgraduate students finished her work, and from here on, more students are finishing their Master in Astronomy and Astrophysics career.

- *AN-111 Introduction to Astronomy (4 credits.)* A general and optional course for students of all university careers. Observations and Models, Solar System, Stars and Interstellar Medium, and Galaxies and Cosmology, are the four units of the academic program developed in this one-semester course, in theoretical classes complemented with some practical and observational activities. Yearly, 5 members of the OACS/UNAH staffs teach to about 500 university students of about 30 careers.

- *Central American Courses in Astronomy and Astrophysics (CURCAA).* These are regional courses where participants have the opportunity to discuss relevant topics on the Sun, solar system, star formation and evolution, interstellar medium, galaxies and cosmology, and observational and data reduction techniques. These courses also promote the interchange of ideas and experiences among staff and students working in Astronomy and Astrophysics. Between 1995 and 2001, six CURCAA have being developed, each one in the Central American countries of Honduras (1995), El Salvador (1996), Guatemala (1997), Panama (1998), Nicaragua (1999) and Costa Rica (2001). In a second cycle, the CURCAA returned to Honduras in 2002, to continue to El Salvador in 2003.

  These CURCAA are programmed within the frame of the Central American Assembly of Astronomers (AAAC, in Spanish), a regional organism already recognized by the International Astronomical Union (IAU) to promote the development of Astronomy though the permanent international cooperation. The AAAC has its own rules and by laws, and sessions are scheduled in periodicals dates, having change their Board every two years.

b) Research. As a results of the level reach at the OACS/UNAH, several research areas have been defined and began to produce some results, to a national and international scale:

- *Education in Astronomy.* Beginning with the Project "An Astronomical Observatory for Central America: a realistic way to strengthening basic space science in developing countries" (Pineda de Carias, 1995), several papers on Education in Astronomy have been published, among which special attention deserves "The Central American Master Program in Astronomy and Astrophysics" (Pineda de Carias, 2001), the first document presented at a General Assembly of the International Astronomical Union, by a Central America astronomer.

- *Astronomical Observations at OACS/UNAH (I).* A project that studies the observational conditions of the site of the OACS, and the type of observational programs that could be performed with the telescope LX200 "René Sagastume Castillo". Following the model designed for this project, 4 avenues are explore: i) Site, in order to characterize the site where the telescope is placed at the campus of the university; ii) Telescope and its accessories, in order to determine the performance of the instruments and facilities already installed; iii) Human Resources, to find out who are the users and in which observational projects they are interested; and iv) Astronomical Observations, to develop different observational programs to find out which one fits best accordingly to the conditions of the site, the instruments already installed and the human resources involved.

- *Astronomical Observations at OACS/UNAH (II).* This is another project looking to study those astronomical events such as eclipses, meteor showers or comets, visible from Honduras territory, before, during and after their occurrence or appearance. Some papers already published in these fields are: "About some measurements done in Honduras during the total solar eclipse of July 11, 1991" (AIP, 1993), and, "About a big fireball seen in Honduras" (Meteoroids, 1998).

- *Dynamics of Planetary Systems.* Evolution of different objects of our Solar System and of other planetary systems around other stars is studied. Currently, there are two areas under development: 1) Minor Objects (asteroids) of the Solar System, an area that after Master in Astronomy and Astrophysics thesis: "About secular perturbations in the outer zone of Saturn" have derived into the study of Instabilities of the Outer Zone of Saturn and Binary Asteroids. 2) Extra solar Planets, currently studying The Possibility of existence of terrestrial planets.

- *Stellar Atmospheres.* The dynamical and thermodynamics structure of circumstellar material in short period binary systems is studied. After having look for a suitable binary in UV data (INES / IUE, Wamsteker et al., 2000), for the identification of specific spectral lines, the application of special methodologies, and after having complete observational campaigns, a study on the dynamic and thermodynamic the circumstellar material in interacting binaries is under development.

c) Outreach. In order to contribute to divulge and to widespread astronomical knowledge, the following outreach projects are under development at OACS/UNAH:

- *Astronomical Ephemeris.* Honduras and Central American Astronomical Ephemeris of the Sun, Moon, planets and special events are prepared and divulge as a monthly publication. Special editions of these ephemeris are prepared for those events that withdraws the general public attention such as eclipses, meteor showers, Sun zenith passage, equinoxes, solstices and others.

- *Academicals Visits to OACS/UNAH.* Under the lemma: "From Honduras, Central America: a window to the Universe", a program of educational and touristy interest is currently developed three day per week. Elementary schools, High schools and university students and teachers, participate of lectures, exhibitions, practical activities and astronomical observations. Yearly, about 4000 students of the different levels of the national educational system enjoy and become beneficiaries of this important project.

- *Astronomical Nights.* Every Friday evening, for about two hours, children, young and grew up people visits the OACS/UNAH to listen special talks intended to popularize astronomy at all levels, and also to allow people to make astronomical observations of the moon, planets, deep sky objects and special astronomical events. Yearly OACS/UNAH receives about 5,000 all age persons, and this program is well known and highly recognized within the country.

- *Solar Activity.* A relation between the background sky intensity, coronal mass ejections and the derived geomagnetic effects is studied. Through bibliographic research about coronography, coronographic observations, coronal mass ejections and other dynamic phenomena in the solar atmosphere and geomagnetic phenomena; and through the statistical analysis that correlates dynamic solar and geomagnetic phenomena, derived from data obtained from MICA (Mirror Coronograph for Argentina); a relation between the background sky intensity as indicator of coronal mass ejections is studied.

- *Maya Archaeoastronomy.* Evidence of astronomical activity among the Mayas, trough the study of dates and orientation of structures and monuments at Copan, Honduras is studied. In this project, astronomers and archaeologists are working together at OACS/UNAH, studding planes, maps, bibliographical documents of the archaeological site of Copan; after choosing the area and the astronomical object under study; the problem is stated and the research methodology; specific structures and monuments are documented, and astronomical observations are done in order to obtain results. Currently there are two areas under development: alignments and orientation of structures and monuments; and, calendar cycles.
- *Remote Sensing.* After the Hurricane Mitch damage in Honduras the need of permanent monitoring of the territory and the need of evaluation of the impact of damage using space technologies was unveiled. After that a Remote Sensing Laboratory was organized and opened at OACS/UNAH. Important part of the Remote Sensing Lab is the Geographical Information System already conformed with satellite images and aerial photographs of Honduras and Central America, for different years. A NASA/CCAD Mesoamerican Biological Corridor Project, Classification and land cover and land use projects, and the Characterization and change detection of protected areas, are some of the projects under development.

2. Some Final Remarks

As final remarks the following must be emphasized:

- With regard to doing research and the training of astronomers for Central America, these are actually the main objectives behind the establishment of the Central American Suyapa Astronomical Observatory. To train, with permanent international cooperation, cohorts of astronomers that will have the responsibility to move Central America further into relevant fields of scientific research. The acquired knowledge, use and applications of astronomical instrumentation and space technology are of particular significance in the furtherance of the general objective of the OACS, and of course, as a contribution for sustainable development of the country and the region. Therefore, in order to strengthen this project, the international community of astronomers may help by providing visiting professors willing to come to Honduras to collaborate as they may be needed;

and with funds for scholarships for graduate students to finish their degrees up to the highest level, while guaranteeing and stimulating them to remain in their own Central America countries.

- The Central America Regional Courses (CURCAA) are excellent opportunities to assemble first line worldwide astronomers with Central American university staff and students interested in the establishment of an astronomical tradition in Central America. Somehow, these events represent a regional chapter of a bigger effort performed by the UN/ESA series of Workshops on Basic Space Science. Because all the national Central American universities have agreed to organize this type of event on an annual basis, in order to continue this effort and reinforce this type of activity, special grants from interested international organizations, institutions, societies and the like could be quite useful. In this way larger numbers of participants from the different countries of Central America will be guaranteed. So far, for all CURCAA, all Central American national universities and other local organizations, the IAU and some other foreign universities and organizations have sponsor participants.
- Further, after having hosted one and participated at most of the UN/ESA Workshops on Basic Space Science, we have learned of the multiple advantages of gathering scientists from different regions of the world in order to reach accurate objectives and goals. Also, after our projects have been presented in such an international forum, our national authorities are made aware of the importance of the fields of astronomy and basic space science to the development of indigenous capabilities. This is especially true after having creating small groups in each of the Central American countries that are jointly looking for mechanisms that could allow them to establish astronomy and astrophysics as professional fields with the permanent international cooperation in their own countries.

References

Pineda de Carias, M. C., 1995, American Institute of Physics. New York, U. S. A., , AIP Conference Proceedings, **320**, 13.

Borovicka, J, Pineda de Carias, M. C., et al. Meteoritics, 1998

Pineda de Carias, M. C., Astronomy for Developing Countries, IAU Special Session at the 24th General Assembly, 2001. A. H. Batten, ed., ASP Conference Series, **SPS**, 69.

Report on the Third UN/ESA Workshop on Basic Space Science, organized in cooperation with the Government of Nigeria and hosted by the University of Nigeria, Nsukka, and the Obafemi Awolowo University, Ile-Ife, Lagos, Nigeria, 18-22 October 1993, United Nations General Assembly Document A/AC.105/592/Add.1.

Report on the Seventh UN/ESA Workshop on Basic Space Science, 1998, Small Astronomical Telescopes and Satellites in Education and Research, hosted by the Observatorio Astronómico

de la Universidad Nacional Autónoma de Honduras, on behalf of the Government of Honduras, United Nations General Assembly Document A/AC.105/682; and *COSPAR Information Bulletin*, **141**, 9.
M. C. Pineda de Carias. Mensaje al Futuro: 10 años de Astronomía Centroamericana en Honduras. OACS/UNAH. 2001. World Wide Web Site available at: *http://www.astro.unah.hondunet.net*
Wamsteker, W., et al., 2000. AP&SS, **273**, 155
Wamsteker, W., and Shustov B.M., 2003, this volume.

Chapter 3.2

THE GEM PROJECT: AN INTERNATIONAL COLLABORATION TO SURVEY GALACTIC RADIATION EMISSION[5]

S. Torres[1], V. Canon[1], R. Casas[1], A. Umana[1], C. Tello[2], T. Villela[2], M. Bersanelli[3], M. Bensadoun[4], G. De Amici[4], M. Limon[4], G. Smoot[4], C. Witebsky[4].

[1]*Observatorio Astronomico, Universidad Nacional de Colombia and Centro Internacional de Fisica, Bogota, Colombia*

sergio.torres@freeflight.com

[2]*INPE/CNPq, Brazil*

[3]*IFCTR-CNR, Milano, Italy*

[4]*Berkeley National Laboratory, UC Berkeley, CA, USA*

Abstract The GEM (Galactic Emission Mapping) project is an international collaboration established with the aim of surveying the full sky at long wavelengths with a multi-frequency radio telescope. A total of 745 hours of observation at 408 MHz were completed from an Equatorial site in Colombia. The observations cover the celestial band $0\ h < \alpha < 24\ h$, and $24° 22' < \delta < +35° 37'$. Preliminary results of this partial survey will be discussed. A review of the instrumental setup and a ~10° resolution sky map at 408 MHz is presented.

1. The GEM Project

Synchrotron radiation from relativistic electrons accelerated by the magnetic field of the Galaxy constitute the main component of diffuse galactic emission at low frequencies (300 MHz to few GHz). At higher frequencies and high galactic latitude free-free (bremsstrahlung) emission from ionized hydrogen starts becoming the dominant

[5] Presented by S. Torres at the UN/ESA Workshop on Basic Space Sciences: From Small Telescopes to Space Missions,Colombo, Sri Lanka 11-13 January 1996. Reprinted from AP&SS, **240**, 225-234 (1996).

W. Wamsteker et al. (eds.), Developing Basic Space Science World-Wide, 113-122.

component (20 - 60 GHz) and beyond 60 GHz interstellar dust emission begins to dominate. A precise measurement and mapping of the diffuse galactic emission at low frequencies can tell us much about the cosmic ray electrons and the dynamics of the Galaxy.

The discovery of cosmic microwave background (CMB) anisotropies (Smoot et al. 1992) and subsequent measurements at large angle scales (Hancock et al. 1994; Ganga et al. 1993) have underlined the importance of an accurate determination of temperature spectral indices ($T_b \propto \nu^{\beta}$) for the synchrotron and free-free mechanisms in the Galaxy. The existing surveys (Reich and Reich 1988; see also Table 1 of Lawson et al. 1987 for earlier work) are insufficient to provide the required accuracy as pointed out by Davies et al. (1996). This fact is due to the large zero level and gain uncertainties (i.e. ± 3 K and 10% for the 408 MHz of Haslam et al. (1982) and ± 0:5 K and 0.5% for the 1420 MHz survey of Reich and Reich (1988) respectively). Except for the Haslam et al. 408 MHz survey, which is a composition of several patches made with four different telescopes, all others have partial sky coverage. The nonuniformity of these data sets and residual striping effects constitute a serious limitations to the quality of the maps. Accurate, multifrequency data are needed in order to extrapolate the galactic emission at higher frequencies (Brandt et al. 1994; Masi et al. 1991), especially due to the fact that the synchrotron spectral index presents variations with galactic latitude (Lawson et al. 1987). Future CMB experiments conducted from satellites such as COBRAS/SAMBA and MAP will carry state of the art detectors but nevertheless will be limited by the accuracy of foreground emission removal. There is a well justified need for a full sky, homogeneous, multi-frequency, and accurately calibrated survey of the galactic radio continuum.

Besides producing crucial information for CMB observations, the GEM maps are scientifically important in themselves. The spectrum of galactic radio emission at long wavelengths is dominated by synchrotron radiation emitted by relativistic cosmic ray electrons accelerated by the large scale galactic magnetic field. The synchrotron power emitted by electrons depends on the electron energy density and the magnetic field intensity (Ginzburg & Syrovatskii 1965). Thus, a survey of the radio emission provides useful information from which the galactic magnetic field and the cosmic ray electron energy spectrum can be studied. Knowledge derived from these studies have special relevance in models of cosmic ray acceleration.

Motivated by the above mentioned need to apply corrections to the galactic contamination present in cosmic microwave background (CMB) maps, an international collaboration was established to measure the galactic continuum emission in the range 408 - 5000 MHz (De Amici et al. 1994). The GEM collaboration was started by groups in Brazil (INPE/CNPq), Colombia (CIF and Observatorio Astronomico), Italy (CNR), USA (Lawrence Berkeley National Lab/UCB), and later joined by Spain (IAC).

2. Experimental Setup

One of the major difficulties faced by a full sky radio survey is the need to achieve accurate calibration. The design strategy consists of a 'portable' radio-telescope that can be moved to sites at different latitudes. Using the same calibrated instrument allows for a consistent merging of patches of the sky taken at different sites. An additional advantage is that moving the instrument at different latitudes allows pointing the main beam at a small angle from the local zenith, thus minimizing the atmospheric absorption seen at large zenith angles (important at the higher frequencies). The GEM telescope is mounted on a rotating base and the pointing of the main beam is kept at a fixed angle from the zenith to keep the atmospheric contribution constant. The mechanical design of the mount system allows for changes in the zenith angle of the antenna.

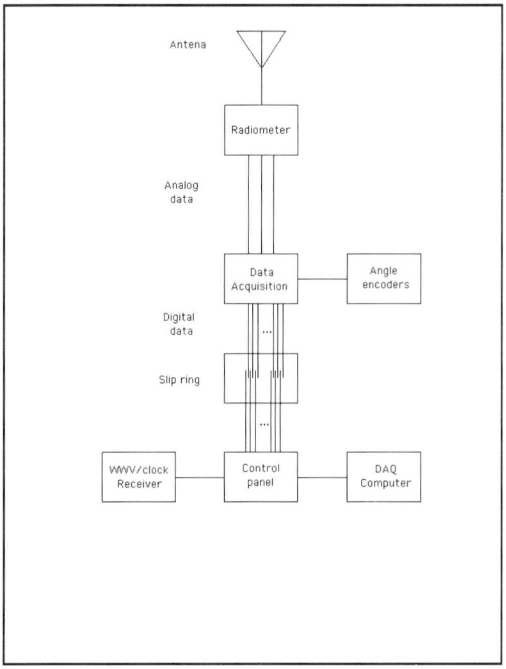

Figure 3.2-1. . Block diagram of the GEM system.

The antenna sidelobes can be checked by observing the same part of the sky at different zenith angles, therefore obtaining a direct measurement of ground contribution. The combined motion of the rotating base and the Earth's rotation results in a swath of the sky seen at each latitude. In principle one could cover the whole sky by collecting data in this mode from three equidistant latitudes (i.e. 60°S, 0° and 60°N). In practice it is desirable to allow for overlaps of the covered regions so as to ensure a self-consistent data set. We have acquired several hundred hours of observation at 408, 1465, 2300 and 5000 MHz from an Equatorial site in Colombia and a Northern site in Bishop, California. The telescope is currently at the IAC Tenerife Observatory in Spain. An overview of the experimental setup and preliminary results of the 408 MHz survey conducted from the Equatorial site will be presented.

A block diagram (figure 3.2-1) shows the main parts that form the GEM system. The main section is made of the parabolic re ector, the feed antenna, the rotating base and the radiometer. Table 1 summarizes the instrumental parameters of GEM some of which have been computed as described below.

2.1 Parabolic reflector, base and feed antenna

GEM uses a Scientific Atlanta 5.5-m parabolic reflector mounted on an alt-azimuth rotating base. The 408, 1465 and 2300 MHz receivers use a prime focus feed: a backfire helix at the low frequencies, and a conical antenna at 2300 MHz. The 5000 MHz receiver unit and conical antennae are mounted at the Gregorian focus.

***Table** 3.2-1.* **GEM parameters**

Parameter	Value
System	
Receiver frequencies	408, 1465, 2300, 5000 MHz
Reflector diameter	5.5 m
Reflector diameter with extension	9.5 m
Postdetection integration time	0.56 s
Base rotation speed (nominal)	1 rpm
408 MHz Receiver	
System temperature	104 ± 6 K
Gain	58 ± 1 K V^{-1}
Band width	28 MHz
Sensitivity	26 mK/integration time
Beam width (FWHM)	11:3°
Baseline susceptibility	-3 K $°C^{-1}$
Gain susceptibility	-2:7 $x10^{-4}$ % $°C^{-1}$

Aluminum panels extend the parabolic reflector surface to a total diameter of 9.5 m. The purpose of this aluminum shield is to minimize diffracted ground emission and to allow us to determine the beam efficiency and loss by covering up with highly reflective opaque material half the 4π solid angle.

The back-fire helix feed antenna (408 MHz) consists of a 9.5 turns made of copper piping of 9.6 mm diameter. The turn length is $0{:}92\lambda$, the spacing between turns is 15.4 cm and the axial length is 148 cm. The feed antenna is sensitive to circularly polarized radiation. The main lobe width of the combined antenna/reflector assembly is obtained using the transit of the sun in front of GEM. A plot of the 408 MHz signal voltage versus the separation angle between GEM's pointing and the sun shows the Gaussian-like shape of the main lobe with FWHM of 11:3°.

The mount assembly rests on a rotating base with a velocity of 1 rpm. An azimuth angle encoder mounted on the rotating axis of the main GEM assembly and a similar encoder on the horizontal axis provide a 0-10 V analog voltage proportional to the angle. The zero angle resulting from this reading is calibrated using the sun signal in the data in combination with the sun ephemerides. These calibration parameters were verified by checking GEM's mechanical orientation with respect to the geographic North employing a theodolite.

2.2 The 408 MHz receiver

The 408 MHz radiometer uses a total power receiver with two RF amplification stages and one DC amplification (x 1000) after detection (figure 3.2-2). A cavity filter ($\Delta\nu$ = 28 MHz) at the front end of the receiver and a tubular filter after RF amplification are used.

The ambient temperature of the receiver is controlled and isolated from the outside temperature by warming the inside of the receiver box with resistance heaters while cooling the inside of a bigger box that encloses the receiver box. Temperature sensors and a regulating circuit keep the operating temperature of the receiver to within ± 0:2 ° C.

The susceptibility of the radiometer output voltage baseline to changes on operation temperature is measured as -3 K/ °C. Thus, within the temperature stability achieved one expects baseline drifts up to 0:6 K which typically occur in time scales of 8 to 9 hours. We correct for temperature induced baseline drifts by using the information from the various temperature sensors and the K/ °C slope quoted above. Gain changes are monitored by injecting a fixed amplitude reference pulse every 45 seconds. The reference noise pulse is generated by a diode and

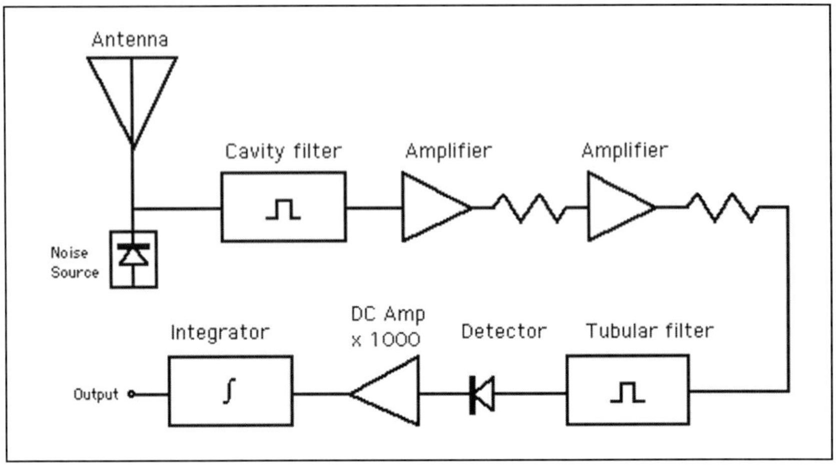

Figure 3.2-2. Diagram of the 408 MHz receiver.

is connected to the input of the first amplifier stage via a directional coupler.

Radio frequency interference (RFI), due to local radio communications or electric discharges in the atmosphere can cause excessive dispersion of the observed signal. An RFI detection circuit signals the presence of such anomalous data. This circuit produces an output voltage V_{sat} proportional to the number of times that the signal average crosses some preset threshold during the integration time. Cutting data above some V_{sat} threshold effectively acts as a low-pass filter on the signal.

2.3 Gain and System Temperature Calibration

Gain, G, and system temperature, T_{sys}, are obtained by recording the radiometer signal voltage as the input is switched from a cold to a hot target. A 50 termination submerged in a bath of liquid Nitrogen (LN) at ambient pressure is used as a reference cold target. A similar termination at room temperature provides the warm reference target. The temperature of the warm target is monitored by a thermocouple while that of the cold target, 75 K, is measured using a reference platinum resistance immersed in the LN. Including the contributions of the connectors and the attenuation and reflection properties of the coaxial cable (at 408 MHz) connecting the termination to the radiometer brings the equivalent LN temperature to 92 ±2 K. The measured parameters are: G = 58 ± 0:9 ±0:2 K V^{-1} , T_{sys} = 104 ± 6 ± 1 K. The first error is systematic, the second statistical. Systematic errors are obtained with the error propagation formulas taking into account the errors in the measurements of target temperatures and signal voltages. The statistical errors quoted come from the standard deviation of the measurements done over a period of 1 hour. The sensitivity of the receiver implied by the measurement of T_{sys} is 26 mK/integration time and is consistent with the rms spread of the antenna temperature when looking at a fixed and known temperature target.

2.4 Data Acquisition System

A 16 bit ADC samples 14 multiplexed analog channels (radiometer signal, angle encoders, temperature sensors, V_{sat} and noise source voltage) and places the digitized information on a serial output. A time stamp and a sequential identifier number ('frame number') is added to the data stream. The time tag is the UT time provided by an external digital clock/receiver synchronized with the WWV station in Denver. Pulses from an internal clock (100 Hz) are divided down to provide the integration time of 0.56 seconds, which effectively defines the basic data rate of 50 bytes/s. The data acquisition (DAQ) unit is mounted in a separate NIM crate box near the receiver whose DC output signal is connected via coaxial cable. The output of the DAQ unit is serial digital data sent to the DAQ computer in the control room through slip rings. The DAQ computer (a Macintosh Performa) performs a routine data quality check and stores the data on disk for future analysis.

3. The Equatorial Site

To complement the observations made or planned from California, Brazil and Spain an Equatorial site seemed to be the natural place to look. Colombia, in particular, is located on the Equator and has high peaks. However, the tropical weather conditions characteristic of equatorial latitudes can place severe restrictions. Site studies (Torres et al. 1992; Hoeneisen et al. 1992) in Colombia showed that there exist at least two dry seasons suitable for radio-astronomy and that there is a large variation of weather conditions at the different regions of the country. Due to the complex terrain there are micro-climates with very dry regions. The selected site in Villa de Leyva (Boyaca, Colombia) is a region characteristic for its dry atmosphere, specially during the dry season (December - April).

The GPS coordinates and altitude of the site are: LAT = 5°37' 7.84'' ±0.59'' N, LONG = 73° 35' 0.53'' ± 0.72'' W, Altitude = 2,173 ±28 m.a.s.l. The errors quoted are statistical while the systematic error is the 3:2'' allowed by DOD.

4. Data Reduction

Every 0:56 seconds the DAQ module sends via serial RS-232 a 14 words frame of data to the on-line computer. The off-line analysis software prepares the data for pixelization and applies cuts as follows: azimuth and elevation angles are calibrated; Julian and Sideral time of each measurement is computed; data points with the sun within 30° of the beam are rejected; noise source pulses are subtracted from the data and used for gain corrections; strong RFI signals are extracted by binning short segments of data in azimuth and rejecting data that shows large deviations from the mean of its corresponding bin; a polynomial fit baseline is found and corrected for temperature induced drifts; finally, the radiometer signal is calibrated and pixelized. Further cleaning of the data is done on the pixelized set by rejecting points that show large deviations with respect to the mean temperature of their corresponding pixel.

A total of 1,116 hours of observation at 408 MHz were completed from the Equatorial site from March 13 to May 13 (1995). Data runs with strong presence of RFI were rejected. Data with elevation angles different than 60° were used for ground contribution studies. The remaining 745 hours of data were analyzed. From the selected dataset 40% is rejected as follows: 10.6% is sun contaminated, 6.3% is RFI contaminated, 3.8% is labeled `anomalous' and is cut at the time of the baseline fit, 6.4% due to temperature sensors out of bounds, 3.5% RFI contaminated as signaled by Vsat, and the remaining goes into the other cuts.

5. The GEM 408 MHz Sky Map

As database structure to store pixelized data we used the skycube pixelization scheme used by COBE (Torres et al. 1989, Torres 1995). This choice is dictated by the availability of several libraries and procedures to handle sky maps in this format and because it makes it easier to compare our maps with others. With the 408 MHz beam size of σ_{beam} ~8:38 x 10^{-2} radians GEM can distinguish ~$(2/\sigma_{beam}) \approx 600$ independent pixels in the sky. Thus, skycube pixelization level 5 with 1536 pixels seems appropriate.

The primary output of the analysis software is a sky map with a 54% sky coverage corresponding to the 60° wide celestial band to which we have access from the Equatorial site. This band covers the sky region 0 h $< \alpha <$ 24 h , - 24° 22' $< \delta <$ +35°37' in right ascension and declination respectively. Figure 3.2-3 is a rendering of this map in an equatorial

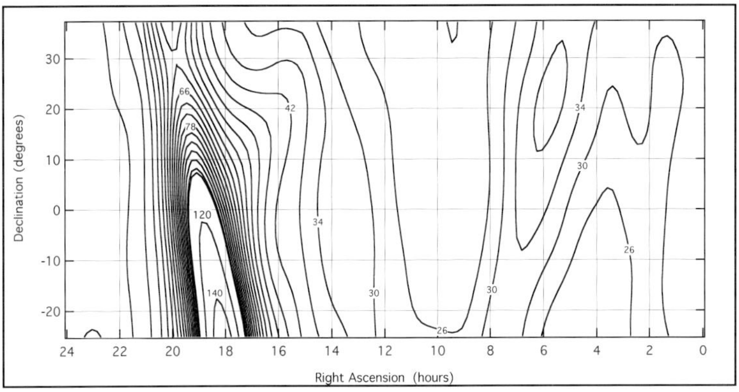

Figure 3.2-3. GEM 408 MHz sky map contour levels in equatorial celestial coordinates. The temperature levels go from 22 to 160 K in constant increments of 4 K up to 100 K where the increment changes to 20 K.

celestial projection. The number of observations per pixel goes from 18 to 13,266 with a mean of 3,576 and a sigma of 2,188.

The standard deviation of the temperature for each pixel is ~2 K, which is somewhat higher than expected from the measured sensitivity, indicating that there remains spurious effects not completely removed from the signal.

The zero level of the calibrated map needs corrections due to all possible known contributions not originating in the Galaxy. The integrated background of faint extragalactic background radio sources at 408 MHz is 3.19 K (Lawson et al. 1987). The CMB temperature is 2.726 ± 0:010 K (Mather et al. 1994), atmospheric emission at 408 MHz is ≤ 1 K. The contribution due to antenna and transmission line insertion loss and sidelobes pickup is estimated using data at several elevations and comparing data taken during the day with data taken at night (difference in ambient temperature ~ 20 °C). The combined contribution of all these factors amounts to 17 _ 8 K and has been taken into account in the map in figure 3.2-3. The preliminary data presented here still has larger uncertainties than what is required for CMB foreground removal. Work towards a more accurate determination of the beam pattern, antenna insertion loss and calibration parameters is in progress. A preliminary comparison with the map of Haslam et al. was done by means of a pixel-to-pixel correlation between the GEM map and the same region of the sky in the Haslam et al. map degraded to the same angular resolution as GEM. The correlation coefficient is 98.3% which is highly significant. A more detailed comparison will be treated in a forthcoming paper.

The tolerable errors in long-wavelength surveys used in CMB work are quite stringent (Brandt et al. 1994). This fact is specially relevant when the `subtraction technique' is used (i.e. measure the foregrounds in the region of the spectrum where they are strong, and extrapolate to the millimeter and sub-millimeter region where the CMB is measured). The rapid propagation of errors in the interpolation does not favor this method. However, a multi-frequency analysis (Masi et al. 1991) or a multi-frequency Wiener filtering analysis (Tegmark & Efstathiou 1995) seem to be a more promising choice to remove foregrounds from CMB maps. Thus GEM multi-frequency data in the range 408 - 5000 MHz combined with IRAS 100 μm, COBE-DMR (31.5, 53 and 90 GHz) and COBE-DIRBE 140 μm data should be sufficient to separate the galactic foreground from CMB maps.

Acknowledgments

This work was supported in part by the Director, Office of Energy Research, Office of High Energy and Nuclear Physics, Division of High Energy Physics of he U.S. Department of Energy under contract No. DEAC03-76SF00098, the National Science Foundation under the International Joint Research NSF 9102295, and Calspace grant No. CS-63-94. Colciencias (projects 222805-007-92, and 2228-05-103-96) provided funds to support GEM operations in Colombia. We owe thanks to many persons who helped make this a successful project. In particular we would like to thank H. Arias, L. M. Cruz, M. Fonsesca, J. Gibson, D. Heine, G. Holgun J. Larsen, D. Maino, E. Mattaini, A. Pantoja, L. Peña, E. Santambrogio, J. Yamada.

References

Brandt, W. N. et al., 1994, Astrophys. J., **424**, 1.
Davies, R. D., et al., 1996, Monthly Not. Roy. Astr. Soc., **278**, 925.
De Amici, G., et al., 1994, AP&SS, **214**, 151.
Ganga, K., et al., 1993, Astrophys. J., **410**, L57.
Ginzburg, V. L. & Syrovatskii, S. I., 1965, Annu. Rev. Astron. Astrophys., **3**, 297.
Hancock, S. et al., 1994, Nature, **367**, 333.
Haslam C.G.T, et al., 1982, Astr. Astroph. Suppl. Ser., **47**, 1.
Hoeneisen, B. et al., 1992, Nuclear Phys. B (Proc. Suppl.), **28B**, 191.
Lawson, K. D., et al., 1987, Monthly Not. Roy. Astr. Soc., **225**, 307.
Masi, S., et al.: 1991, Astrophys. J., **366**, L51.
Mather, J. C., et al., 1994, Astrophys. J., **420**, 439.
Reich, P. and Reich W., 1988, Astron. Astrophys., **74**, 7.
Smoot, G. F., et al.: 1992, Astrophys. J., **396**, L1.
Tegmark, M. & Efstathiou, G., 1995, MPI-PhT/95-62 (preprint).
Torres, S., et al., 1989, in Data Analysis in Astronomy III, Eds V. di Gesu, L. Scarsi, and M.C. Maccarone) **40**, 319.
Torres, S.: 1995, AP&SS, **228**, 309.
Torres, S., et al., 1993, Rev. Col. de Fisica, **25**, 23.

Chapter 3.3

THE GALACTIC EMISSION MAPPING (GEM) PROJECT: SUMMARY AND RESULTS

S. Torres
Centro Internacional de Física
Bogotá, Columbia
verada@earthlink.net

Abstract The results of the activities under the GEM Project are described in the context of the more recent results of the modern missions to study the Cosmic Background Radiation. A summary of the papers and thesis based on the Gem Project is also given.

Introduction

With the results coming from recent Cosmic Background Radiation (CBR) experiments, large-scale surveys and deep sky observations the age of precision cosmology has been heralded. The physics of the early universe links the mechanism that originates the large structure of the universe that we see today with observable features of the CBR angular power spectrum. Density fluctuations at the time when the CBR photons decouple (t_d = 300,000 years) leave a characteristic imprint on the CBR angular power spectrum. CBR anisotropies at large angular scale (> 1°) are sensitive to interactions of the radiation field with the gravitational potential on the surface of last scattering, thus providing access to the primordial spectrum of fluctuations. Smaller angular scales (< 1°) on the other hand result from inhomongeneities influenced by gravitational instability growth, which in turn are very sensitive to the major cosmological parameters: baryonic matter density, dark matter density, curvature, Hubble constant and optical depth. Clearly, accurate measurements of CBR at all angular scales are a well established mechanism to pin down the details of the standard cosmological model.

W. Wamsteker et al. (eds.), Developing Basic Space Science World-Wide, 123-129.

The lofty goal of unmasking the exact value of the cosmological parameters with CBR observations cannot be achieved without paying close attention to potential contamination of these observations by foreground emission, such as those originating locally in the Milky Way. Depending on the angular scale and the band of the electromagnetic spectrum chosen for observations, the galactic emission could be the dominant contribution to the signal. For the range of frequencies relevant to CBR studies the galactic emission is always present and has to be dealt with either by creating a template to mask out regions of the sky dominated by galactic signal or by modeling the galactic emission and subtracting it out of the CBR maps. In this age of precision cosmology the ability to account for and separate galactic foregrounds from CBR maps has become a crucial task. When current and future CBR experiments (i.e. WMAP, Plank, etc) attempt to measure the polarization of the CBR, galactic contamination may be the source of the largest systematic error limiting the accuracy of the results.

The Galactic Emission Mapping (GEM) project originated from the recognition that accurate multi-frequency galactic emission models were a necessity for the analysis of CBR data and at the same time the existing galactic emission measurements at the time were not adequate to accomplish this task [15, 19]. A clear understanding of the galactic emission mechanisms in the radio, infrared and microwave regions of the electromagnetic spectrum is not only a necessary condition in the analysis of CBR data but in itself is an important source of data for studies of our galaxy. Galactic dynamics and morphology, the temperature density of the interstellar medium, the energy spectrum of cosmic ray electrons and the intensity of the magnetic field of the galaxy are some of the areas that can be studied with the GEM surveys.

The main sources of galactic emission are: a) synchrotron radiation from relativistic electrons spiraling under the influence of the magnetic field of the galaxy; b) thermal bremstrahlung emission inside hydrogen clouds and c) thermal radiation from dust along the plane of the galaxy. Each one of these sources has a characteristic and different spectral shape (parameterized with an spectral index) allowing for a separation into components at a given frequency. GEM measurements focus on the low frequency part of the spectrum (408 MHz, 1465 MHz, 2.3 GHz, 5 GHz and 10 GHz). These observations can be complemented with the CBR maps themselves (i.e. from COBE, WMAP, Plank, etc), which show the galaxy at higher frequencies. The goal of the GEM project is to obtain a full sky survey (spatial template) of galactic emission at the low frequencies mentioned above and an accurate determination of the spectral indices including spatial variations.

More accurate measurements of the spectral indices of the galactic emission components, allow for the reduction of the systematic errors affecting the determination of cosmological parameters from CBR observations. It will also be possible to perform a more precise determination of the baseline in point source surveys whose contribution affects the measurements of anisotropies at small angular scales. With respect to the large angle scale anisotropies an accurate estimation of the quadrupole component due to galactic contamination can help solve the unexpectedly low value of the CBR quadrupole measured by COBE (10_{-4} μK) and WMAP (8 ± 2 μK). Galactic contamination is the leading uncertainty in the measured CBR quadrupole, thus improvements in the estimation of the galactic contribution to this component may point to hidden interesting physical effects, such as curvature effects, non-standard metrics, global rotation, etc.

1. The Project

The GEM radio-telescope consists of a 5.5 m diameter parabolic reflector with receivers at 408 MHz, 1465 MHz, 2.3 GHz, 5 GHz and 10 GHz mounted on a moving platform (~1 rpm) that allows to map bands of 60° from a network of observing sites spread in latitude for full sky coverage. The experimental approach and the goals of the project are described in papers [1, 6, 7, 8, 9, 10, 19 and 21].

The GEM collaboration began forming in 1990 when the first proposals were prepared. The collaboration is composed by the Lawrence Berkeley National Laboratory of the University of California, USA, the Centro Internacional de Física (CIF) and the Observatorio Nacional of the Universidad Nacional de Colombia, Bogotá, Colombia, the Instituto Nacional de Pesquisas Espaciais (INPE), São José dos Campos, Brazil, the University of Milan, Milan, Italy, University of Rome, Rome, Italy, and the Instituto Astronómico de las Canarias, Tenerife, Spain. These groups have been active during different phases of the experiment.

A prototype system was assembled and field-tested during December 1991 at the South Pole near the Amundsen-Scott Research Station. From these observations it was possible to obtain calibrated measurements of the sky temperature along directions of fixed pointing angle or 'drift' scans and 'nodding' scans. The results of these measurements are presented in [20].

During the period November 1993 through November 1994 the completed GEM system was assembled in the White Mountain Research Station, Owens Valley, California (+37° 22′ latitude and 1,250 m). Most of the observation time was devoted to diagnostics and calibration tests of the equipment. Data acquisition at 408 MHz and 1465 MHz was achieved and a partial 1465 MHz map covering the region $0^h < \alpha < 24^h$ and $+7° 10' < \delta < +67° 30'$ was completed. The results of the White Mountain acquisition period constitute the main topic of the PhD thesis work in [22]. Observations at 408 MHz from the White Mountain site had to be suspended due to radio frequency interference emanating from a meteorological station in the local airport.

GEM observations from equatorial latitudes were conducted from January through June 1995. For this campaign a site in a desert (+5° 38′ latitude and 2,173 m) near the town of Villa de Leyva, Colombia was selected [11]. A total of 1,618 hours of observation were completed as follows: 1,116 hours at 408 MHz, 131 hours at 1465 MHz, 231 hours at 2.3 GHz, and 140 hours at 5 GHz. A calibrated partial sky map at 408 MHz including a selection of 745 hours of observation covering the celestial band $0^h < \alpha < 24^h$, and $-24° 22' < \delta < +35° 37'$ was produced and is discussed in papers [2, 4, 5, 7, 8, 14, 16, 17, and 18]. The results of the analysis of the 2.3 GHz data including a partial map were presented in [13]. Observation with the 1465 MHz receiver was aborted due to strong radio frequency interference from a geo-stationary satellite. Two graduate thesis [23, 24] and two undergraduate thesis [25, 26] were produced from the GEM observations in Colombia.

Observations of the southern sky are being conducted since 1998 from the site in Cachoeira Paulista, SP, Brazil (-22° 41′ latitude and 500 m). The data acquisition campaign from this site has completed 1,356 hours of observation at 1465 MHz and 818 hours at 2.3 GHz taken during the time period July 1998 – November 1999. Partial maps have been obtained at these frequencies. These results have been presented in [3,12]. There are plans to take measurements from the Antarctic Brazilian Station.

The GEM results obtained thus far satisfy the goal of achieving sky brightness temperature measurements with absolute calibration of the zero level of the map to better than 0.1 K and an accuracy of the gain level to better than 3%. The requirement of internally consistent maps and full sky coverage have been achieved by using the same equipment from sites at different latitudes from the South Pole, to Brazil, to Colombia, to White Mountain in California. The major technical problem encountered is the presence of man-made radio interference. The project has provided a great opportunity for scientists and students in Latin America to participate in an important international scientific collaboration.

2. GEM Publications

2.1 Publications in Journals

1. C. Tello, T. Villela, G.F. Smoot, M. Bersanelli, N. Figueiredo, G. De Amici, M. Bensadoun., C.A. Wuensche, and S. Torres, "Spillover and diffraction sidelobe contamination in a double-shielded expriment for mapping Galactic synchrotron emission", *Astronomy & Astrophysics Supplement Series*, **145**, 495-508 (2000).
2. S. Torres, "The UN/ESA Workshop on Basic Space Science in Colombia, 1992: What has been achieved since then?", *COSPAR Bulletin*, **No. 144**, pp. 13-15, April 1999.
3. C. Tello, T. Villela, A. M. Alves, L. G. Arantes, G. Smoot, G. Deamici, S. Torres, M. Bersanelli, M. Bensadoun, A. Wuensche, N. Figueiredo, "A radio continuum survey of the Sky at 1465 MHz from 53 degrees South to 78 degrees north", *Revista Mexicana de Astronomía y Astrofísica Serie de Conferencias* (in press 1998).
4. V. Cañón, S. Torres, "Medición de la Radiación Sincrotrón en la Galaxia", *Revista Momento, Universidad Nacional*, (1998).
5. R. Casas, S. Torres, "Corrección de mapas de Radiación Cósmica de Fondo por contaminación galáctica y estimación del cuadripolo cosmológico", *Revista Momento, Universidad Nacional*, (1998).
6. C. Tello, T. Villela, C.A. Wuensche, N. Figueiredo, S. Torres, M. Bersanelli, M. Bensadoun, M. Limon, G. De Amici, G. Smoot, "Diffraction analysis of a double-shielded antenna in the Fraunhofer and Fresnel regimes", *Radio Science*, **34**, No. 3, May/June 1999.
7. S. Torres, "Cosmología y Modelos Galácticos con el Proyecto GEM", Boletín del Observatorio Astronómico, (1997).
8. S. Torres, V. Cañón, R. Casas, A. Umaña, C. Tello, T. Villela, M. Bersanelli, M. Bensadoun, G. De Amici, M. Limon, G. Smoot, C. Witebsky, "The GEM project: An international collaboration to survey galactic radiation emission", astro-ph/9605060, *Astrophysics and Space Science*, **240**, 225-234 (1996).
9. De Amici, G., Bensadoun, M., Limon, M., Smoot, G., Bersanelli, M., "A new instrument to map the absolute brightness of the sky at low radio frequencies", *Astroph. Lett Comm.*, **32**, 153 (1995).
10. G. De Amici, S. Torres, M. Bensadoun, G. Dall'Oglio, M. Limon, G. Smoot, G. Sironi, T. Villela, C. Witebsky, "A research program to map the galactic emission at low frequencies", *Astrophysics and Space Science*, **214**, pp. 151-160 (1994).
11. S. Torres, G. Smoot, G. De Amici, M. Becerra, M. Chaux, J. Gómez, A. Umaña, "Site Evaluation and RFI Measurements in Colombia for a Galactic Radio Survey", *Revista Colombiana de Física*, **25**, pp. 23-30 (1993).

2.2 Publications in Conference Proceedings

1. C. Tello, T. Villela, G.F. Smoot, S. Torres, M. Bersanelli, "A new radio continuum survey of the sky at 1465 MHz between declinations –52° and +68°", in *New Cosmological Data and the Values of the Fundamental Parameters*, International Astronomical Union, Symposium no. **201**, Manchester, England, Aug. 2000.
2. S. Torres, V. Cañón, R. Casas, A. Umaña, C. Tello, T. Villela, M. Bersanelli, G. De Amici, G. Smoot, "Long wavelength sky surveys with the GEM radio telescope", *Astrophysics and Space Science*, in *Seventh UN/ESA Workshop on Basic Space Sciences*, Eds. H. Haubold, M. C. Pineda, Tegucigalpa, Honduras, June 16-20 (1997)
3. S. Torres, T. Villela, G.De Amici, G. Smoot, M. Bersanelli, C. Tello, A. Umaña, "A Survey of the Galactic Emission at 408 from an Equatorial site", XXI Reunião Anual da Sociedade AstronômicaBrasileira, Sao Loureno, MG, Brasil, Jul 31-Agosto, 1996, Boletin da SAB, **16**, No 1, 99 (1996)
4. S. Cortiglioni, M.R. Attolini, N. Mandolesi, E. Palazzi, S. Montebugnoli, A. Orfei, M. Bersanelli and G. De Amici, "The importance of accurate galactic observations at radio-frequency for CMB experiments", *Mem. Soc. Astr. It.*, **66**, 84 (1995).
5. A. Umaña y S. Torres, "El Proyecto GEM de Observación de la Galaxia", en *De la Astronomía a la Cosmología: estudios y resultados recientes, Primera Escuela Nacional de Astrofísica*, Ed. S. Torres, Bogotá, Nov 28 -1 Dic. 1994, pp. 169-176
6. L. Chacón y S. Torres, "Interferencia en observaciones radioastronómicas" en *De la Astronomía a la Cosmología: estudios y resultados recientes, Primera Escuela Nacional de Astrofísica*, Ed. S. Torres, Bogotá, Nov 28 -1 Dic. 1994, pp. 177-184
7. T. Villela, A. M. Alves, C. Tello, G. Smoot, G. De Amici, S. Torres, A. Umaña, M. Bersanelli, "Resultados Preliminares do Projeto GEM em 408 MHz" en XX Reuniao Anual da Sociedade Astronômica Brasileira, Campos do Jordao, Sao Paulo, Brasil, Agosto 1 al 4, 1994, **14**, No. 1, p. 100 (1994)
8. G. De Amici, G. Smoot, M. Bensadoun, M. Limon, W. Vinje, C. Witebsky, S. Torres, A. Umaña, M. Becerra, "Mapping the Absolute Brightness of the Sky at Low Frequencies" in *Back to the Galaxy*, Maryland, USA, Octubre 1992, edts. S.S. Holt and F. Verter, AIP Conference Proceedings Series, **278**, pp. 206-209 (1993)
9. De Amici, G., Smoot, G., Bensadoun, M, Bersanelli, M., Levin, S., Limon, M., and Vinje, W., "A new program to map the absolute temperature of the sky at low frequencies", in G. Chincarini et al. (edts): *Observational Cosmology*, ASP Conference Series (ASP, San Francisco) **51**, 527 (1993)
10. G.F. Smoot, G. De Amici, M. Limon, S. Torres and T. Villela, "A research program to map the absolute sky brightness at low frequencies," in *5th Symposium on Pan-American Collaboration in Experimental Physics*, August 17-21, 1992, Cartagena, Colombia.

2.3 Doctoral and Undergraduate Thesis

1. "Um experimento para medir o brilho total do céu em comprimentos de onda centimétricos", Camilo Tello, doctoral thesis, Instituto Nacional de Pesquisas Espaciais, São José dos Campos, Brazil, 1997, INPE-7036-TDI/664.
2. Medición de la Radiación Sincrotrón en la Galaxia", Victor Cañón, master thesis, Universidad Nacional, Bogotá, Colombia, October 1997.
3. "Corrección de mapas de Radiación Cósmica de Fondo por contaminación galáctica y estimación del cuadripolo cosmológico", Rigoberto Casas, master thesis, Universidad Nacional, Bogotá, Colombia, Octuber 1997.
4. "Medición desde Colombia de la Radiación Difusa de la Galaxia a 408 MHz", Andrés Umaña , undergraduate thesis, Universidad de los Andes, Bogotá, Colombia, August 1995.
5. "Estudio de la Interferencia en Señales Medidas por un Radio Telescopio", Liliana P. Chacón, undergraduate thesis, Universidad de los Andes, Bogotá, Colombia, August 1994.

Chapter 3.4

A CENTRE OF ASTRONOMY FOR PARAGUAY: A QUEST FOR A MODERATE-SIZED TELESCOPE

A E. Troche-Boggino†
Universidad Nacional de Asunción
Facultad Politécnica
San Lorenzo, Paraguay
atroche@pol.com.py

Introduction

Young people from every nation of the world have dreams about outer space and all that it contains. When these dreams become realized, many bright horizons are open to them. In more pragmatic words, technology transfer should help a nation or society progress, provided there exists an appropriate environment. Opportunities for technology transfer should be taken advantage of and from that springs urges of creativity, enabling a nation to develop its own proper technology. This is especially important as the new millennium nears; human beings need to find new paths opened for their future. Our planet needs more sensitive inhabitants concerned about the environment, development and social affairs.

Today, astronomical databases are available on CD-ROMs, through the Internet and so on. There are no restrictions for accessing them and prices are low. Also, image processing methods sometimes can reach resolutions as good as those obtained through high-resolution and other very expensive astronomical instruments. But other types of database resources, like those from remote sensing and optical or radar satellites views of land, are still very expensive and restrictive. But data analysis methods used in astronomy and other fields are often similar. So, astronomy might provide excellent training in the use of advanced data

† Shortly after the completion of this paper Professor Alexis E. Troche-Boggino passed away.

W. Wamsteker et al. (eds.), Developing Basic Space Science World-Wide, 131-136.

analysis techniques often not available in other applied fields. Young students should have the opportunity to be trained in astronomy and learn about these data analysis techniques as well as have some experience through the use of moderate-sized telescopes and its parts (CCD, photometers and spectroscopes). They must know how to acquire data as much as to analyze and record them.

The telescope is essential to stimulating observations, for data acquisition and to interact with other astronomical centers abroad. A telescope should be seen as a kind of tractor for interest in space science. Some of these astronomy students would like to go further. They might conduct research through astronomical observations and use data acquired from databases. Other trained students should be ready to use their knowledge in applied fields like those of remote sensing, meteorology, computer science and communications. Astronomy can be a stimulus for training future scientists and high level technicians, not only for space science but also in many related fields.

Unfortunately, because of a lack of such opportunities in developing countries, many of these bright young people opt for professions in fields far from science. Though the United Nations has 185 Member States, only 100 have professional astronomers and astronomical amateurs working groups and only 60 support their astronomical community and scientific interests through membership in the International Astronomical Union (IAU), which promotes astronomy research, education and astronomers exchange. Many countries lack an astronomical observatory, and Paraguay is no exception. The United Nations, through its Office for Outer Space Affairs, has been seeking cooperation from developed countries in providing developing countries with moderate-sized telescopes. A possible offshoot of this idea is to develop a network of observatories that are useful for research on the sun, binary and variable stars and small solar system bodies throughout the world.

Paraguay has such support from the United Nations and has received a donated telescope (e.g. Doncel et al., 2003) from the Government of Japan, which, in cooperation with leading astronomers from the National Astronomical Observatory of Japan, is particularly supportive in establishing astronomical facilities in developing countries around the world.

1. Buenaventura Suarez S.J.: A Pioneer Astronomer in Paraguay

There is an important historical reason for having an astronomical observatory in Paraguay: Buenaventura Suarez. Suarez was a pioneer astronomer, the first native astronomer from the southern regions of South America, and he built a sort of astronomical observatory in the 18th century. He was born in Santa Fe in 1678 and studied at Cordoba, both cities now located in Argentina. He did most of his work in San Cosme y Damian, one of 30 Jesuit communities for Guarani Indians in the Great Province of Paraguay, until his death in 1750. With the help of local artisans, Suarez built various astronomical instruments, including some Kepler-type refractors with lenses polished from local crystalline rocks; sundials; a quadrant with degrees divided into minutes; and a pendulum clock divided into minutes and seconds.

In 1743, the Jesuit Order provided him with telescopes and two Martirion clocks imported from England. The telescopes had focal lengths ranging from 2.2 to 6.5 m. For 13 years, Suarez accurately observed eclipses of Jupiter's satellites. He also observed eclipses of the sun and the Moon. He corresponded with N. Grammatici at Amberg, N.L' Isle at St. Petersburg, I. Koegler at Beijing, Pedro Peralta in Lima and O. Celsius at Upsala. He made determinations of longitudes (as meridian differences) and latitudes of San Cosme y Damian and all other Jesuit towns.

In 1739, Buenaventura Suarez readied his calculations for his book "Lunario de un Siglo", a kind of astronomical calendar of one century containing the phases of the Moon, solar and lunar eclipses, church festivities and geographic coordinates of 70 cities. The first edition of "Lunario de un Siglo" was printed in 1743 and reissued four times until 1856. By the order of King Carlos III, Jesuit priests and brothers were expelled from Spain and all its colonies around 1767. Time and military adventures took their toll. All of Suarez's instruments were lost, with the exception of a sundial at San Cosme y Damian and now serves as a lonely testament to this exceptional man.

Suarez's life is an example of a self-made astronomer in colonial Paraguay able to apply rules for computations and build appropriate instruments. He obtained data from his own observations. He published this data in a book and wrote reports to other scientists around the world. Furthermore, he worked with natives as a missionary priest, and also worked as an artisan making bells.

2. Efforts at Teaching Astronomy in Paraguay

- IAU Visiting Lecturers Programme (VLP)

From 1988-94, visiting professors from Argentina, Mexico and Italy conducted six astronomy courses, mainly at the Universidad Nacional de Asuncion, on General Astronomy and Astrophysics (Else Recillas,

Mexico); Radio Astronomy (Maria Cristina Martin and Carlos A. Alan, with practical lectures from two radio-electronic engineers, all from JAR, Argentina); Galactic and Extragalactic Astronomy (Josi Lums Sersic, Argentina), Astronomical Instrumentation (Armando Arellano Ferro, Mexico) and Stellar Oscillations with some data analysis (Michele Bossi, Italy). About three dozen physics and engineering students from the Universidad Nacional de Asuncion and Universidad Catolica de Asuncion participated in these courses. Three of these students obtained grants to study and do research abroad and two others received grants to participate at the IAU School for Young Astronomers in Bello Horizonte, Brazil, in July 1994.

- The Total Solar Eclipse of 3 November 1994

A group of professors, astronomers and students from Meisei University in Tokyo visited Paraguay in order to observe the total solar eclipse. They came under the leadership of Eijiro Hiei and N. Takahashi. Two days before the solar eclipse an international forum was held at the Universidad Nacional de Asuncions campus in San Lorenzo. For the observation of the eclipse some former VLP physics students joined the Japanese observers in Chaco.

Mr. Hiei suggested that the Japanese Government, through the O.D.A. programme, be solicited to donate a moderate-sized telescope and its peripherals for an astronomical observatory. President of the Republic of Paraguay, Juan Carlos Wasmosy, who observed the eclipse at one of the best places in Paraguay, also promised to help with the construction of the national observatory on that occasion.

- Visit of the IAU Commission 46 "Teaching of Astronomy"

John R. Percy, then-President of IAU Commission 46, visited Paraguay for a week early in August 1997. He delivered four public talks related to astronomical teaching and the need for a Centre for Astronomy for Paraguay. He mentioned the success of the Central American Astronomical Project in Honduras (Pineda de Carias, 2003). The goal is to let young people learn about and train in space science at all levels of education.

He invited everyone involved in astronomy to work together to make the Centre for Astronomy work. He also did his best to obtain support for the observatory from the Universidad Nacional de Asuncion administrators and contacted Japanese astronomers for support from the Cultural Grant Aid of the Japanese Government for a donation of an astronomical telescope and its parts. Paraguay has now received, through the cooperative efforts of the Japanese Government and scientists, the telescope and its peripheral parts necessary to give life to the proposed

Center for Astronomy. There is a committee to support this project comprising the National Astronomical Observatory of Japan at Tokyo, the IAU, and the United Nations Office for Outer Space Affairs. The members are: M. Kitamura (Japan), E. Hiei (Japan), B. Hidayat (Indonesia), J. Sahade (Argentina), H.J. Haubold (United Nations), W. Wamsteker (European Space Agency) and others. This telescope is now fully in operations and first results have been presented by Doncel et al. (2003).

3. Some Plans for the Use of the Requested Telescope

Teachers and students in physics and astronomy, engineering, meteorology and geography from the Universidad Nacional de Asuncion and other universities, have now the opportunity for training in the use of the telescope and its peripherals instruments. Also, teachers and younger students from the primary and secondary levels will have opportunities to receive basic training in astronomy. Astronomy is a good stimulus for these high school students in deciding to pursue science and technology professions. A library and useful database and data analysis techniques are also required. A personal computer was donated with the help of IAU funds, some books and magazines are being received.

It is very important to develop cooperation among amateur groups for finding resources to improve the activities of the Centre for Astronomy. Interaction with astronomy club members is essential for the life of the Centre. Interested amateurs must also receive training and telescope time.

The exchange of astronomers and students from Paraguay, Japan and abroad is another goal. Joint observations with astronomers from Paraguay and other countries would be complementary for certain observations that deserve different latitudes and time. The Facultad Politecnica of the Universidad Nacional de Asuncion has contributed to the construction of the observatory, office space and an auditorium for the Centre (see figure 3.4-1). The location of the observatory has been chosen at the UNA Campus in San Lorenzo. The telescope is used mostly to observe of bodies of the solar system, variable and binary stars and be open for other projects. Astronomy is an international academic discipline. The results of acquired knowledge do not belong to a specific country, but constitute a cultural resource to be shared by all of humanity.

References

Doncel, F.A., Troche Boggino, A.E., Noguchi, T., 2003, this volume.
Haubold H.J. and Wamsteker, W. 1998, *Space Technology*, **18**, 149.
Pineda de Carias, M.C., 2003, this volume.

Chapter 3.5

THE URUGUAYAN AUTOMATED AND ROBOTIC TELESCOPE "B U S C A"

G. Tancredi[1], A.Sosa[1, 2], E. Acosta[2], A. Ceretta[2], E. Joliet[1], M. Ruétalo[3] and F. Bonsignore[2]

[1]*Departamento de Astronomia, Instituto de Fisica, Facultad de Ciencias*
Iuga 4225, Montevideo 11400, Uruguay
gonzalo@fisica.edu.uy

[2]*Obnservatorio Astronomico "Los Molinos", Ministerio de Educacion y Cultura*
Camino de los Molinos 5769, 12400 Montevideo, Uruguay
oalm@fisica.edu.uy

[3]*Department of Astronomy & Astrophysics and Canadian Institure for Theoretical Astrophysics*
University of Toronto, Canada

Abstract The efforts to discover NEOs have been concentrated up to now in the Northern Hemisphere where there are already 6 big NEO surveys functioning. The Observatorio Astronómico "Los Molinos" obtained a grant to install a new observatory dedicated to the NEO survey in the countryside of Uruguay (South America). The new telescope has started operations in mid 2002. The name of this program is "Búsqueda Uruguaya de Supernovas, Cometas y Asteroides - BUSCA".

Introduction

The search for Near Earth Objects (NEOs) has been concentrated in the northern hemisphere. Six dedicated NEO surveys programs are already in place: 4 in the Southwestern USA, one in Hawaii and one in Japan. None of these can reach declinations south of -30°; therefore more than 25% of the celestial sphere is not covered by any project (see figure 3.5.1). Several scientific and political groups have recognized this North-South asymmetry and they have appealed for a prompt solution by installing new survey telescopes in the southern hemisphere.

W. Wamsteker et al. (eds.), Developing Basic Space Science World-Wide, 137-150.

Recommendations have been made by the Spaceguard Foundation, the IAU- Working Group on Near-Earth Objects, and the UK Task Force in NEOs.

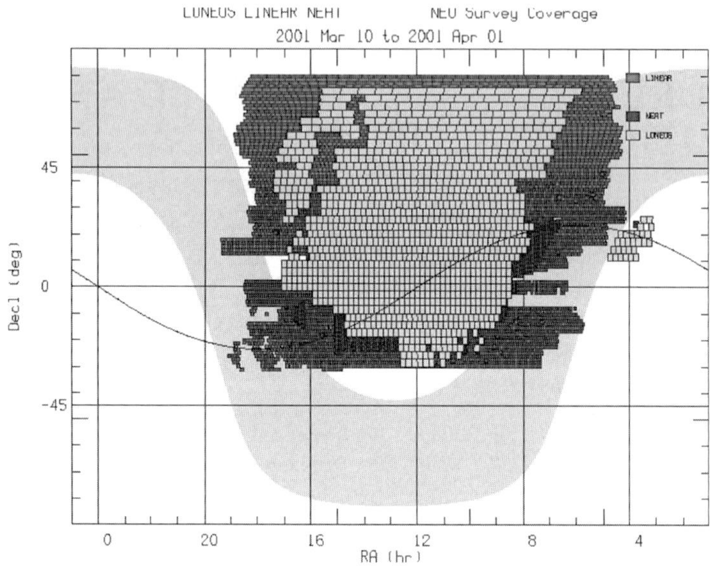

Figure 3.5-1. Sky coverage of the big surveys functioning in March 2001. Note the sharp cut at declination -30° (Created by the Lowell's Asteroid Observing Service - Sky Coverage of NEO Searches, Version 1.2 - *http://asteroid.lowell.edu/*).

1. Objectives and Methodology

Our main objective is the search for NEOs in the southern sky as well as follow-up observations to determine accurate orbits. With a larger number of known NEOs and their orbits, we can improve the extrapolation of their motion into the future. To search for moving objects we require a dedicated telescope and a CCD camera with a wide field of view. We have started our survey with a telescope in the lower size range of those already operating. Software for automatic control of the telescopes has been installed. Follow-up observations of the discovered objects will be done from other telescopes of our own institute as well through collaboration with colleagues of the South American Spaceguard Association, with telescopes in Argentina, Brazil, Paraguay and Uruguay.

This new survey in the southern hemisphere and the follow-up capability will contribute considerably to reach the goal of locating the complete population of Earth-approaching asteroids larger than about 1 kilometer in diameter.

2. The Research Team

The institutes involved in the project are: the Dept. Astronomy (Fac. of Sciences) and the Observatorio Astronómico "Los Molinos" - OALM (Minister of Education and Culture). Their main research topic is the study of minor bodies of the Solar System. Our group has been involved in several searches for asteroids and comets using telescopes from the European Southern Observatory (Tancredi and Lindgren, 1994; Lagerkvist et al., 1996; Hernius et al., 1997; Lindgren et al., 1997) and the Cerro Tololo Interamerican Observatory (Tancredi and Sosa, 1998).

The OALM also has a strong commitment in outreach activities. More than 10.000 persons visit the Observatory every year, most of them are primary and secondary students. Some hundred people visit our monthly open house. Also, several amateur groups have their instruments in the Observatory campus. The press frequently requests us information about astronomical events.

3. The Present Status

The National Research Council of Uruguay (CONICYT) granted OALM a project to install a telescope to search for NEOs. The total amount of the award is US$ 27,000. This money was used to buy the 46cm (f/2.8) telescope (Centurion 18" by Astroworks, http://www.astroworks.com) shown in figure 3.5-3. With further support from our home institutions (Universidad de la República - Uruguay and the Minister of Education and Culture) we acquired a Personal Computer (PC) and the necessary software. One of the members of our group (Dr. T. Gallardo) also obtained a grant from The Planetary Society to buy a CCD.

Regrettably, the ST9 does not fully cover the available focal plane of the Centurion telescope, but due budget limitations, this was the largest CCD available. Detectors as large as 30x30 mm could be installed without considerable image distortion.

The characteristics of the system are presented in Table 3.5-1. In the future the telescope will be re-located in a dark area (see figure 3.5.2) in the countryside, 200 km from Montevideo at the Tourist Ranch Posada "La Laguna":

Coordinates: Lat.: -34°20'01" ;Long.: -54°42'44"; Alt: 240m.

Figure 3.5-2. The future location of the BUSCA telescope at the Posada "La Laguna. On the right side the map of Uruguay and on the left side the nighttime view of light distribution in Uruguay.

Figure 3.5-3 The OALM Centurion 18" Tel. at current location.

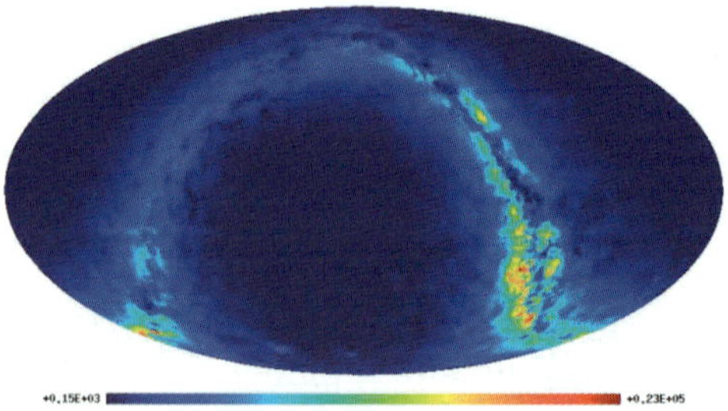

Figure 3.5-4 The distribution of the number of stars in the sky brighter than R mag =16. The horizontal bar gives the scale in number of stars per square.

***Table** 3.5-1.* **Characteristics of Instrument at OALM**

Telescope	Centurion 18" by Astroworks
Primary Mirror	46cm
Field corrector in primary focus	
Focal Ratio	f/2.8
Focal Distance	1.27m
Plate Scale	161"/mm
CCD Camera	ST-9E de SBig (USA)
Chip	Kodak KAF-0261E thick, front-illuminated
	512x512 pixels
Pixel size	20x20 μm
Field of view	28′x28′
Limiting Magnitude	~19

The Government of the Province of Maldonado will support for the building construction, which will be finished by the end of 2003. We have obtained some further support from the National Telecommunications Company (ANTEL) and the National Energy Company (UTE) to provide telecommunications and energy. A weather station to monitor wind, temperature and humidity as well as clouds, instantaneous rain, and lightning detectors will be included. At the current time we are operating the telescope at the site of the OALM (15 km north from down-town Montevideo in a semi-rural area) where we are testing the software, the hardware and have initiated our survey observations (figure 3.5.3).

The telescope and the CCD are controlled with the following software: 1) *Astronomer's Control Program* (ACP) (http://acp2.dc3.com) to control the telescope and the dome, 2) *MaxIm DL/CCD* (http://www.cyanogen.com/maximccd.html) to acquire and pre-process CCD images; and 3) *Pinpoint* (http://pinpoint.dc3.com) to detect moving objects as well as stars of varying brightness.

We concentrate our survey efforts in the region not covered by the northern surveys, i.e. declinations southern than -30 deg.

The telescope will be fully controlled from our home institute in Montevideo. Every afternoon we submit the jobs for the night and the controller system will decide whether to open the dome and start the observational routines depending on weather conditions. Four frames of each field will be taken separated by half an hour. The software for detection of moving objects will analyze the images and produce a report. Next morning the report of the discoveries as well as the discovery images will be sent back to Montevideo. After analyzing the information we submit the discoveries to the Minor Planet center (MPC).

Although the CCD is a blue-enhanced chip, the peak QE barely reaches 60% in the visible part of the spectrum. We take 45 sec. exposures and 4 images per frame. Allowing for time to download the images and move the telescope, in two hours we can cover 30x0.22 = 6.6 deg ^{2}. In a typical night of 8 hours, we cover ~26 deg ^{2}. The available sky south of –30° in declination in any night is ~5000 deg ^{2}. In a month, we will be able to cover an area on the order of 300 deg ^{2}, a tiny fraction of the available sky.

With a larger CCD (e.g. 1024x1024 pixels of 24μm) we would increase the efficiency of our southern survey by a factor of ~3.

4. Development of Software

In parallel with the hardware improvements we are working in several areas related to the development of software for our project. These will be discussed below.

a) The Observing Plans and image acquisition

An observing plan is an ASCII text file with several keywords that are interpreted by an ACP2 script. The keywords set parameters for the image acquisition (exposure time, delay between images, etc.) and the coordinates of the regions to be observed.

In order to select a region we consider the following criteria:

- The regions should have δ< -30° and be close to the meridian. The regions are randomly generated following a distribution with $\cos^3\delta$ and $\cos^3$RA and, at the time of observation, must be more than 30° above the horizon and an elongation from the Moon greater than 60°.
- A new region should not overlap with previous observed regions.
- The regions should have a number star density less than 10% of the maximum star density in the sky (see figure 3.5-4).

This later point deserves a more detail explanation. In order to know the expected number density of stars for a given direction of the sky, we have done a pixelization of the sky following the HEALPIX software (http://www.eso.org/science/healpix/). The sky was divided in $12{*}8^2$ = 49152 regions of equal size (0.83x0.83 deg^2). Using the more than 500x10^6 stars of the USNO-A2.0 catalog, we compute the number of stars in each pixel as a function of limiting magnitude. Figure 3.5-4 shows an all-sky map of the number density of stars. We adjust a spherical harmonic function to the data and compute the coefficients, supplying an analytical function that gives us the number of stars per square degree for a given limiting magnitude in any given direction of the sky. A web page service has been created to provide this information: http://www.fisica.edu.uy/~joliet/cgi-bin/xcounts.cgi.

The ASCII text file with the Observing Plan is generated by an Excel template with severalVBScript macros.

The regions to be observed at any time are arranged as a square mosaic of 5x5 regions centered on the selected position. The frame centers are separated by 24 arcmin in δ and 24 arcmin in RA*cosδ allowing a small overlap between adjacent regions.

The image acquisition is done with the VBScript *AcquireImages.vbs* written for the ACP2 software, modified to be able to restart an aborted plan.

b) The image pre-processing

Before analyzing the images to detect the moving objects we must do a pre-processing of them. The following steps have been incorporated in the automatic pre-processing of a set of images from an observing run:

- List the images where the astrometric plate solution fails
- In case of failure, try a new plate solution with new parameters with the *SolvePlate.js* script of PinPoint.
- For each set of 4 images of the same sky-region, we select the best 3, according to a quality coefficient. The quality coefficient is a function of the number of stars in the image, the mean area occupied by the stars and the distance of each frame to the common center.
- Creation of a Super FlatField or Night-Flat from 50 images taken during the night, and the associated corrections for all images.

c) The detection of moving objects

Our specific hardware configuration (e.g. the sub-sampling of the PSF - our pixel corresponds to 3.26" - and therefore, faint objects occupy only 4-5 pixels) and the very unstable weather conditions at the site, with sudden changes requiring to abort the observing routine, introduce special problems for the application of the standard procedures.

The detection of moving objects is done with *PinPoint*. The scripts in the standard version impose several limitations for our survey and introduce a large number of false detection. Hereby, we outline the functions performed by the program as well as our improvements:

- Number of images used to look for moving objects

It analyzes 3 images of the same sky-region taken at different times. Future improvements foresee to use 4 images correlating all possible combinations of 3 images.

- Detection thresholds

This lists all the sources detected in the images. The sources are recognized in the standard version on the basis of the integrated brightness over the star area, the Signal/Noise ratio, and the radius of the source should be larger than a preset minimum. To address the problems associated with hot pixels, cosmic rays and faint stars that occupies very few pixels we decide to use the total area of the source rather than the radius as a selection criteria. Also the closeness to bright stars generate spurious identifications, this problem is solved by rejecting detection close to saturated stars. Finally, the soft detects in some cases more than one star in a pixel, we then keep only the brightest star detected.

- Identification of isolated sources

Correlating the positions of the objects on each image where common position objects are rejected as fixed objects. If the positional difference is significant, an object is included in a sublist of isolated objects.

The list of detected objects is created in the standard version using the same limiting brightness threshold in the 3 images. Stars close to the detection limit could be in an image just above the threshold and in others just below this limit. To avoid this problem, we create for every image two lists of detected objects: one with a reliable detection limit and another one with a lower detection limit where there are a lot of spurious objects. Every "reliable" list is compared with the two other lists corresponding to the two other images but with the lower detection thresholds. We are then able to reject a lot of stars close to the detection limit.

The astrometric error is a function of the location of the star in the image, we then make the position tolerance limit a monotonic growing function of the location respect to the center of the image.

- Identification of objects in rectilinear motion

Every isolated object in the first list is correlated with all the objects in the second list in increasing order of distances and then with those in the third list. A final check is introduced through a regression analysis (assuming linear motion) of the derived motion for each individual sublist. Those candidates with a value of the regression coefficient less than a given value are accepted as a valid identification.

Nevertheless, we have faced several false identifications due to problems like unusual motions for some detected objects, e.g.:

1. Objects with displacements only in Declination, generally close to saturated stars. Since the chip readout is along columns, the "bleeding" of saturated stars produce the apparition of spurious objects along the columns. Therefore objects with $\Delta RA \approx 0$ are rejected.
2. Objects with prograde and retrograde motions out of plausible limits. The region of plausible velocity for objects in heliocentric motion can be better analyzed in ecliptic coordinates. We have computed the velocities for the first 50.000 numbered asteroids for several dates and

plot them for different elongation ranges. Since we are doing an all sky-survey, it is not enough to consider the velocities close to opposition. In figure 3.5-5 we present plots of velocities in longitude (*dl* x *cos b*) vs. velocity in latitude (*db*), for different elongation ranges. Motions faster than 1 deg/day in retrograde motion and than 2 deg/day in prograde motion are highly unusual for asteroids, and objects falling beyond this range are rejected.

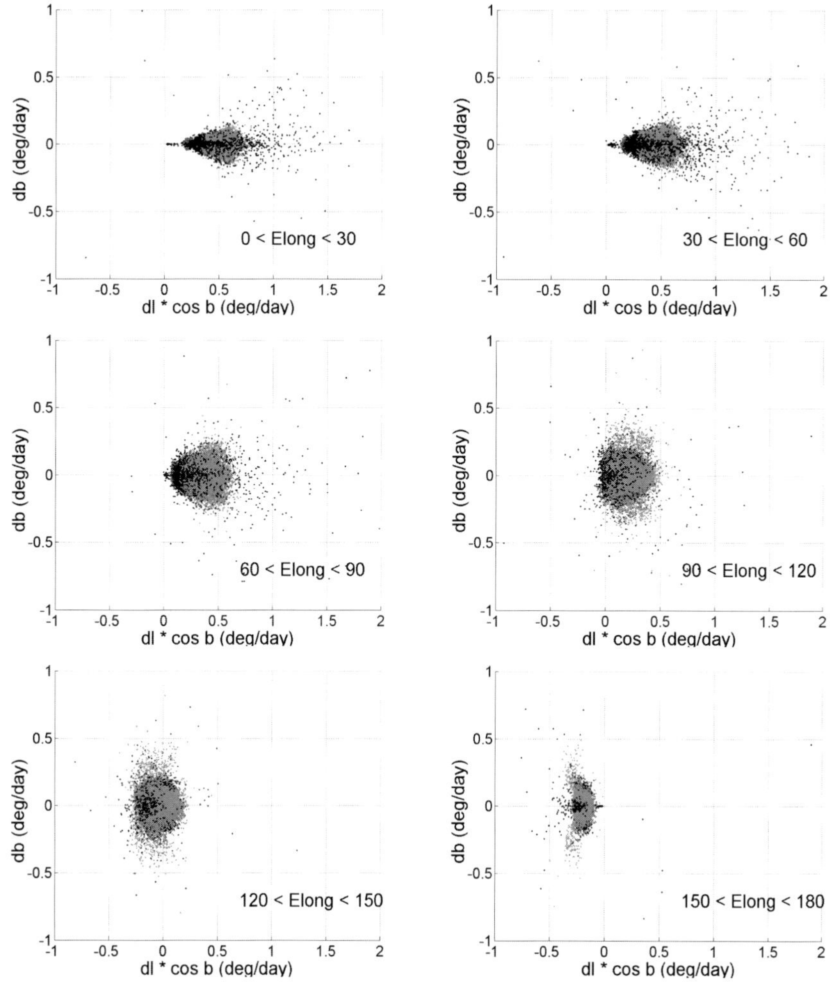

Figure 3.5-5. Six different plots of the ecliptic longitude velocities (dl x cos b) vs. the ecliptic latitude velocities (db) for different ranges in elongation.

- Final report

The final report is an ASCII file in MPC format plus some additional information about the equatorial velocities of the candidate discoveries.

In interactive mode this analysis of a set of 3 images takes approximately a minute. Considering the number of fields covered in a good night (some 100 fields to be analyzed), this is a very time consuming process. On the other hand, if we run this in background mode, the only output is the ASCII final report. To be able to run the detection algorithm in background mode and verify the results reported in a short time, we develop a blinking check for all the detected objects allowing the inspection of only the detected objects on all relevant images. We generate new images, copies of the original ones, with the discovery candidates encircled with a yellow line

d) Automatic detection of trails

Historically, near-Earth asteroids were discovered by the recognition of the trails they left in photographic plates. Present CCD-based surveys make use of automatic detection techniques. Various survey teams have implemented software similar to the one described above. Only the Spacewatch survey has implemented a detection method that includes automatic trail detection, but this is only effective in identifying bright trailed images of fast moving nearby objects with trails longer than about 10 pixels and peak signals with $S/N > 4$ (Scotti, 1993). Other methods have been suggested to automate the detection of fainter streaks, but either they report many false streaks or fail to detect many real ones. Milani et.al. (1995) propose the use of a different algorithm to detect trails with low signal to noise ratios (even below $S/N \approx 1$), but due to its computational complexity it is too slow.

We have developed a new approach using the Hough Transform and an algorithm based on it capable of lowering the automated streak detection limits to trails with peak signals with $S/N \approx 1\text{-}2$ and shorter computational time.

The Hough Transform

The Hough Transform is a robust algorithm for detecting curves with an analytical representation in images and estimating the parameters that represent them. Introduced by Hough (1962), it has become a standard tool in many tasks within the domain of artificial vision. Useful references can be found in the review article published by Leavers (1993). The simplest and most straightforward application of the Hough Transform is the detection of straight lines. A straight line can be parameterized in different ways, but in any case it will be defined by two parameters (e.g. slope and ordinate intercept). These leads to a 2-dimentional parameter space in which each point of the image is projected into a curve. Each point (x,y) of the image is projected into a straight line defined by $c=-x\ m+y$.

The trail detection algorithm

In a digital implementation of the Hough Transform the parameter space will be represented by an accumulator array. A set of points in the image belonging to a curve will generate a peak in the accumulator.

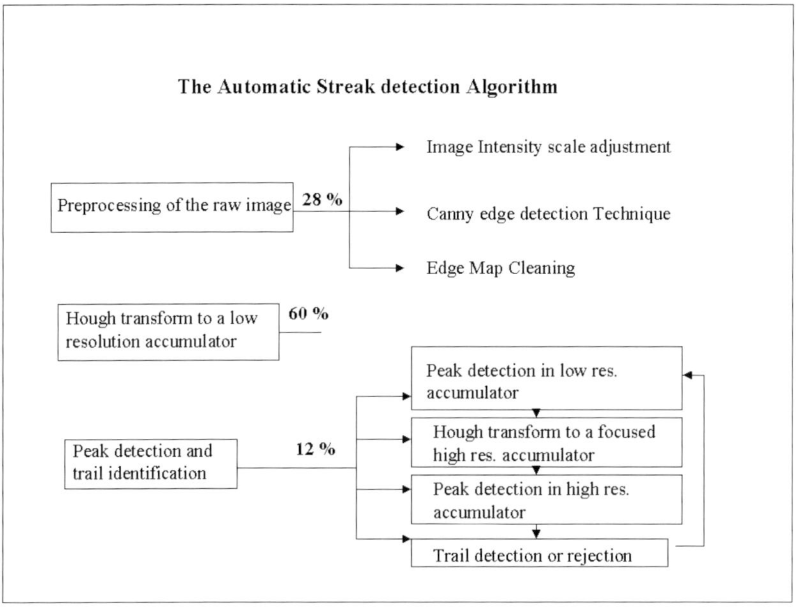

Figure 3.5-6. Flow chart of the automatic trail detection algorithm

The basic structure of our streak detection algorithm is shown in figure 3.5-6. In figure 3.5-7a we present a section of an image taken by Spacewatch telescope showing a long trail. The first stage is the preprocessing of the image. In it we first change the intensity scale in order to enhance those pixels with values near the noise level. After that, the "Canny" edge detection technique is applied (a Gaussian filter followed by a gradient filter and the ``non-maximum suppression" cleaning technique). Finally, we clean the edge map eliminating the edges produced by noise using an algorithm that considers the local density of edge points (figure 3.5-7b).

The preprocessed image is then transformed to a low-resolution accumulator (figure 3.5-7c) using the double parameterization of straight lines. To reduce the effects of the noise in the positions of the edge points, we increment the cells of the accumulator using a Gaussian influence function (see e.g. Ballester, 1994).

The last part of the algorithm involves the detection of peaks in the low-resolution accumulator and the identification of the corresponding object in the image. Once a peak has been detected in the low-resolution accumulator, the edge map is transformed again, but to a smaller, higher resolution accumulator centered in the detected peak (inset of figure 3.5-7c). The parameters can be then estimated with higher accuracy detecting the peak in the latter. Then, only a narrow band of the image is considered, and it is placed over the straight line with parameters corresponding to the detected peak. Finally, only if a long straight edge is found within the band, the detection is reported (figure 3.5-7d). Otherwise, the same procedure is applied to the next accumulation peak in the low-resolution accumulator and so on.

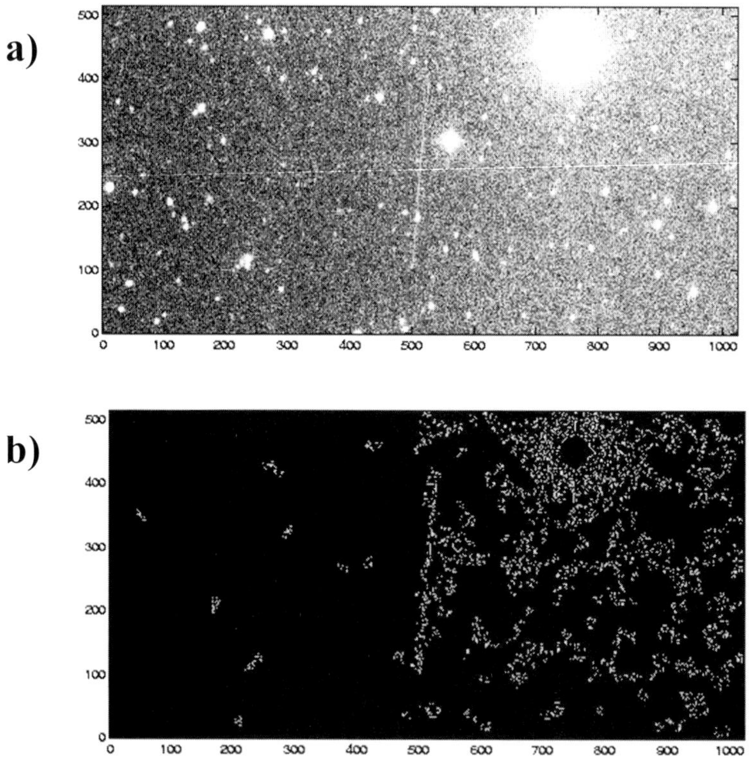

Figure 3.5-7 Different steps of the trail detection algorithm (see also text).

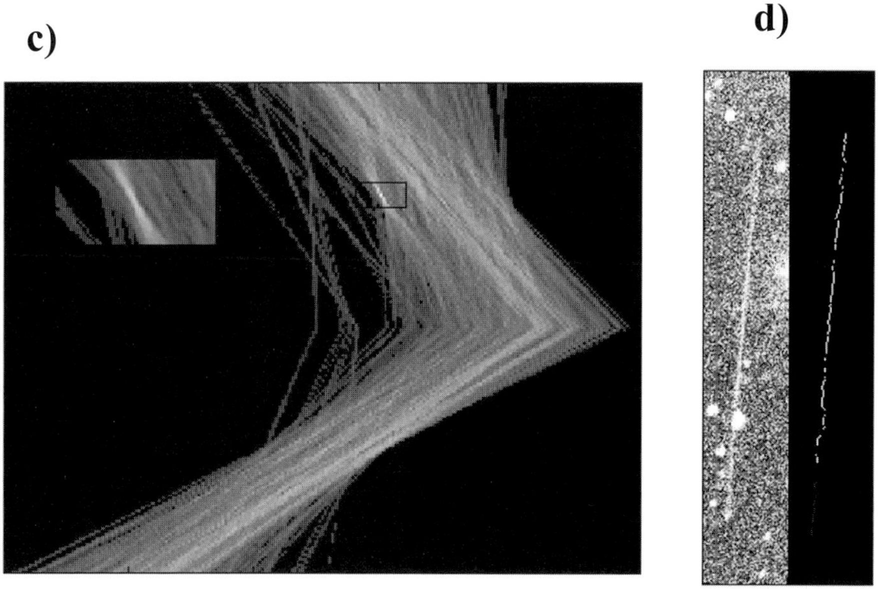

Figure 3.5-7. Different steps of the trail detection algorithm (see also text)

The algorithm has been successfully applied for a set of images taken by Spacewatch. We are now working in a public version of the software that can read images in FITS format.

e) Educational software in Spanish to work with asteroid surveys data

The Japanese Spaceguard Association in the framework of the International Schools Observatory promoted by the British Council has developed an Educational software "Asteroid Catcher B-612". Groups of students can work with real images taken from the NEO survey telescopes of the Bisei Spaceguard Center and visually blink them and look for asteroids. The software is very powerful and quite similar to the professional software in use in the different surveys. We have translated the manuals of the software into Spanish and we plan to distribute it among interested groups of students in the secondary level and in science clubs. We will also offer our own images in the public domain.

5. First Asteroid from Uruguay

We started testing the telescope and software in April 2002. In the night of April, 18 we discovered the first asteroid from Uruguay. The Minor Planet Center assigned the provisional name: 2002HA9. It was a main belt asteroid. Since then many more new objects have been discovered. We also recovered two short-period comets: P/1992 Q1

=2002 Q4 (Brewington) (IAUC 7961 and M.P.E.C. 2002-Q41, 2002) and 79P/Du Toit-Hartley (M.P.E.C. 2003-E32, 2003). We have also continued the follow-up observations of known objects both with the Centurion 18" telescope and the 35cm telescope of the OALM. During 2001 and 2002 we have prepared 342 astrometric reports on asteroids and 207 on comets for the Minor Planet Circulars, many concerned NEO type objects. These reports lead us to occupy the 4th place among the South American groups in number of asteroid reports and the 1st place for comets. Nevertheless, as mentioned above we are just starting the full operation of the project. We are confident that in the near future the our contributions will greatly increase.

Acknowledgements

The BUSCA project has been supported by the CONICYT (Uruguay), The Planetary Society, Universidad de la República and PEDECIBA (Uruguay), Raúl Zabala and his family (owners of Estancia "Posada La Laguna"), the Government of the Department of Maldonado, ANTEL and UTE (Uruguay). We greatly appreciated the support of all this institutions. Emmanuel Joliet developed the software for star counting under the advice of G.T. Marcelo Ruetalo developed the trail detection algorithm under the advice of G.T. Federico Bonsignore did the translation of the ASTCAT manual. The images taken from Spacewatch were kindly provided by D. Rabinowitz. We thank Andrea Sánchez for a careful reading of the manuscript. During the course of this project, the support given by our families have been crucial to its success. We are very grateful to them.

References

Ballester, P., 1994, Astron. Astrophys., **286**, 1011.

Hernius, O,et al., 1997, Astron & Astroph **318**, 631.

Hough, P.V.C., 1962, "Method and Means for Recognising Complex Patterns", U.S. Patent N°3069654.

Lagerkvist, C.-I., Hernius, O., Lindgren, M. and Tancredi, G., 1996, Earth, Moon & Planets, 73, 267.

Leavers, V.F., 1993, CVGIP:Image Understanding, **58:2**, 250

Lindgren, M., Tancredi, G. ,Lagerkvist, C.-I. and Hernius, O. (1997), Astron & Astroph. Suppl. Ser., **118**, 293.

Ruetalo, M. and Tancredi, G., 1996 Rev. Mexicana Astron. Astrof., Serie de Conf., **Vol. 4**, 117.

Tancredi. G and Lindgren, M., 1994, Icarus, 107, 311.

Tancredi, G. and Sosa, A., 1998, Rev. Mexicana de Astron y Astrof. Serie de Conferencias, **Vol. 11**, 17.

Chapter 3.6

ASTRONOMY AT THE UNIVERSIDAD DE SONORA

A. Sánchez-Ibarra
Universidad de Sonora
P.O. Box 5-088
Hermosillo Sonora, 83190 Mexico
asanchez@cosmos.astro.uson.mx

Abstract Astronomy has been a long tradition in Mexico since the Mayas culture. However, most of the development of this science has been concentrated in the center-south region of this country. The creation of the Astronomy Area at the Universidad de Sonora, permitted since 1990 a different perspective for the development of this science, not only related with the scientific work, but also with the participation in education and public outreach, as also develop facilities, as a fundamental part of the future for this science. I present here a description of the work made and the projects in development for the near future.

Introduction

Although the interest in Astronomy was present in different people since the foundation of the Universidad de Sonora at Hermosillo in 1942, it was not until 1989 that a systematic effort began to work on this discipline. People from the Center for Research in Physics, CIF-US, explored this in the form of an Area of Astronomy.

The Area began its work in January 1, 1990, although the official dedication was in February 15, 1990. With the participation of three astronomers, Julio Saucedo-Morales, Manuel Corona-Galindo, and Antonio Sanchez-Ibarra, the main objectives of the Area were established:

a) To make research in Astronomy and Space Sciences.
b) To develop observational facilities.
c) To participate at all the educational levels.
d) To develop programs for public outreach.

W. Wamsteker et al. (eds.), Developing Basic Space Science World-Wide, 151-157.

Something to remark here is that there is no priority order on these goals: each one is as important as the others. So, we have the education not as something complementary to our work, but as something very important.

I will try to resume in this paper the activity made for each goal and the plans we have for the near future. However, is interesting to remark the main ways which have permitted us to reach the results we have, considering that officially only two people has been the staff of the Astronomy Area, and that our annual budget is mostly symbolic. Those are:

a) Trying to make an effective work, as far as possible of bureaucracy.
b) Giving personal financial support to the project.
c) Combining creativity and action.
d) With the support of many students and people that agree with our philosophy.

This way, a very interesting history has developed, in addition to the work itself that has been done during almost twelve years.

1. Scientific Works

The research work at the Astronomy Area has been centered in two fields: solar physics (A. Sanchez) and extragalactic astronomy (J. Saucedo). In last year, 2002, also Cosmology projects were included (A. Lipovka).

The work in solar physics has been on the topic of Coronal Holes activity. Working mostly with the He I data of National Solar Observatory at Kitt Peak, we made a Catalogue of Coronal Holes (CHs), published as the report UAG-102 of the SGDC in 1992. The catalogue also includes data from the OSO satellites and Skylab. The data obtained on CHs permitted us later to study the nature and relationship of CHs with other solar phenomena, as also with the sunspot solar cycle. Eight papers more made between 1992 and 1998 describe how CHs rotate, develop and interact with other solar phenomena. Also, the study of CHs was extended to the work on SXT Yohkoh images and EIT SOHO images. Currently we are working in the problem of visualization of CHs through the analysis of images at several wavelengths. The result of this research would permit later to make CHs Atlas with more precision. Also, especial work have been made during the total solar eclipse of July 11, 1991, the annular solar eclipse of May 10, 1994, several partial solar eclipses, and the transit of Mercury on the solar disk in November 15, 1999.

2. Facilities

At the beginning of the Astronomy Area in 1990, we had only a 25 cm catadioptric telescope.

Knowing that a 41 cm reflector telescope was inside boxes at the Ecological Center of Sonora, we proposed to that institution to install the telescope and put it in operation in exchange to use in training students. We signed the agreement and in two months the telescope was installed with the first light ceremony in March 20, 1990. Since the last 12 years, this observatory has been open for scholar groups and public observations, giving a national record for operation of a public observatory. Also we have used the telescope for several stellar observations and the training of Astronomy Area students.

At the end of 1990, we requested to CIF-US a classroom to be converted in a solar observatory. We received the classroom in 1991 and through three years we completed the "Estacion de Observacion Solar" (Solar Observation Station), EOS. A hole was made in the roof to install a 25 cm heliostat in an equatorial mount. That heliostat sends the solar beam inside the room and is received by another 25 cm flat mirror. That mirror reflected the solar beam to three solar telescopes fixed on a table. Two 6 cm refractors project the white light solar image on a screen. A 12.5 cm catadioptric telescope has worked with a H-Alpha DayStar filter or a Calcium filter, or both through a beam splitter and CCD video cameras. The signals from the cameras go to VHS-VCR to record on tape and to PC computers with video card to get images of video at MPEG format. In November 2000 we began live broadcast of the solar image through Internet. The data collected at EOS is the wider in the country. Some years, such as 1995, we were able to observe 352 days. In other years, we usually observe more than 320 days at year. Currently, EOS is in charge of the continuum and Calcium observation.

In April 1996 we organize and held the meeting "The Solar Cycle: Results and future work", with the participation of 25 astronomers from 7 countries.

The work in extragalactic astronomy has been focused in the nature and relationship of isolated elliptic galaxies. Morphologic and brightness studies of these galaxies are giving interesting results on the distribution of galaxies and nature of themselves. We have included other observations as such of Comets Hyakutake and Hale-Bopp, the encounter of Comet Shoemaker-Levy 9 .with Jupiter in 1994, and planetary occultations.

Figure 3.6-1 Building of the Astronomy Area and prototype of Observatory Carl Sagan at the University campus.

Figure 3.6-2 The first of three solar telescopes of Observatory "Carl Sagan".

In January 31, 1996, we presented to the University authorities a project to build a new observatory in the mountain "Cerro Azul", 200 km north of Hermosillo and with an elevation of 2480 m above sea level. The project would have to be feasible not only to be made, but also in maintenance. Through evolution in the first three years, the "Carl Sagan" Observatory was designed as a battery of four telescopes: a 41 cm stellar telescope with four 15 cm solar telescopes attached in a fork mount (see figure 2.5-1 and 2). The stellar telescope would have a direct CCD camera to work in search of Supernovae type 1a. The solar telescopes would be equipped with continuum, H-Alpha, Calcium and He I filters and direct CCD cameras. Especially the He I telescope will be used to observe Coronal Holes, the main research subject of the Astronomy Area. The telescopes would be in a 5 m diameter, 10 m high building in three levels and a 5 m dome, completely operated with solar-energy and observing by remote control both from a Base names "Complejo Astronomico Jose Martinez Rocha", at Magdalena de Kino, a little town 36 km from the mountain, as also from the University campus at Hermosillo. The complete cost of the Observatory is estimated in US$ 250,000.

In summer of 1999 the foundation of the prototype building was made at the University campus. With a special and cheap design, the building was made in 2000 and also the dome was installed. In April 2001 the first of three solar telescopes was installed and in May 2 we saw the first light. Something remarkable is that the H-Alpha image from this telescope is live broadcasted through Internet during all the observation time (http://cosmos.astro.uson.mx/webtv/index.htm).

In 2003 the work on the top of "Cerro Azul" will begin. Meanwhile, the other telescopes will be installed at mid- 2003 at the prototype building here at Hermosillo with also the development of the control and communications systems. The Observatory "Carl Sagan" could be completed in 2004 at the mountain. Different to any other astronomical institution, the Observatory "Carl Sagan" will broadcast all the observations, both solar as stellar, through Internet. On this way, any researcher will be able to use the data. Also at least the 40% of observation time will be devoted to educational programs. In addition to participate in the research program of the Area, a menu of small astronomical projects for students and teachers at all educational levels will be available. So, any student or teacher will be able to request observation time to participate in both.

In 2002, a Planetarium was added to the Base in Magdalena and it is now in construction. It will be dedicated by summer of 2003.

3. Education

The Astronomy Area participated in all the educational levels, from kindergarten to superior education. The main work is:

- Update scholar programs in Astronomy and Cosmography.
- Included Astrophysics courses at the Physics career.
- To create workshops and seminars for teachers at all educational levels.
- To offer the Basic Astronomy Course for everybody.
- To have continuous advisory for teacher and students of all educational levels.

At the beginning of 2002, we started the Program "Constellation", focused to promote the creation of planetariums, based in the development of a very cheap projector made by Saul Grijalva Varillas, a young Systems Engineer. The cost of the complete projector, with similar capabilities of the typical Planetarium projectors, is only US$3100. The Astronomy Area team worked out in the design of the complete Planetarium (building, 8" telescope, sound system, computers, exposition gallery) for only US$ 120,000. That without a commercial vision, but with a strong educative purpose.

The Astronomy Area has been expanding their work in education through Internet. Several WebPages are concerned with this subject and in July 1, 2002, we started the live webcast through ASTRO-USON WebTV Education (http://cosmos.astro.uson.mx/webtv/choose.htm). Also, at the end of that year, we offered the Basic Astronomy Course through this system with 42 virtual students from all our country, South-America and Spain. This program is continuous from February-may and September December each year.

Additional to the Basic Astronomy Course that will be offered twice a year, a Diplomate (degree program) in Basic Astronomy will start during 2003. Also, during this year the Astronomy Area will continue working in the model to offer the Career in Astronomy Via Internet. That would be offered mainly to Latin-America people, and would permit to have directly the career in Astronomy, because in Mexico there is only career in Physics.

4. Public Outreach

A wide program for public outreach has been active at the Astronomy Area since its creation. I may mentioned:

- An hour radio program in Astronomy since 1997.
- A half-hour radio program on Science News weekly since 2000.
- Participations in radio and TV programs as requested.

- A weekly lectures program in Astronomy for everybody since 1996.
- Lectures at schools as requested.
- Public observations in school and little towns, as also during important astronomical phenomena such as eclipses and comets.
- A weekly article for newspaper since 1990.

a) Articles for magazines as requested.

5. The Future

The Astronomy Area has been participating in the Basic Space Sciences Workshops of UN/ESA since 2001. We hope that our activities will encourage people and groups from countries in development to realize the feasibility, even under difficult financial conditions, to make important progress, do important scientific research and develop facilities, such as the "Carl Sagan" Observatory and the Planetariums. Also we hope to help the increase the communication and participation between individuals, groups and institutions over the whole of Latin America with the creation of our coordination website RECABAS (http://cosmos.astro.uson.mx/un-esa.country.htm), the Regional Coordination of Activities on Basic Space Science for America, as recommended at the 11th United Nations/European Space Agency Workshop on Basic Space Science held at Cordoba, Argentina, last year. We are strongly committed not only to science, but also to society and to the scientific education. On this basis, we will continue maintaining our programs and trying to expand it.

Chapter 3.7

SPACE SCIENCES IN LATIN AMERICA

S. Torres
Centro Internacional de Física
Bogotá, Columbia
verada@earthlink.net

Introduction

For institutions and individuals managing the funding and formulation of development strategies and priorities, basic space sciences is perhaps one of the most difficult areas to tackle. On one hand space sciences, both basic and applied, promise to be of great potential for applications that are urgently needed in developing countries. Monitoring of land use and atmospheric conditions, telecommunications, remote sensing applications, etc. are only a few of the areas in need of expertise and participation in developing countries. On the other hand, space technology is by its nature an expensive endeavor, which on the face of competing priorities in countries with seemingly insurmountable budgetary constraints, becomes relegated to the category of luxurious activities.

For the scientist interested in developing space sciences, his/her task becomes even more difficult because given the very well known obstacles that scientists have to face in developing countries one finds that the activities of space scientists, often times, are considered 'out of place' even by their fellow scientists working on other fields. It is not uncommon for scientists from developing countries to express that the 'useful' sciences should be funded and practiced in their countries and the 'not-so-useful' sciences should be left for the developed countries. Space scientists in particular face this mentality and thus the concern for survival becomes imminent. What can be done to change this situation? Is there any hope for developing countries to see a consolidated scientific

W. Wamsteker et al. (eds.), Developing Basic Space Science World-Wide, 159-175.

space community capable of generating and using its own technology one day?.

The term 'basic space sciences' is used here to refer to research related to observations of the celestial sphere in the full range of the electromagnetic spectrum (Haubold and Wamsteker, 2003). This broad definition includes the areas of astronomy, astrophysics, cosmic rays, cosmology, solar-terrestrial interactions, and planetary and atmospheric studies.

Basic space sciences in developing countries share with the other scientific fields similar problems. However, in space sciences the known difficulties become amplified due to the fact that space and the stars are seen as a subject too remote to be considered by policy makers pressed to find rapid solutions to their country's problems. Thus, a close examination of the issues involved in developing space sciences in developing countries serves as evidence of the problems that all our scientists are facing. Possible solutions, recommendations and strategies emanating from this exercise will also apply to other fields in basic sciences. The problems that most afflict the development of space sciences in poor countries are well known and pose the hardest challenge for scientists and policy makers as well. Insufficient capital, lack of a scientific tradition, brain drain, lack of governmental support, inadequate infrastructure, competition, and an ever more growing gap with developed nations, have been the conditions of these countries. As a consequence of the bleak panorama that we observe, scientists have inherited an immense responsibility and their task has increased to also include lobbing, policymaking, and outreach activities.

A review of past experiences and the current status of space research activities on the part of developing countries reveals that the strategy to follow is one in which the initiative of individuals to start research activities must be complemented with a coordinated plan at the national and international level. One could differentiate two models of development: the first in which the actions of a small group of individuals driven by a common scientific interest results in the formation of collaborations towards the achievement of very specific goals (the 'bottom-up' model). The second consists of projects that can be conducted only as a result of international agreements and development plans that need governmental support (the 'top-down' model). The right solution, as argued here, is a judicious combination of both strategies.

An exhaustive review of the status of basic space sciences in developing countries is a daunting task that is not the scope of this article. However, one can single out few examples to illustrate the virtues and difficulties with the two development strategies mentioned above (for a complete treatment of the subject the reader is referred to the proceedings of UNISPACE 82, Chipman 1982).

1. The Scientific Development of Latin American Countries

Based on science development indicators space sciences in Latin America (LA) follow the same pattern as research and development (R&D) in general. It is therefore instructive to review the current status and trends of science in LA countries. This review relies on the input and output indicators commonly used to evaluate scientific development: gross expenditure on R&D (GERD), personnel working in R&D, and bibliometric data. The sources used to elaborate this review are primarily the UNESCO report (see UNESCO2001), the National Science Foundation (see NSF 2002), the Ibero-American Network on Science and Technology Indicators (see RICYT 2003), and the Institute of Scientific Information (see ISI 2003).

A note about indicators: the above-mentioned science indicators have limitations and are subject to bias. By any means it is claimed here that the state of scientific development of a whole nation can be encapsulated in such a simplistic set of metrics. They are however useful to indicate the relative development among countries. In absolute terms it does not make sense to compare for instance the science budget of a poor LA country with that of the US. Beyond the obvious disparity little insight can be gained from such exercise. In the analysis below, absolute comparisons have been used only to contrast and to put into context some of the indicators. By using the US to compare indices it is not suggested that the US is the best model for developing countries to follow. Also, when interpreting relative indices (i.e. GERD relative to gross domestic product (GDP) or number of publications relative to number of researchers) much caution must be used, as results could be misleading. Take the case for instance of a very poor country with GERD below the mean for its group and a number of researchers not exceeding few tens (not an atypical situation).

Lets assume that the country in question has a very prolific researcher producing tens of publications a year (not atypical either). The number of publications per researcher in that country could surpass developed countries by a large margin. This result of course has nothing to do with the country as a whole. Notwithstanding the aforementioned shortcomings, lets take a look at what can be discern about the state of science in LA.

A most striking characteristic of the scientific development of LA countries is the vast disparity of development among its member nations. The level of development goes from countries where the existing infrastructure is negligible, the poverty level is alarmingly high, and the investment in R&D is minimal (in some of these countries the total R&D budget does not buy a fraction of a research telescope) to countries such as Brazil which have developed and launched their own satellites. This state of affairs is revealed by the large variance in the scientific indicators for the whole region and for this reason it is best to divide the LA countries in separate groups based on the level of scientific development based on the metrics mentioned above. Latin America (excluding the Caribbean countries) is made up of 24 countries with a total population exceeding 500 million people and a geographic area of more than 20 million square kilometers. LA countries can be grouped in three clear levels of scientific development. The first group (G1) made by Argentina, Brazil and Mexico is the most advanced. In these countries scientific research is well established, research groups have been consolidated, the number of researchers is greater than 25,000 and the GERD has consistently been in the range US$1 -4billion per year, publication levels have over passed the 2,000 publications per year mark. Brazil, Mexico, and Argentina have respectively about 3000, 2200 and 2000 PhDs involved in physics research (Moran-Lopez 2000). A second group (G2) made by Chile, Colombia, Cuba and Venezuela has a more recent growth in researchers and R&D spending. Countries in this group have a population of researchers in the range 5,000 to 7,000; 100 to 500 physics PhDs; a GERD of US$ 100 - 400 million per year and publication levels averaging 400 per year. The remaining LA countries (Belize, Bolivia, Costa Rica, Dominican Republic, Ecuador, El Salvador, French Guiana, Guatemala, Guyana, Haiti, Honduras, Nicaragua, Panama, Paraguay, Peru, Suriname, and Uruguay) belong to the third group (G3). In the latter group the level of development ranges from countries that do not even appear in the radar screen of scientific development to countries like Panama, El Salvador, and Costa Rica, where small research groups have been formed in the last decade.

Unlike industrialized countries where the investment money in R&D is provided mostly by private industry (in the US 68.4% or US$ 181 billion of R&D funding for 2000) in LA countries R&D funding originates almost in its entirety from government sources which are overwhelmed by other urgent needs and by outside pressures to adjust budgets and to repay foreign debts.

In absolute terms the combined GERD of all LA countries was US$ 9.4 billion in 2000. In the same year the United States alone spent 28 times that sum in R&D. Of all the GERD in LA it is in G1 countries where most of this spending takes place (86.1%) while for G2 countries it is 12% and for the G3 group it is only 1.9%. Brazil by far is the LA country with the highest GERD (US$ 3.5 billion in 2000) by contrast the average GERD of G2 countries is US$ 280 million and US$ 22 million for G3 countries. The combined expenditure in R&D of the 17 countries in the G3 group is comparable to the cost of a single telescope, the Gemini South telescope installed in Chile. The R&D expenditure of the US in 2000 was 12,200 times larger than the average GERD of G3 countries the same year. The wide gap between the less developed and more advanced countries in LA is clearly evident. The trends in spending during the decade of the 90s are favorable for the whole region (a 53.6% increase) with G2 countries seeing the largest increase (80%). This trend has reversed in the last 3 years. Because of budgetary mechanics the practice in most countries is to allocate R&D budgets based on a percent of GDP level, however GDP growth has been dismal or negative in some cases. The distribution of R&D budget by discipline for LA countries shows an allocation to exact sciences of 27% in average with a minimum of 7% (Mexico) to a maximum of 36% (Panama). Among LA countries the difference in GERD relative to GDP is not nearly as large as the difference in absolute expenditure numbers. The GERD relative to GDP of the country was in average (2000) 0.58%, 0.40% and 0.20% for G1, G2 and G3 respectively (The US GERD was 2.66% in 2000). In many of the LA nations R&D spending has not been growing in proportion to the GDP. In Bolivia, Chile, Colombia, Ecuador, Panama, and Uruguay the R&D budget relative to the GDP has decreased in the period 1996-2000, and in other countries where the relative GERD has increased the actual dollar amount of R&D investment has decreased due to an overall decline in the economy of these countries.

Has the investment in R&D produced results? Lets look at the output indicators: number of researchers and publications. Of the 142,000 researchers in Latin America (as of 2000) 78.4% are from the G1 group, 15.7% from G2 and the remaining 5.9 % from G3 countries. The US has 13.7 times more researchers than all of the LA countries combined. The country with the largest number of researchers per capita is Argentina with 713 per million, the average number of researchers per million population by group is 420 for G1, 286 for G2 and 170 for G3. To contrast these numbers, the researchers in the US are 52 times the average number of researchers in G1 countries, 350 in G2 and 1,635 times in G3 countries. Trends of number of researchers since 1996 show a positive growth for all LA countries (2.4%) except for Bolivia, Ecuador, and Peru where researcher population has decreased. The G2 group saw the largest increase (4.5%) while the G3 the smallest (0.3%). Colombia shows the largest percent increase (20.7%) in the number of researchers during this period. With 5,163 graduate scholarship recipients during the 1996-2000 period Colombia also shows a strong emphasis in investment in education of researchers. The number of publications is another indicator commonly used to measure scientific output. The primary source for this metric is the Science Citation Index (managed by ISI) who compiles bibliographic information. From the more than 70,000 scientific journals that are published worldwide only roughly 3,300 make it to the SCI. Only journals (mostly in English language) that follow a strict set of criteria are accepted to the SCI, which highlights the potential bias against developing countries resulting from a count of publications on SCI journals when it is known that many researchers in those countries publish their work in non-SCI journals (Gibbs 1995). A comparison of the number of publications by country derived from ISI data with those published by UNESCO (which uses many other bibliographic databases) reveals a consistent underestimation by a factor of 2 for the ISI numbers for developing countries (however in relative terms the bibliographic ISI and UNESCO indicators seem to scale, thus retaining the relative weight).

Of all the scientific publications in the world in 1999 2.26% came from LA countries. If one uses the number of publications as a proxy to measure scientific production it is then possible to say that the production of science results from LA countries is only 2.26% of the world production. Although a small proportion, it is nevertheless an encouraging indicator since the number of publications from LA more than double in the last decade. It was 1.18% in 1986. However it is also important to note that these results comes mostly from the G1 group which accounts for 82% of all the LA publications. The contribution of G2 and G3 countries to the LA publications is 14.5% and 3.5% respectively.

Another encouraging trend is that the total number of publications per researcher in LA is comparable with that of developed countries. This indicator has been used as a measure of relative performance (May 1997). While in the US the ratio of publications per 100 researchers per year is 8.4 in LA the same number is 8.3 and some countries have publications per researcher larger than the US: Brazil 10.3 and Chile 12.2 (these figures are derived from ISI 1999 data, when using UNESCO and RICYT numbers the results are consistently higher by a factor of 2). By G1, G2 and G3 the numbers are 8.6, 7.4 and 3.6 respectively.

2. Astronomy and Astrophysics in Latin America

Astronomy and astrophysics has been an important area of research in Latin America and an activity with deep roots in Central and South America (Sahade 1990). Brazil and Argentina have developed space technology to levels comparable with industrialized countries and Chile is host of an array of the best astronomical observatories in the world. Local scientists, however, have been only marginally involved with those large facilities. Only recently Chile has passed a law demanding access to 10% of time to all international telescopes in Chilean territory. Chile has insufficient number of astronomers to fill in this time at the moment but the measure will surely strengthen the local scientific community.

In the region physicists have been an important group in establishing research activities (Moran-Lopez 2000). Most of the countries in groups G1 and G2 have physical societies that have helped foster academic activities. The Argentine Association of Physics was founded in 1944 and the Mexican Physical Society in 1950. Undergraduate programs in physics first appeared in the 1940s, starting with Mexico, Argentina and Brazil and graduate programs appeared only in the 1950s. Doctoral programs in astronomy are offered only in Argentina, Brazil, and Mexico. The National University in Colombia offers PhD degree in physics and Master degree in Astronomy, coordinated by the National Observatory. The G1, G2 and G3 country classification applied to science in general is also valid for basic space sciences (with the exception perhaps of Chile where the strength of astronomy relative to the other sciences is higher than in other countries).

An indicator that reflects the relative importance of astronomy and astrophysics in Latin America is the membership in international space science organizations. Of the Latin American countries only Argentina, Brazil and Mexico are members of COSPAR (Committee for Space Research). The membership to the International Astronomical Union (IAU) is shown in the table below.

Table 3.7-1. **IAU Membership**

Country	Year	Members
Brazil	1961	125
Argentina	1927	101
Mexico	1921	90
Chile	1957	46
Venezuela	1953	13
Cuba	2000	5
Uruguay	1970	3
Central America	1998	2
Bolivia	1997	0
Peru	1988	1
Colombia	1967	1
Ecuador	Not member	1
Paraguay	Not member	1

In terms of scientific production, how does basic space science fare in Latin America? To get an idea of the relative strength of research in space sciences one can look at the total number of publications on these subjects and compare with the total number of publications in all fields of research. The astro-ph archive (see astro-ph reference) is a good tool to achieve this measurement.

The on-line astro-ph archive is the preferred means by researchers of making manuscripts of articles available to a large audience before the article is actually published. There are obvious shortcomings (some astro-ph articles are not published at all, or published in non indexed journals, etc) nevertheless it is a useful metric to evaluate the relative strengths in space sciences research among countries because it is expected that the number of papers submitted to astro-ph is correlated with the research activity in the subject. Of all the astro-ph papers coming from LA countries (at least one author from an institution in Latin America) 70.3% originated in G1 countries, 28.7% from G2 countries and 1% from G3 countries. Relative to the total number of publications of the country Chile has the largest proportion of space science papers with 2.8%, followed by Mexico with 2% and Brazil with 0.7%. The latter results are underestimated as a result of astro-ph not being able to capture all of the basic space science papers.

A study by SRI International (Ailes 1988) based on publication trends finds that for LA as a whole, astronomy represents 3.4 percent of its total active research areas, which is above the US or world average. The same study points out that the weight of astronomy with respect to other areas of research is very high for Chile (9.9%), following Mexico (4.3%) and Brazil (3.5%). The difference of the astro-ph derived numbers and the SRI study results can be accounted for by the reasons stated above (underestimation of astro-ph numbers) and the fact that since the SRI study (1988) the relative contribution to publications coming from other disciplines has increased.

3. Regional Projects

There are a large number of successful projects that started as a small group's initiative. The local presence of a strong research team is a driving force and an important element in the establishment of research groups in developing countries. The impact to the advancement of the local scientific community in Bolivia and Argentina as a result of the installation of the Chacaltaya cosmic ray station in Bolivia and the cosmic ray laboratories in Argentina are two clear examples. The case of Argentina and Bolivia help to stress once more the importance of looking at the geographic attributes in order to attract international collaborators. Astronomical observations of variable starts and gamma ray burst afterglows, environmental and bio-diversity studies, environmental geo-chemistry, solar physics, physics of the upper atmosphere, ionospheric physics, galactic studies, the south atlantic anomaly, equatorial electrojets, the ozone hole, cosmic ray moon neutrinos, etc. are examples of subjects with special geographic requirements that developing countries can offer (Hoeneissen et. al. 1992). However, geographic attributes alone are not enough. The presence of international researchers without a local educated scientific community ready to absorb and participate will not benefit the region. On the contrary, a strong local group could attract mutually beneficial collaborations with the international community. The case of the cosmic ray group in Argentina formed in 1949 around the activities promoted by Estrella Mathov and Juan Roeder is an illuminating example (Roeder 2003). This group attracted the attention of the scientific luminaries conducting cosmic ray research in Europe at the time, which resulted in a collaboration from which fundamental contributions to field were made. Today Argentina is the host of the Auger project, the largest and one of the most important cosmic ray experiments done thus far in the world.

Several international scientific organizations have flourished in Latin America that have accelerated the consolidation of research groups in the region by promoting and providing efficient contacts with the international scientific community. The Latin American Center for Physics (CLAF) founded in 1962 in Brazil has organized workshops and scholarship programs. The Centro Internacional de Física (CIF) in

Colombia founded in 1985 has promoted the formation of research groups with the participation of countries in the region and scientists from developed countries. The areas of research most developed by CIF are bio-physics, high-energy physics, astrophysics, opto-electronics, microprocessors, and biotechnology. FELASOFI, the Federation of Latin American Physical Societies founded in 1984 promotes scientific exchange among Latin American physicists. The Society of Latin American Experts in Remote Sensing (SELPER) with basis currently in Argentina and rotating among its members countries coordinates workshops and other activities for experts in the subject.

The leading role of an international organization such as the United Nations (UN) is a key factor in the coordination and the formation of international collaborations tending to open space activities by developing countries. Given the economic and social pressures that these countries have to deal with, it is almost inconceivable that one single country can afford to develop its own space technology (this fact applies to industrialized countries as well). An international coordinated effort by a large group of developing countries, however, can result in a successful space mission. This is where the actions of the UN and other international bodies can have a decisive impact. In 1958 the UN formally recognized the need and the potential for international cooperation in the field of space sciences thus establishing the ad hoc Committee on the Peaceful Uses of Outer Space (COPUOS). In 1991 the UN, in cooperation with the European Space Agency, began a program to provide basic space science education for the benefit of developing countries (Haubold et al. 1995). As a result of this program a series of Basic Space Science Workshops have taken place. Three of the Workshops have been in the LA region: Costa Rica and Colombia (1992), Honduras (1997) and Argentina (2002). The UN/ESA Workshops on basic space sciences are held in developing countries and their format is designed so that besides scientific talks of very high quality there is ample time for discussions on the main issues faced by scientists from the region. The interaction of scientists from the region with their peers in more scientifically advanced countries results in the identification of key problems and recommendations that should be adopted to ameliorate the situation. A conclusion that has came out of these workshops is that even though the different regions do not share the same conditions and level of development, there are similar problems faced by the scientists of developing countries. Due to this fact, the recommendations emanating from discussions of scientists from the different regions where the UN/ESA Workshops have taken place apply in general equally well to all of the regions. It has been recommended, for example, to build first- rate educational and research capabilities for internationally recognized scientific research by building the necessary computational infrastructure to allow access and use of space science data archives. The participants of the Workshops recognized the need to establish local electronic communications and access to the international computer networks. The possibilities of exploiting local and geographic attributes such as latitude, high peaks, climate and bio-diversity, has also been pointed out as an important element of great potential.

Thus far an attempt to present a general overview of the state of development of basic space sciences in the Latin American region has been done. The remaining of this article highlights the most important projects and activities in individual countries.

4. Brazil

Of the Latin American countries the most advanced in space technology and basic research is Brazil. The country has an advanced infrastructure for space activities, with an ambitious agenda that includes the construction of four scientific satellites (SCD-1, SCD-2, SSP-1 and SSP-2), and the construction with China of a remote sensing satellite. Research in space sciences is coordinated and funded through the civilian National Institute of Space Research (INPE) with headquarters in Sao Jose dos Campos in Sao Paulo. The INPE institute formally created in 1971 grew from the Commission of Space Activities (GOCNAE) established in 1961. The first projects conducted by INPE involved atmospheric studies with the launch of ionospheric probes. By mid 1970 the main projects conducted by INPE consisted of the MESA project for the reception and interpretation of images from meteorological satellites, the SERE project to use remote sensing technology from aircraft and satellites to be used in surveys of natural resources, and project SACI a geostationary communications satellite used to improve the education system. The end of the 70s decade was the beginning of the development of the first fully developed space technology project MECB (Complete Brazilian Space Mission).

The first results of the MECB program began appearing in the 90s. The first Brazilian satellite, SCD-1 is placed in orbit in 1993. In 1998 SCD-2 is successfully launched. In 1994 the CBERS-1 satellite developed in collaboration with China was launched from the Taiyuan base in China in 1999. Brazil also participates in the ISS.

Brazil has a 2.5% participation in the Gemini Observatory and more than 30% (together with the University of Virginia and the NOAO - National Optical Astronomical Observatory) in a 4-meter telescope currently in final stage of construction in Cerro Pachón (Chile) in the same place where the Gemini South telescope is located.

5. Argentina

The largest telescope in operation in Argentina is the 2.15-meter of the El Leoncito Astronomical Park in the Sanjuanina Mountains established in 1986. The University of Córdoba Observatory with a 1.52 meter reflector is the only observatory in LA that has functioned uninterrupted since its foundation in 1871. Argentina has a 2.5% participation in the Gemini Observatory (see report on Chile below).

Space science in Argentina is carried out by a civil organization and became firmly establish in 1992 with the SAC-B (Satélite de Aplicaciones Científicas) mission. The SAC-B was an international effort with the participation of Argentina, Brazil, US and Italy. Brazil contributed with its installations to control the satellite; Italy provided the solar panels of the spacecraft and one of its instruments. The scientific aim of the mission was to achieve observations of the bright, brief transient emissions of astronomical objects in the range 5 eV - 1 MeV. On board the SAC-B satellite there was an Argentine experiment of solar physics and a high energy astrophysics experiment of Penn State University. Unfortunately the SAC-B was lost during launch from NASA Wallop base in 1996 when the launcher could not release the experimental platform.

After the SAC-B accident the National Space Affairs Commission (CONAE) built the SAC-A satellite to demonstrate and test new technologies for future missions including newly developed solar panels. The NASA Shuttle placed SAC-A in orbit on December 12, 1998. The Argentine SAC-C satellite was placed in orbit on November 21, 2000 by a Delta II 7320 rocket launched from the Vandenberg Base in California.

The payload of SAC-C includes instruments developed by CONAE, NASA, the Italian Space Agency (ASI), CNES and the Danish Space Agency. The final tests were done in Brazil. The SAC-C together with LANDSAT 7, EO-1 and TERRA forms the "morning constellation" of satellites. The construction of these satellites and many other technological developments are conducted by INVAP a consortium of the National Atomic Energy Commission and the government of the province of Río Negro.

CONAE plans to build two new satellites: the SAOCOM as part of the SIASGE project of the Italian Space Agency for emergency management, and SAC-D with the participation of NASA. One of the scientific goals of SAC-D would be to measure the salinity of the ocean. CONAE's ground station is located 40 km from the city of Córdoba where several antennae are being operated, including one of 13 m diameter.

Argentina was chosen as the site to deploy the south component of the Auger Project, a large area ultra-high energy cosmic ray experiment operating in conjunction with a matching site in the northern hemisphere (Utah, in the US). The construction of the austral observatory started in 2000. Prior to that, in 1995, an international collaboration had been formed including more than 200 scientists and technicians from 55 institutions in 16 countries. The Pierre Auger project studies the highest energy cosmic rays known in nature (> 10eV) from a giant observatory spanning an area of 3,000 square kilometers located east of the Andes Mountains in the Province of Mendoza, Argentina. The expected arrival rate at these energies is only 1 particle per square kilometer per year (hence the large area requirements). The other distinctive feature, besides the exceptional size of the observatory, is the hybrid nature of the detectors. The experimental technique uses 24 fluorescence detector telescopes and 1,600 surface water tank detectors for Cerenkov radiation. The hybrid technology provides a large number of events with less systematic detection uncertainties and the best possible information about the primary particle types. The goal is to observe close to 1000 events during a 20 year period. The first stage of construction and installation of the first detectors was completed in January 2002. Since then 30 detectors have been in operation and more than 70 air showers have been seen in hybrid mode confirming that the equipment operated within the design parameters (Etchegoyen et al., 2003).

6. Mexico

In Mexico the Autonomous National University (UNAM) has taken the lead in coordinating and promoting research and activities in space sciences. In January 1990 the UNAM established the University Program for Space Research and Development (PUIDE), comprising four areas: basic and applied research, aerospace engineering, teaching and space law. The UNAM together with the National Institute of Astrophysics, Optics, and Electronics (INAOE) participates in the 10-meter telescope project of the Instituto Astronómico de las Canarias (GRANTECAN). The Astronomical Institute of the UNAM has a 2.10-meter telescope installed in San Pedro Martir.

The INAOE, managed by the University of Sonora, in association with the University of Massachusetts at Amherst and with the financial and institutional help of the government of the state of Puebla is working on the construction of the Large Millimeter Telescope (LMT), an antenna 50 meter in diameter which is expected to start operations in 2004. LMT is the largest single dish being built. The telescope will be located on top of Sierra Negra at an altitude of 4,640 meters, 270 kilometers southeast of Mexico City. The US$ 86 million radio telescope will be used by astronomers to detect and study galaxies in all stages of evolution, measuring spectral lines from gases and radiation emitted from dust.

7. Chile

Because of its ideal geographic conditions (dry weather, high altitudes, south latitudes) Chile offers the best place for astronomical observations of the southern sky. This fact has attracted the attention of the international community of astronomers and has converted the country in the home of several of the best observatories in the world. A review of the large astronomical observatories in Chile follows. Chile is participating in the Gemini project consisting of a pair of 8-meter telescopes in Hawaii (Mauna Kea at 4,200 meters altitude) and Chile. The project, managed by the US Associated Universities for Research in Astronomy (AURA), has a cost of US$ 184 million. The Gemini telescopes will provide unique, high quality coverage of the entire sky in the infrared and optical, spectral regions. The contribution to the project by country is US (50%), United Kingdom (25%), Canada (15%), Brazil (2.5%), Argentina (2.5%), Australia (5%). Chilean astronomers have automatic rights to 10% of the observation time in Gemini South and to any other international telescope in Chile.

The Cerro Tololo Inter-American Observatory opened in 1962 with support of the National Science Foundation. The site is located at 2,200 meters near the city of Vicuña. The most potent instrument of the observatory is a 4-meter telescope.

The European Southern Observatory (ESO) opened in 1969 on top of Cerro La Silla includes a 3.5-meter telescope. The infrastructure has expanded to Paranal in the north of Chile where 4 new8-meter telescopes have been installed. These can be used independently or as an interferometer array (including the use of smaller telescopes).

The Carnegie Institution of Washington has an observatory in Cerro Las Campanas operating since 1976. The site is equipped with 3 telescopes including a 2.5 meter (the largest) and one owned by the University of Toronto. The Department of Terrestial Magnetism of the Carnegie Institution of Washington has cooperation with the Maipu Radioastronomy Station located 30 kilometers southwest of Santiago.

The University of Chile inherited the instruments (including a Maksutov astrograph) left by Russian astronomers of the Pulkovo Observatory when they departed at the time the military regime took power in Chile.

The NSF and the ESO have signed an agreement to build the Atacama Large Millimeter Array (ALMA) in Chajnante in the Atacama desert in Chile at 5,000 meters altitude. The US$ 620 ALMA array will be used to study the formation of structures in the early universe. The ALMA interferometer consists of 64 movable 12-meter antennae with an extension of up to 14 kilometers. Completion of construction in the Chilean desert is scheduled for 2011. The government of Japan through the Cultural Grant Aid has delivered in 1999 two 45-centimeter reflector telescopes to the National University of Asuncion, Paraguay and to the Observatory of Cerro Calán, Chile. Results of research on eclipsing variables have been already presented by researchers in Paraguay using one of the Japanese telescopes (Doncel Invernizzi et al., 2003).

Other smaller groups of astronomers operating in Chile include the Astrophysics Group of the Chilean Catholic University in Santiago with a 91-cm reflector at Cerro San Cristóbal and the Isaac Newton Institute in Santiago.

8. Colombia

The National Astronomical Observatory (OAN) built in 1803 is the oldest observatory in the continent. The OAN is part of the National University in Bogotá and has a small team of researchers with strong emphasis in the areas of Cepheid variables, cluster membership, AGN, stellar physics and cosmology. The OAN offers a Masters degree in Astronomy and has a 3-meter refractor, a 2.25- meter telescope and several other smaller telescopes. In 1995 the OAN installed the GEM radio telescope in the town of Villa de Leyva (120 kilometers east of Bogotá). GEM radio maps of the galaxy at 408 MHz and 2.3 GHz were completed from data taken at the Villa de Leyva site during the first half of 1995 (Torres, 2003).

9. Bolivia

Cosmic ray physics is a research field with deep roots in Latin America. It was through experiments conducted in the Chacaltaya Cosmic Ray Laboratory in Bolivia by the Brazilian physicist Cesar Lattes that the identities of the particles in pion decay were determined. The Chacaltaya Laboratory, one of the oldest international laboratories in the region, is associated with the University of San Andres and was created in 1952. Scientists from Brazil, Japan, Argentina, England, India, Italy and the US have conducted leading edge cosmic ray experiments in Chacaltaya.

10. Concluding Remarks

There is no single magic recipe to follow in order to improve the conditions of researchers in space sciences in developing countries. The problem is a complex one and has long and historical roots that are not easy to change. One can however distinguish some key areas that need particular attention and can bring immediate change if applied vigorously. These can be summarized as follows: a) improving or creating computing and network infrastructure, b) supporting training programs and postgraduate studies for young researchers, c) creating incentives for good researchers so that they stay in their countries, thus helping to reduce brain drain, d) implementing educational programs for the public in space sciences in order to increase awareness and public support, and e) increasing international cooperation in research projects.

Of all the points mentioned above, improving the computer and network infrastructure is perhaps one whose impact can have the most immediate results. Nowadays there exist extensive data archives on space sciences readily available at virtually no cost to any one with INTERNET access. Space missions of millions of dollars such as WMAP, COBE, ROSAT, IRAS, IUE, HST etc. have made public releases of their data. The same data is available to a researcher in Germany or Nigeria provided that there is access to a computer network. Subsequent to the first results published by the COBE collaboration astronomers throughout the world started using the released data resulting in more than 160 papers produced. Certainly the re-analysis papers did not have associated with them the US$ 400 million price tag of the COBE project. For the first time it is possible to participate in "big science" on a limited budget. Computer networks also allow the access to electronic mail and electronic publications with no delay, solving the traditional problem of isolation and obsolete libraries in developing countries.

An additional feature offered by computer networks is the possibility to have CPU power at a remote site, obviating the problem of having to buy expensive machines to install locally.

In summary, the task of the scientist is to bring to the public attention and to the attention of the politician the need to develop adequate conditions for research. To achieve this goal researchers needs to invoke the cooperation of the relevant local and international organizations. On the other hand, governments should never abandon support of the small research teams that show promise. The combined efforts of individual scientists and the actions of governments and international organizations could bring a change to the bleak situation faced by researchers in a world where market interests and social emergencies dictate where public spending goes. One must not forget that the joy of exploring the universe and the excitement of finding new discoveries belongs to the human culture as a whole, not just to a privileged group of countries.

Acknowledgments

Special thanks to Jorge Sahade who provided large amounts of the information presented here.

References

Ailes, C. P., et al., 1988, Technical note STPP-TN-3164-4, SRI International (Washington), 307pp. (astro-ph, *http://arXiv.org/; http://babbage.sissa.it/archive/astro-ph*)

Chipman, R., 1982, "The world in space - United Nations. A survey of space activities and issues", Ed. Ralph Chipman, Prentice Hall (New Jersey), 689pp.

Doncel Invernizzi, F.A., et al., 2003, this volume.

Etchegoyen, A., et.al., 2003, AP&SS, *in press.*

Gibbs, W., 1995, "Lost Science in the Third World", Scientific American, August, 1995, 76.

Haubold, H.J., et al., 1995, ESA Bulletin **81**, 18.

Haubold,H.J., and Wamsteker,W., 2003, this volume.

Hoeneissen, B., Torres, S., Violini, G., 1992, Nuclear Physics. B., (proc suppl.), **28B**, 191. (*www.isinet.com*)

May, R. M., 1997, Science, **275**, 793.

Moran-Lopez, J. L., 2000, "Physics in Latin America Comes of Age", Physics Today, October, 2000, p. 38. (NSF, (2002). "Science and Engineering Indicators - 2002", National Science Board, NSB-02-1)

2003, Red Iberoamericana de Indicadores de Ciencia y Tecnologia. *http://www.ricyt.edu.ar/.*

Roeder, J. G., 2003, "Early Cosmic-Ray Research in Argentina", Physics Today, January 2003, p. 32.

Sahade, J., 1992, in Status and Problems of Science in Latin America and the Caribbean, Eds. M.L. Guardiola, J.L. Villaveces, and G. Violini, (Guadalupe: Bogotá), 225.

Torres, S., 2003, this volume.

UNESCO, 2001, "The State of Science and Technology in the World".

SECTION:

4. SUB-SAHARAN AFRICA

Chapter 4.1

ASTRONOMY IN SOUTH AFRICA

B. Warner
University of Cape Town
Department of Astronomy
Rondebosch, South Africa
warner@physci.utc.ac.za

Abstract This note reviews those aspects of astronomy in South Africa that may be useful for the support of astronomical development on the rest of the African continent.

1. Infrastructure

Optical and infrared astronomy are largely consolidated at the Sutherland site of the South African Astronomical Observatory (SAAO), about 350 km from Cape Town. The administrative and technical headquarters of this institution occupy the buildings of the old Royal Observatory (founded in 1821) in Cape Town. In Cape Town there is a twin 18/24 inch (0.45/0.6m) refractor and a 18-inch (0.45m) photometric reflector that are no longer in active use but are operational.

At Sutherland the 74-inch (1.8m Radcliffe) reflector is equipped for spectroscopy, CCD imaging and IR photometry. Newtonian and coudé foci are available but not commonly used. Also at Sutherland there are 40-inch, 30- inch and 20-inch telescopes (approx. 1m, 0.75m and 0.5m) used primarily for CCD imaging (40-inch), optical and infrared photometry (30-inch) and UVBRI standard photometry (20-inch). An automatic photoelectric telescope (30-inch aperture) is in operation. Other telescopes, owned and managed by overseas institutions, also exist at Sutherland. Observing conditions at Sutherland provide about 50 % photometric time.

The 10-m class telescope known as SALT (Martinez, 2003) is scheduled for completion at the end of 2004. It will be available for use by suitably qualified users throughout Africa.

W. Wamsteker et al. (eds.), Developing Basic Space Science World-Wide, 179-181.

The SAAO is a National Facility, funded through the National Research Foundation (NRF). Extensive use of the SAAO has been made by overseas users during the past three decades. Many international collaborative projects have been and are being undertaken. Applications for observing time may be made to the Director, P.O. Box 9, Observatory 7935, South Africa. The SAAO also supports together with the French Observatoire Midi-Pyrénées to the publication of the Newsletter"African Skies/Cieux Africains" under the auspices of the Working Group on Space Sciences in Africa (Martinez, 2000).

Radio astronomy is concentrated at Hartebeesthoek Radio Astronomical Facility (HartRAO) sixty kilometers north-west of Johannesburg. This is a National Facility, equipped with a 26m equatorially mounted radio dish (evolved from a NASA Deep Space Station) suitable for centimetric and longer wavelength studies and operates in the wavelength range from 2.5cm to 18 cm. Radio, spectroscopic, VLBI, survey and pulsar-monitoring studies are undertaken, including international collaborative studies. Variable sources and radio-continuum mapping are also included as main field of interest. Applications for observing time or collaborative studies may be made to The Director, HartRAO, P.O.Box 443, Krugersdorpf 1740, South Africa.

2. Education

Astronomical education at the tertiary level is available at several universities, the principal of which are:

1. University of Cape Town (UCT), which has provided undergraduate courses in the Department of Astronomy since 1970 and B.Sc.(Hons), M.Sc. and Ph.D. programmes. The research programmes make use of the SAAO facilities. The Applied Mathematics Department runs a cosmology and general relativity programme (under Prof. G.R. Ellis). The Physics Department has research programmes in Big Bang physics. A new programme - the National Astronomical and Space Science Programme (NASSP) has started in 2003, based at UCT. This brings lecturers from several Universities within South Africa to provide intensive courses at Honours and Masters degree levels;
2. Rhodes University (Grahamstown), Physics and Electronics Department specializes in research in radio astronomy. The personnel are the principal academic users of HartRAO;
3. University of Potchefstroom, Department of Physics, specializes in ground-based gamma-ray astronomy, cosmic-ray, heliosphere and high-energy astrophysics research;

4. University of South Africa (Department of Mathematics, Applied Mathematics and Astronomy, P.O.Box 392, Pretoria 0001, South Africa) is a correspondence university with about 150,000 students of which 87% are from South Africa,11.5% from the rest of Africa and the remainder from all parts of the world. They have about 90 first-year astronomy students, 10 second-year and 3 third-year students, on average. In addition, B.Sc. (Astronomy), B.Sc. (Honours), M.Sc. and Ph.D. courses are offered;
5. Astronomy courses and supervision are also offered at the University of the Witwatersrand, the University of the Orange Free State (Bloemfontein) and the University of Natal (Durban).

An advantage of tertiary education in South Africa is that fees and living costs are much lower than the equivalents in Europe or North America. Degree standards are maintained at international levels (by the use of, for example, overseas examiners for higher degrees).

The Universities, libraries and scientific institutions in South Africa are connected by an academic network, which itself is connected to Internet.

Other astronomical activities include major planetaria at the South African Museum (Cape Town) and the University of the Witwatersrand, which process about 100,000 schoolchildren per year, as well as large numbers of adults. The Astronomical Society of Southern Africa is a largely amateur organization with significant professional membership.

References

Martinez, P., 2000, Bull. On Teaching of Astronomy in the Asia-Pacific Region, **16**, 6.
Martinez, P., 2003, this volume.

Chapter 4.2

THE SOUTHERN AFRICAN LARGE TELESCOPE (SALT): A 10 METRE CLASS SPECTROSCOPIC SURVEY TELESCOPE

P. Martínez
South African Astronomical Observatory
P.O. Box 9, Observatory 7935, South Africa
peter@saao.ac.za

Abstract Over the next five years, South Africa and its international partners will construct the Southern African Large Telescope (SALT), a 10-metre optical/near-infrared spectroscopic survey telescope. SALT is based on a design that costs about 20% of a conventional telescope, while allowing about 70% of the science capability of the more costly conventional design. This paper discusses the main design elements of SALT and gives a brief overview of the anticipated suite of first-light instruments. The paper also traces the arguments used to justify the construction of such a facility in Africa. The collateral benefits expected from this project are discussed, along with suggestions of how to realize these benefits. SALT has the potential to promote the development of astronomy on the African continent. The paper ends with a description of the African Network for Astronomy Education and Research, an initiative currently in the pilot phase.

Introduction

Since the early 1970s, the major facility for optical/infrared astronomy in sub-Saharan Africa has been the South African Astronomical Observatory (SAAO) in Sutherland. By exploiting advances in detector quantum efficiency and computing power, SAAO has made significant contributions to modern southern-hemisphere astronomy. The largest of the SAAO telescopes is the 1.9-m Radcliffe reflector, which is over 50 years old. South African astronomers realized that in order to remain internationally competitive in ground-based astronomy well into the 21st Century, they required a telescope in the 8- to 10-metre class. Since a conventional telescope of this aperture was

W. Wamsteker et al. (eds.), Developing Basic Space Science World-Wide, 183-194.

clearly unaffordable, it was decided to adopt the design of the Hobby-Eberly Telescope (HET), built in Texas by a consortium of American and German astronomers. This telescope represents a radical departure from conventional telescope design, and offers much of the functionality of a conventional telescope at a fraction of the cost. The decision to build SALT as a copy of the HET was welcomed by the HET collaboration, as they had been considering the construction of a southern-hemisphere counterpart of HET. The South African astronomers thus went into partnership with the some of the partners in the HET collaboration for the construction of SALT.

Even at the reduced cost of the HET design, South Africa still required additional partners. However, with SAAO's well-developed infrastructure in Sutherland, it seemed likely that South Africa would succeed in attracting international partners to join a 10-m telescope project in the Southern Hemisphere. In 1998, the South African Government announced that it would commit 50% of the funds required to construct SALT and to operate it for ten years, provided that South African astronomers could find foreign partners to raise the balance of the funds. By November 1999, commitments had been secured from a number of international partners, and the project was allowed to proceed.

As of August 2001, the SALT partnership included South Africa, the HET Board, Poland, Rutgers University, Göttingen University, the University of Wisconsin-Madison, Carnegie Mellon University, the University of North Carolina, Canterbury University (New Zealand), Armagh Observatory (Northern Ireland), and a consortium of universities in the United Kingdom comprising the University of Central Lancashire, Keele University, Nottingham University and Southampton University. The National Research Foundation of South Africa is the largest shareholder, although its initial 50% stake in the project has dwindled to 40% because the South African rand has weakened against the United States dollar since the Foundation's commitment of R50 million in cash and a non-cash contribution equivalent to $1.76 million.

Figure 4.2-1 A computer-generated rendition of the Southern African Large Telescope (image courtesy of the SALT Board).

Figure 4.2-2. Architects' model of the SALT Visitor Centre currently under planning.

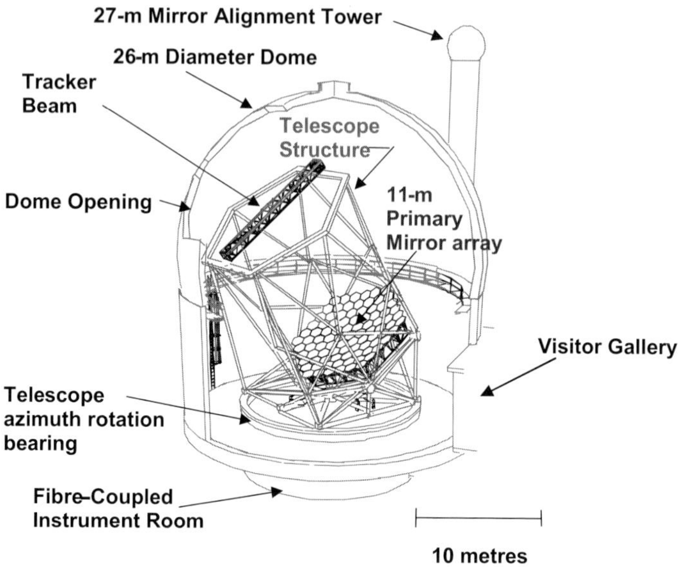

Figure 4.2-3. Schematic view of the SALT facility (image courtesy of the HET Board).

The total funding required for the SALT project is \$30 million over ten years. This comprises \$18 million for the telescope system, \$4.5 million for scientific instrumentation and \$7.5 million for the first ten years of operation. As of this writing (August 2001), SALT is about 90% funded for construction and the first ten years of operations, but the instrumentation is not fully funded. The SALT consortium is therefore not yet closed, and prospective additional partners are in the process of raising funds to join SALT on a 'first come, first served' basis.

How does South Africa justify building one of the world's largest telescopes when it has so many other pressing needs? This paper outlines the ideas that have emerged (and continue to emerge) in response to this question. Section 2 contains a brief description of the design of SALT. Section 3 discusses SALT science goals and instrumentation options. Section 4 discusses current plans for a programme of collateral benefits to be implemented during the construction and operation phases of SALT. Section 5 discusses SALT as an African facility.

1. The Design of SALT

SALT is based on the design of the Hobby-Eberly Telescope (HET), which was built at McDonald Observatory, in Texas. For a description of the HET project, refer to the paper by Ramsey et al. (1998) and references therein. Briefly, the HET design concept is as follows. The telescope has a segmented spherical primary mirror array 11 metres across. The mirror array comprises 91 hexagonal mirrors, each 1 metre across. This mirror array is supported in an open space frame structure inclined at a fixed tilt (37 degrees, in the case of SALT). Each mirror segment is individually oriented by a computer-controlled mirror mount. The mounts are required to align all the segments to an accuracy of 0.06 arcsec RMS. The telescope structure moves only in azimuth (not elevation), and only when slewing. Rotation in azimuth is accomplished by means of floating the telescope structure on an air cushion (figure 4.2-3). During an observation, the telescope structure remains stationary, and tracking motions are accomplished by a tracker beam that moves across the top hexagon of the telescope structure.

In one full rotation, the telescope sweeps out a 12-degree-wide annulus on the sky, centred on the telescope's latitude. Astronomical objects are accessible to this telescope only when passing through this 'annulus of visibility'. The amount of time it takes an object to cross the annulus of visibility is declination-dependent. At the worst extreme, near the celestial equator, objects can be tracked for about 48 minutes, while at the southern extreme of the annulus of visibility, objects can be tracked for up to 2.4 hr. This feature of the telescope's operation dictates that observations must be carried out in a queue-scheduled mode. That is, unlike a conventional telescope, which is dedicated to a single programme for a whole night, SALT will typically observe targets from many different programmes during the same night. The fixed tilt of the telescope simplifies the structure considerably since the gravity load on all structural members and on the mirror array is constant. Likewise, the use of a spherical primary means that all 91 mirror segments are identically figured. Together, these cost-saving features allow access to 70% of the sky accessible to a fully steerable conventional telescope at 20% of the cost.

SALT will operate in the wavelength region of 0.3 to 2.5 microns. The thermal background increases rapidly redward of 2.5 microns, and the telescope is not efficient at those longer wavelengths. Moreover, the altitude of Sutherland (1,757m) makes it uncompetitive in the thermal infra-red, compared to other high-altitude sites.

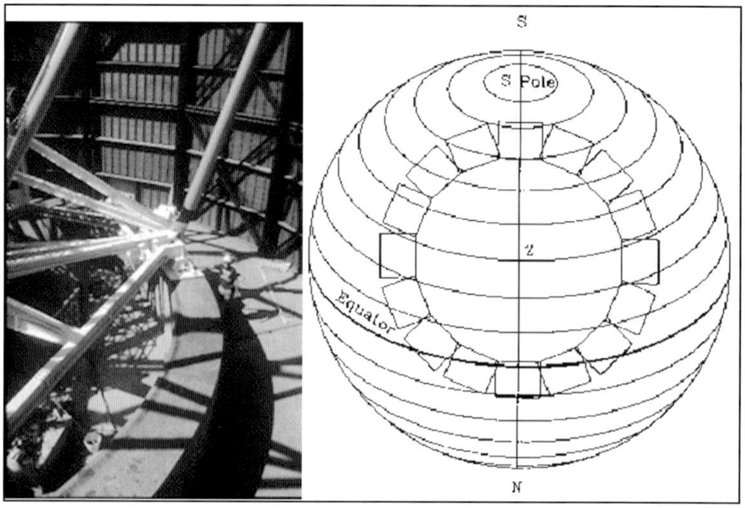

Figure 4.2-4. The telescope structure for SALT rests on a concrete ring as seen in this HET image (left). The footpads (one of which appears in the centre of the image) contain airbags, which are inflated to allow the telescope to rotate on an air cushion. During an observation, the airbags are deflated and the telescope structure rests on the ring wall. Owing to the fixed-tilt design of the structure, the telescope sweeps out an 'annulus of visibility' 12° wide on the sky (right), covering the declination range $+12° \leq \delta \leq -76°$. Targets are observable only while crossing this annulus of visibility. Images courtesy of the HET Board.

One of SALT's most complex subsystems is the tracker beam, which moves across the top hexagon of the telescope structure to accomplish tracking. This tracker beam weighs 1.5 tons and must be capable of 10 degrees of freedom to move accurately on a spherical focal surface with a linear precision of 6 microns. The tracker beam will support the Prime Focus Instrument Platform (PFIP), a package housing the spherical aberration corrector, atmospheric dispersion corrector, exit pupil baffle system, acquisition/guide charge coupled device (CCD) camera, a Prime Focus Imaging Spectrograph (PFIS), and a fibre feed to medium- and high-resolution spectrographs in an environmentally controlled chamber under the telescope. For a more detailed discussion of SALT instrumentation plans, refer to the paper by Buckley et al. (2000).

SALT is a copy of the HET, but not an identical copy. The HET Board is a partner in the SALT project, and the SALT project team is benefiting from the hindsight of the team that built the HET. The tilt of SALT has been changed from HET's 35-degree tilt to 37 degrees to allow SALT to access most of the Small Magellanic Cloud. SALT will also incorporate edge-sensors to facilitate alignment of the primary mirror array, a modification currently being applied to the HET as well.

The most significant departure from the HET design is in the manner in which spherical aberration is corrected in SALT. An improved Spherical Aberration Corrector design by O'Donoghue (2000) promises to deliver substantially better-quality images, an 8-arcmin field (double that of the HET), somewhat reduced vignetting, and an increased pupil size (pending implications of redesigning the top hexagon). Potentially more efficient protected silver coatings offering much improved blue performance (λ < 400 nm) are being investigated. A different primary mirror segment alignment system, probably using Shack-Hartman and possibly even an artificial Rayleigh beacon, is under consideration. The design of the SALT dome uses louvres to provide more natural ventilation and assumes a more aggressive attitude to heat sources in the dome. The prime focus instrument platform is planned from the outset to have a larger mass/volume envelope. The possible use of a cooled moving baffle to decrease the infrared background is also under discussion. The differences between SALT and HET are discussed in more detail by Stobie et al. (2000).

The SALT project team was appointed early in 2000 and the ground-breaking took place on September 1, 2000. As of August 2001, contracts have already been placed for the telescope structure, dome drives and controls, the mirror support truss, mirror segments and the polishing of these segments. The civil works commenced in earnest early in 2001. The first polished segments are expected to be installed in the last quarter of 2002. Commissioning is expected to take place in 2004. The construction of SALT may be viewed on the SALT web site (http://www.salt.ac.za/).

2. SALT Science and Instrumentation

SALT is primarily a spectroscopic survey telescope with a spectroscopic capability over the wavelength range 0.35 to 2.0μm. It will be most competitive scientifically when used on astronomical targets uniformly distributed on the sky or clustered on a scale of a few arc minutes. Because the telescope will be queue scheduled it is particularly effective in time variability studies on time-scales of a day or longer. The telescope will be able to address topical problems in a wide area of astrophysics, inter alia:

- Study of the early Universe and cosmology
  Hubble medium deep survey, luminosity functions
  X-ray Multi-mirror Mission (XMM) and *Chandra* surveys
- Synoptic observations of variability
  gamma ray bursters, support for satellite observations (e.g. *SWIFT)*
  stellar variability (e.g. accretion, proto-stellar activity)

- Stellar Populations in the Milky Way and nearby galaxies
- Magellanic Clouds, nearby dwarf galaxies
  galaxies with different evolution, mergers – interactions – cannibalism
- Dynamical studies of stellar systems
  globular clusters and galaxies
  dark matter searches
- Planetary searches
  discovery of planets around other stars

The choice of first-light instruments for SALT should aim to maximize these primary science goals. Current instrumentation plans call for a Prime Focus Imaging Spectrograph (PFIS), and a fibre feed to medium- and high-resolution spectrographs in an environmentally controlled chamber under the telescope. It is likely that these instruments will have will have multi-object capability and dual visible/near-infrared beams extending to at least the H-band (~1.7μm). For a more detailed discussion of SALT instrumentation plans, refer to the paper by Buckley et al. (2000).

3. The SALT Collateral Benefits Plan

SALT will enable South African scientists to remain internationally competitive in astronomy well into the 21st century. However, SALT has an importance to South Africa (and Africa) far beyond its astronomical research mission. SALT will provide distinct and substantial benefits in the development of people, technology, and the economy. The *SALT Collateral Benefits Plan* is designed to maximize the benefits from the investment of public funds in the construction and operation of SALT. There are three main thrusts to this plan: industrial empowerment, educational empowerment, and public outreach. The Collateral Benefits Plan is being drafted in consultation with stakeholders from government, industry and education. The discussion in the remainder of this section draws heavily on the discussion document "SALT Collateral Benefits Plan" (1999).

3.1 SALT Industrial Empowerment

The SALT project is being planned in such a manner as to maximize the industrial pay-offs from the development of components, subsystems, equipment, and labour for the construction and provision of SALT. Of the total of $18 million that is budgeted for the telescope, 57% is planned to be contracted to South African firms. Thus, for the $8 million that South Africa is contributing towards the telescope construction, $10.5 million will be spent in South Africa – a net inflow of $2.5 million.

The SALT procurement policy places high emphasis on empowerment and capacity building of people, firms and companies from previously disadvantaged communities in South Africa. Foreign companies contracted to work on SALT are being requested to host South African technical personnel at the contractor's facility to collaborate in the design and manufacturing efforts. Careful selection of personnel will maximise this benefit. About two thirds of the labourers involved in the construction of SALT have been recruited from the nearby town of Sutherland, providing work in an economically depressed area.

The result of these activities will be industrial collaborations joining South African and foreign companies in new ways to increase technical capabilities. This will result in technology transfer in the areas of optics, electronics, micro-positioning, dynamic control systems, precision sensing, and others. These companies will be able to compete for additional business within and outside of South Africa.

3.2 SALT Educational Empowerment

One of the biggest educational challenges confronting South Africa is to broaden the science, education and technology-training base, particularly in the black community. SALT can play a significant role in meeting this challenge by providing educational and training opportunities for astronomers, physicists, computer scientists, and engineers during the construction (5-yr) and operational (>25-yr) phases of the project. This will specifically involve student participation in the SALT project team, student placement with SALT contractors, student participation in collaborations with SALT partners, and student involvement in instrumentation development. Since these students all have to be registered at their own academic institutions, this will promote closer ties between SAAO and those institutions, and empower them in new technologies.

The involvement of students in all phases of the SALT project will equip them with the experience of participation in a large, complex, high-technology project, an experience that will be carried with them into later careers. Knowledge of the technology, interactions and techniques for development of such projects is vital, as South Africa must address ever more ambitious technical undertakings. Experience with SALT will provide confidence in using SALT's constituent technologies in future projects.

Another component of this educational thrust is to promote the development of astronomy at the historically black universities in South Africa, where astronomy has not featured in the past. This will have the advantage of increasing the numbers of scientists and engineers participating in the construction and use of SALT, and will also make SALT more accessible to the wider community in South Africa. The achievement of research success in astronomy at these institutions will have long-term spin-offs in terms of encouraging success in other fields at those universities by stimulating both students and faculty. It is also worth noting that the students trained in astronomy will be equipped with analytic, computing and problem-solving skills that will make them highly sought-after in technology-intensive industries in Africa.

3.3 SALT Education and Outreach Opportunities

The South African government recognizes that scientific literacy is the key to ensuring the technological competitiveness of the nation. Unfortunately, the majority of South Africans have had very limited access to science, engineering and technology education, training and careers in the past, with the result that the work force is largely unskilled in this vital area. South Africa requires a sustained campaign to promote awareness and understanding of science, engineering and technology at all levels in society. SALT will contribute to this process by stimulating excitement about science and technology in general.

A careful programme of educational and public outreach benefits is being compiled in collaboration with other stakeholders to capitalize on the interest generated by SALT. This programme includes activities such as training of educators in astronomy, a 'Starbus' science roadshow for outreach to rural communities, and development of educational resources (Rijsdijk 2001). To accommodate the large numbers of visitors expected to visit SALT, a visitor centre will be constructed in Sutherland. The visitor centre will contain interactive hands-on displays on SALT, astronomy, and other scientific aspects related to the Sutherland/Karoo region. A collateral benefit of these activities will be the economic development of the communities in the vicinity of SALT.

4. SALT as and African Facility

Capability in the space sciences will be a vital factor in the development of Africa in the 21st Century. The applied space sciences are underpinned by the basic space sciences, such as astronomy and astrophysics. At present there is a dearth of large-scale astronomical facilities in Africa, and African astronomers are forced to develop their careers elsewhere. This is reflected in the membership of the International Astronomical Union (IAU): currently, only four out of 53 African states are members of the IAU. Isolation, and lack of access to world-class facilities, are the two factors that retard the development of basic space science in Africa. The creation of new large-scale facilities like SALT will contribute to encouraging highly trained professionals to remain in their own cultural and political environment. As the development agenda of the region precludes the construction of another facility comparable to SALT in the foreseeable future, this project was conceptualised from the start as one that would serve the needs of the region. Hence the appellation *Southern African* Large Telescope. Steps are already being taken to develop an African user community for SALT by the time that the telescope is commissioned in 2005.

In order to promote the development of astronomy in Africa, one has to provide adequate opportunities for African astronomers to practice their science in Africa, rather than pursuing careers elsewhere. The advent of the Internet has revolutionized science by giving scientists everywhere unprecedented access to information and data. Although most African scientists do not enjoy the same standards of connectivity as their European or North American colleagues, many now have access to electronic mail, and limited access to the World Wide Web. This access provides the foundation upon which to build an "African Network for Education and Research in Astronomy". Such a network would allow astronomers from a variety of African countries to access instruments at one the best astronomical sites on the continent, without crippling investments in their own countries.

The Working Group on Space Sciences in Africa (Martinez 2000) and the South African Astronomical Observatory recently conducted a pilot project for an African Network for Education and Research in Astronomy. This pilot operated under the aegis of the United Nations Educational, Scientific and Cultural Organization (UNESCO)'s Pilot African Academic Exchange Programme. The South African Astronomical Observatory acted as the host institution for three visiting Fellows from Ethiopia, Uganda and Zambia. These Fellows worked for 6 months at SAAO, during which they developed research skills in astronomy, as well as the personal acquaintances so important in scientific collaboration. They also collaborated on the development of educational resources to be used upon returning to their home institutions.

The programme for the fellowship was devised to ensure the long-term goal of establishing a sustainable network after the fellows return to their home institutions. The scientific programme was developed to be topical, yet accessible to physicists with little or no background in astrophysical techniques. The focus of their work was on pulsating variable stars. This was chosen because of its scientific relevance, the availability of leading scientists in this field at SAAO, modest computing hardware and software requirements, and modest data storage and data transfer requirements. The Fellows gained considerable experience in observing with a variety of telescopes, ranging in size from 0.5 m to 1.9 m.

An important consideration in setting up this pilot project was to structure the network in such a way that the fellows may collaborate with each other fruitfully after returning home. To facilitate this, the fellows were provided with computers, printers and uninterruptible power supplies. The computers are loaded with a common set of software tools and teaching resources. Some measure of remote access (via e-mail) to the 0.75-m robotic telescope at SAAO is planned. In this way the fellows will continue to obtain new data of excellent quality for their own research projects. If this pilot project is successful, it will be repeated periodically and the network will grow with time. In this way, there could be a number of potential users of SALT by the time the telescope is commissioned in 2005.

Acknowledgements

The author gratefully acknowledges generous financial support from the United Nations for participation in the Tenth United Nations/European Space Agency Workshop on Basic Space Science.

References

Buckley, D.A.H., O'Donoghue, D.O., Sessions, N.J., Nordsieck, K.H., 2000, in Optical and IR Telescope Instrumentation and Detectors, Eds. Masanori Iye and Alan F. Moorwood, Proc. SPIE **4008**, 72.

Ed. Alan Batten, Astronomical Society of the Pacific, 117. SALT Collateral Benefits Plan, 1999, South African Astronomical Observatory/National Research Foundation.

Martinez, P., 2000, Bulletin on Teaching of Astronomy in the Asia-Pacific Region, **16**, 6 (See also *http://www.saao.ac.za/~wgssa*).

O'Donoghue, D.O., 2000, in Optical Design, Materials, Fabrication, and Maintenance, ed. Philippe Dierickx, Proc. SPIE **4003**, 363.

Ramsey, L.W. et al., 1998, in Advanced Technology Optical/IR Telescopes VI, Proc. SPIE **3352**, 34.

Rijsdijk, C. 2001, in Astronomy for Developing Countries,

Stobie, R.S., Meiring, K., Buckley, D.A.H. 2000, in Optical Design, Materials, Fabrication, and Maintenance, ed. Philippe Dierickx, Proc. SPIE **4003**, 355.

Chapter 4.3

AN INTER-AFRICAN ASTRONOMICAL OBSERVATORY AND SCIENCE PARK ON THE GAMSBERG IN NAMIBIA?

H. Elsässer†
Max Planck Institute for Astronomy
Heidelberg, Germany

Abstract This paper concerns astronomical research in the southern hemisphere and concerns in particular the current situation in this hemisphere and presents a conceivable development for the African continent.

Editors note: This paper was originally written before the inception of the Southern African Large Telescope (SALT see also Martinez, 2003). However the visionary nature and motivation for plans for the long term future of Astronomy in Africa on a scale extending beyond the national boundaries deserves to be reported in this book. Some editing has been done to assure that the ideas expressed are maintained in the context of recent developments.

Introduction

Today, high-performance observatories for the exploration of the southern sky are mainly found in Chile. On the La Silla and Paranal mountains, the European Southern Observatory (ESO) has at its disposal a series of well-equipped telescopes with apertures of up to 8 meters. The Very Large Telescope Interferometer is already in early test operations. Of similar capacity are the observatories in Cerro Tololo of AURA, Las Campanas (Carnegie Institute) and the Gemini-South associated with its multi-national consortium. All these observatories profit from the favourable climate of the boundary between the Chilean coastal regions and the Atacama desert, at altitudes well above 2000 meters. Further improvements in local capacity are under way with the construction the Atacama Large Millimeter Array (ALMA) Observatory in cooperation

†Shortly after the completion of this paper Prof. Elsässer passed away

W. Wamsteker et al. (eds.), Developing Basic Space Science World-Wide, 195-201.

between ESO and the US National Science Foundation (NSF). In Argentina the Observatory El Leoncito (CASLEO) has been in operation since 1983 with a 2.15 m telescope at the continental side of the Andes mountain range (http://www.casleo.gov.ar/).

In Australia, in addition to the installations at the Mount Stromlo Observatory *(Editor's note: the Mt.Stromlo observatory was devastated by the major Canberra bushfires in January 2003)*, is the Anglo-Australian Observatory on Siding Spring Mountain, which has made important contributions to astronomy of the southern sky by means of its 4-m class telescope and a powerful Schmidt telescope.

Compared to South America and Australia, Africa was until recently somewhat behind. The South African Astronomical Observatory (SAAO) in Capetown with its observatory in Sutherland is a well-known institute with worldwide reputation and of great tradition and abundant first class results (http://www.saao.ac.za/). Construction has started on the expansion of the Sutherland site with the Southern African Large Telescope (SALT) as an international Project (Martinez, 2003). The SALT is expected to be commissioned in 2005 and foreseen to be an inter-African institute. The non-recurrent investments already made in the Sutherland site contributed to the decision the locate SALT at Sutherland.

In general, due to the considerable initial costs required for the operation of a modern, competitive observatory, it is logical to think of common efforts within a framework of international cooperation. Moreover, an inter-African institute could create ideal preconditions for scientific research across many national borders.

1. The Gamsberg and the "Max-Planck-Institut für Astronomie"

In 1968, the "Max-Planck-Gesellschaft zur Förderung der Wissenschaften" (Max Planck Association for the Promotion of the Sciences) established the "Institut für Astronomie" (Max Planck Institute for Astronomy, MPIA) with the objective enable German astronomers to keep up with the advances in international research in the field of optical observational astronomy. In addition to the Institute's headquarters in Heidelberg, the construction of two observatories was planned. The first is the Hispano-German Observatory at Calar Alto in Southern Spain. This is today equipped with five telescopes, including one with an aperture of 3.5 meters in service since 1985. The second observatory was foreseen to be dedicated to astronomical observations of the southern sky and to be similarly equipped with high-performance instruments

(Elsässer, 1985). In our search for a suitable site, the south-western part of Africa was pinpointed for consideration (Elsässer et al., 1971). As a rule, both in the southern as well as in the northern hemispheres, the most favourable climatic conditions are found in the range of the subtropical high-pressure belts in the range of approximately 20° to 40° terrestrial latitude. From the data collected at an earlier German observatory near Windhoek, a lot was known about the local weather conditions for that area (Schulze,1965).

In addition to a large number of clear nights, the site of a modern observatory requires a mountain with an elevation of at least 2000 meters above sea level, which should also project some 100 meters above its surroundings to keep the ground haze layers below the level of the observatory and maintain the benefits of a dry atmosphere of high transparency above. Another important aspect is that of "seeing", i.e. the effects of atmospheric turbulence. Turbulent air impairs the sharpness of the stellar images and reduces both the spatial resolution capacity as well as the limiting magnitudes that can be reached with the telescopes. Good seeing can be expected if the atmosphere is heated up as little as possible from the ground over the course of a day, and if it adapts quickly to the thermal condition of the free atmosphere after sundown.

A particularly appealing location appeared to be the 2350 m high Gamsberg mountain, a table mountain at the edge of the Namibian desert (see figure 4.3-1). Its plateau is 2.5 km long, with a portion of the table almost 1 km wide, and covers an area of 240 ha. In the eastward, inland direction, it rises by 500 m above the surroundings. In the westward direction, towards the Atlantic Ocean, the height difference is considerably greater. From Windhoek (international airport), the mountain can be reached in two to three hours via one of the country's main highways (Windhoek - Swakopmund).

In 1970, a station for test measurements was installed on the mountain, the results of which surpassed our expectations. The annual distribution of cloud cover is complementary to that of the Chilean La Silla and Cerro Tololo region. The season of low cloudiness in Namibia corresponds to longer winter nights from April to October. The yearly total of clear nights for observation is almost identical. The high purity of the Gamsberg atmosphere is not only supported by our photoelectrical extinction measurements, but also suggested by the impressive horizontal visibility extending, as a rule, for more than 100 km as far as the mountains of Windhoek. Air humidity (important for IR Observations) is frequently so low that it can not be measured with standard hygrometers. Seeing was registered in two different ways:

1. using photographic star trails, and

2. a photoelectric method (a knife edge following the image).

Figure 4.3-1. The Gamsberg (2350m above sea level) in Namibia; seen from the north-east.

Figure 4.3-2. The Telescope structure of one of the H.E.S.S telescopes at the foot of the Gamsberg seen in the background (Courtesy HESS collaboration).

The first method is easy to operate, while the second one is more accurate for small image-motions. For a detailed comparison with the photoelectric measurements made on the Gamsberg and La Silla over 100 nights each in 1971 to 1973, I refer to Birkle et al.(1976). Their results are shown in figure 4.3-3, and show that image-motion amplitudes of approximately 0.5 seconds of arc are most frequent in both places. This implies that the average observing conditions are very good.

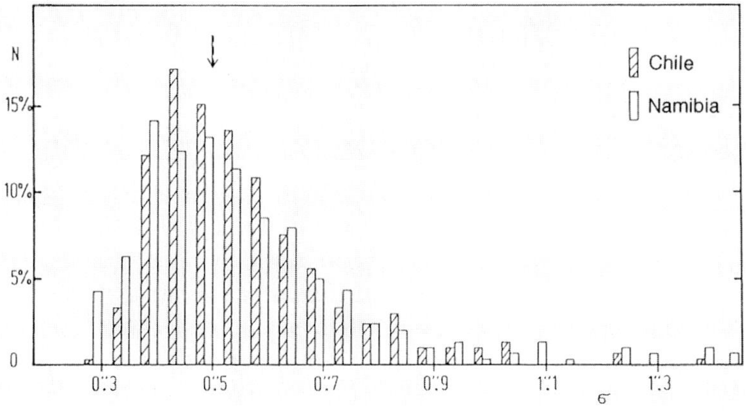

Figure 4.3-3. Seeing comparison between La Silla (Chile) and Gamsberg (Namibia) based on photoelectrical measurements; N frequency of the image motion σ in terms of arcseconds (Birkle et al., 1976)

However, especially the smaller values around 0.3 arcseconds were observed more often in Namibia than in Chile, a point which strongly reinforces the excellent quality of the Gamsberg.

Another important property of the at Gamsberg is complete absence of artificial light sky background. Settlements with nocturnal illumination are widely spread and so remote that light contamination of the night sky, a serious problem at most primary observing sites today, is practically zero.

On the basis of our first positive impressions, the plateau of the Gamsberg was purchased from its former owner, and since 1970 it has been property of the "Max-Planck-Gesellschaft". Today, one has to use a four-wheel drive vehicle to ascend the steep and bending road, which would need to be remade to allow realistic access for a science park. Over the years, a small observatory with a 50-cm telescope and several barracks for technical equipment and staff accommodation have been installed.

The experience gathered from extensive astronomical programs confirm the results of the initial test measurements and reinforces our conviction that the quality of the Gamsberg site is equivalent to the major observatory sites associated with the Atacama desert in Northern Chile. It is certainly one of the few excellent places left on Earth. The quality of the Gamsberg site should be recognized and preserved as a potential site for the future. This is not only important for the world astronomical community, but also for future generations of African scientists, which will, hopefully, be the result of the efforts described in this book. They need to be able to develop their own facilities on the continent.

2. A new start

Will there be chances for future scientific exploitation of the Gamsberg? The MPIA has made efforts to initiate a developments with this objective in mind. MPIA would like to work with others interested in the current efforts to encourage international cooperation with the objective of creating an inter-African Center (Elsässer, 2000).

Other basic space science disciplines are interested in this exceptional site. In a multi-national effort the High Energy Spectroscopic System (H.E.S.S. see figure 4.3-3) is being constructed and expected to be in full operation with 4 telescopes in 2004 at the foot of the Gamsberg plateau. This is dedicated to the observations of cosmic γ-rays in the 100 GeV range and will benefit very much from the excellent atmospheric conditions (http://www.mpi-hd.mpg.de/hfm/HESS/HESS.html)

Meteorology and atmospheric research could also benefit from the Gamsberg. It could e.g. be a suitable terrestrial basis for the continuous observation of solar ultraviolet radiation, monitoring the evolution of the ozone hole. The Gamsberg has long been on the mind of botanists and geologists. The flora on its plateau is unique to the local conditions. In would therefore be a logical idea to make the astronomical observatory part of a science park, open to other disciplines as well. The large extend of the table plateau makes this quite feasible. Such a center with an international aura, would undoubtedly be a great attraction and be very beneficial to the development of the country in the broadest sense.

3. Cost and Organization

A feasibility study on the development of the mountain and the expected cost of the required infrastructure has been made. It was estimated that a well-developed access road to the Gamsberg plateau as well as the necessary basic supply of electricity and water would require a total of approximately US$ 6 million.

About US$ 30 million is the present cost for the construction of a 4-m class telescope, including the dome and building. Accommodation for technical equipment and staff would require an investment in the range of US$ 2 to 3 million. This would represent a minimum infrastructure for a first stage from which further scientific activities could be developed. The annual running costs for this first stage would amount to some US$ 2-3 million.

Of course, the organizational form and the charter of such installation would need to be developed and negotiated at an international level. The purpose of this report was to attract attention to the project of an inter-African science park in Namibia. In the view of the author, the objective and chance to create a "Center of Scientific Excellence" on the basis of international cooperation in the southern part of the African continent should not be lost.

References

Birkle, K., et al., 1976,.Astron. and Astroph., **46**, 397.
Elsässer, H., Neckel, Th., Birkle, K., 1971, Mitteilungen der AstronomischenGesellschaft, **30**, 75.
Elsässer, H., 1985, Weltall im Wandel. Die neue Astronomie (Stuttgart: DVA), pp. 275.
Elsässer, H., 2000, Sterne und Weltraum, **No. 2/3**, 121.
Martinez, P., 2003, this volume.
Schulze, A., 1965, Climate of South Africa (South Africa: Weather Bureau).

Chapter 4.4

CURRENT DEVELOPMENTS IN BASIC SPACE SCIENCE IN NIGERIA

P.N. Okeke
University of Nigeria
Space Research Center
Department of Physics and Astronomy
Nsukka, Nigeria
misunn@aol.com

Abstract Astronomy is important to developing African countries. In this paper, a brief review of the situation of astronomical research in Africa before 1991 is given. During that period only South Africa and Egypt were carrying out observational research in astronomy. In other African countries astronomy research was in its infancy, except the University of Nigeria Space Research Centre (UNNSRC) in theoretical areas. A summary of the important recommendations for Africa at the United Nations/ European Space Agency (UN/ESA) series of workshops on basic space science were itemized to help identify those which have now been accomplished. Additionally, UNNSRC has now embarked on further observational programmes through the establishment of strong collaborative ventures with two observatories in South Africa, the Hartesbeesthoek Radio Astronomical Observatory (Hart RAO) and the South African Astronomical Observatory (SAAO). UNNSRC has also made permanent arrangements with HartRAO, SAAO, and the Jodrell Bank for collaborations in data analysis. A new interest in astronomy appears to have awakened in Nigeria with three more universities joining this area of basic space science. It is recommended that the time has come for all African countries to contribute towards a common facility such as the Southern African Large Telescope (SALT. The efforts of UN/ESA which resulted in tremendous achievements are commended.

Introduction - Is Astronomy Important for Africa?

Borrowing some of the arguments put forward by Stobie (1994), astronomy is widely seen as a major growth point in the physical sciences and therefore will play an equally indispensable role in the development of African science as has been the case in other

W. Wamsteker et al. (eds.), Developing Basic Space Science World-Wide, 203-209.

industrialized nations. There is an urgent need to strengthen the basic space sciences in Africa as a foundation for technological progress (Abiodun and Odingo, 1983). It is known from studies in other countries that contact with astronomy at an early age excites young minds and acts as a catalyst in encouraging students to follow careers in science and technology. Astronomy can be used to teach physical principles at all levels. For example, in explaining modern science and technology to the general public and promoting the interest of young people in a science career. By providing young graduates with exciting examples of the applications of physical principles and training at postgraduate levels where projects of real scientific or technological value are coupled with the development of a wide range of scientific and engineering skills.

Astronomy can thus do much to raise the general level of scientific awareness in Africa and draw young minds towards a career in physical science or associated areas of technology. Countries that see science as an essential part of their future wealth and well-being, participate actively in the development of astronomy. Any modern observatory requires not only astronomers but skilled engineers and technicians in electronics, mechanics, optics, computers and software in order to function. It requires advanced industrial capabilities and precision engineering to set up an astronomical instrument. Industrial capabilities acquired through the fabrication and instrumentation of an astronomical telescope could prove invaluable to companies developing hi-tech products. In addition to all this, it is important to note that some parts of Africa by virtue of their geographic position, excellent climate and good atmospheric conditions (e.g. Elsässer, 2003) satisfy most of the pre-requisites for setting up future astronomical observatories. Since astronomy involves international collaboration, it is important that all African countries should share to work towards the establishment of a front-line research astronomical telescope in the most suitable part of Africa as a central facility for all researchers. This instrument will help to complement observational results from other parts of the world.

1. The UN/ESA Workshops 1991-1994: Situation in Africa

This section of the paper attempts to review the situation of astronomy/basic space research and education in Africa during the period 1991-1994. This corresponds to the period during which UN/ESA organized four of the series. of workshops in various developing countries (Haubold and Wamsteker, 2003). At these workshops it was found that a high level of astronomy research/education exists in only two of the African countries namely, South Africa and Egypt. Details of astronomical research in these two countries can be found in Warner (2003), Martinez (2003) and Hassan (2003).

Before 1992/93, the situation in Nigeria and other African countries was as follows. In Nigeria, only one university in the country - the University of Nigeria Space Research Centre - was making contributions in theoretical research in high energy astrophysics and cosmology, but not in observational astronomy. However, relevant research is being done in some other areas of space research such as geomagnetism, geodesy, meteorology, remote sensing and communication physics by at least six universities in the country; these include: Universities of Obafemi Awolowo, Nigeria, Lagos, Ibadan, AhAmadu Bello and Rivers State. The Basic Space Science Workshop held in Nigeria also identified a small number of astronomers in some other countries in the African region, including Algeria, Saudi Arabia, Kenya, Malawi, Morocco, Tunisia, and Republic of Congo. However, based on replies the questionnaires distributed to all African countries, which unfortunately may not have reached all concerned, it was found that other African countries are carrying out research mainly in other areas of space science such as atmospheric physics, remote sensing, meteorology and geodesy. The level of research being carried out in these areas, was not assessed. But it was clear is that astronomy education/research in these countries, if any, was still in its infancy. Based on the above, it must be emphasized that there is an urgent need to form an "Association of African Astronomers" which could meet annually or biennially to discuss astronomical problems in Africa. This would not only foster inter-continental collaborations but also would enable African astronomers to know the nature of astronomical research going on in the African continent (see also Martinez, 2000).

2. Recommendations Emanating from the Workshops

The UN/ESA workshops at a number of meetings arrived at some of the following important recommendations:

It was recognised:

- That global joint observations and multilateral collaborations are indispensable for both ground-based and space-based astronomical research;
- That most of the African countries have no viable nor front-line observational facilities, and since astronomical research/education is still in its infancy in Africa. It was strongly recommended that the most effective way of getting the African countries to catch up with recent developments in astronomical research is to encourage them to

be involved in intercontinental and international collaborations with more advanced countries;

- That researchers and post-graduate students from African countries should be encouraged to learn and use observational facilities in more advanced countries;
- That researchers and post-graduate students should learn data handling techniques, analysis and interpretations. It will then be possible for them to access copious data from various observatories which they can handle at their home base assuming that they have acquired suitable powerful computers. In this way they could contribute immensely to one important aspect of astronomy research - data analysis and interpretation. To assist in this, some international organizations have in 1994 already donated thirty computers systems to six developing countries;
- That institutions in African countries need to be linked to the world through the Internet. This would ensure close contact between researchers in different parts of the world and, additionally, it would permit access to remote databases and computer power from any part of the world. This would drastically reduce the isolation factor.
- That since South Africa has the premier observatories for optical and infrared astronomy as well as radio astronomy on the African continent, South Africa is in a unique position to act as the focus for the development of basic space science in Africa. The initiatives already taken in South Africa (Martinez, 2003) in this context, are very important.
- That the Solar Seismology project between Nigeria, the United States (Arizona), the Russian Federation and China should be strongly pushed to get the project firmly off the ground;
- The importance of the support of advanced countries and organisations to assist researchers from African countries with travel funds and subsistence allowances to attend international meetings and conferences;
- That short-term fellowships to African astronomers to participate in observational programs at observatories in advanced countries are of great importance. The efforts of international professional organizations like the IAU present already an important support.
- That some countries have promised to send their preprints and reprints including their extra back-volumes of journals to identified African astronomical institutions.

3. The Impact of UN/ESA Workshops on Astronomy in Nigeria

Based on the above recommendations and results, I will try here to explain how the UN/ESA workshops have dramatically improved the level and direction of astronomical research in Nigeria as a whole and UNNSRC in particular which is currently incorporating meaningful observational programs into their theoretical work in astronomy. Additionally, the number of astronomical institutions in the country has significantly increased in the past years. Moreover, a new interest in astronomy appears to have awakened throughout the country as many students are now anxious to have courses offered in astronomy at both graduate and post-graduate levels. During the UN/ESA Workshop held in Nigeria in 1993 (Onuora., 1999; Haubold and Onuora, 1995), based on meetings and discussions held with the two Directors of the South African astronomical facilities, it became clear that in order for Nigeria to initiate observational programs in astronomy, for the time being (pending the availability of national observational facilities) the first four items of the UN/ESA workshops recommendations listed above should be adopted. The details of our collaborative efforts and other developments are as follows:

1. *New Collaborative Efforts with SAAO and HartRAO*
   The SAAO in Cape Town and the HartRAO have both initiated collaborative efforts with the University of Nigeria Space Research Centre (UNNSRC). This collaboration was enhanced with the appointment, in 1995, of the present author, as an external board member of the National Astronomical Facilities Board by the Foundation for Research Development (FRD) of South Africa, furthering the collaboration with both the board members and the President of FRD and had their full support. The President of FRD in fact will also willingly extend the offer of collaboration to other qualified astronomers from other African countries. Various collaborative programs were initiated on optical study of cataclysmic and pulsating variable stars; Radio observations of pulsars.
   P.N.Okeke and A.Ubachukwu were formerly theorists, expanding their experience in observational and data handling techniques, they will be in a position to guide their post-graduate students on their return to Nigeria. Ubachukwu's travel fund was paid by the IAU and subsistence by HartRAO. Okeke's travel fund was provided by FRDand subsistence by SAAO. The Nigerian Government plans to sponsor future observational programs of PhD students at SAAO and HartRAO respectively.

2. *Astronomy In Other Universities In Nigeria.*
Before 1992, astronomy existed only in the University of Nigeria, but by 1996 three more universities have introduced astronomy in their university curriculum. Below are the list of universities and names of staff in charge:
Imo State Univ.- Prof.C.E Akujor - a former PhD astronomer from UNN.
He is an observer and uses facilities at Jodrell Bank and VLA;
Namdi Azikiwe Univ. - Dr. G. Anene - a former PhD astronomer from UNN, with additional specialized training at one of the Indian observatories, and
Obafemi Awolowo Univ. Prof. E. Balogun is currently introducing astronomy in this university.
3. *Plans for Data Handling Collaboration.*
Plans are under way to assist in the analysis of data obtained from three of the following observatories: SAAO, HartRAO, and Jodrell Bank. For this purpose arrangements are being made to install the five computers donated by ESA to UNNSRC, as soon as local formalities are cleared. Plans to purchase at least one additional powerful computer capable of analyzing the CCD images are also underway as well as arrangements to acquire the relevant software required for the various projects are being made. The UUNSRC also continues to mount astronomy popularization programs, through newspaper articles, radio, student quiz and TV programs.
4. *Plans For A Common Facility For Africa - SALT.*
For this it is best to refer to the paper by Martinez (2003) in this book.
5. *The creation of a Nigerian Space Agency*
In 2001 the Government of Nigeria created the National Space research & Development Agency (NASDRA) to advance Nigerian indigenous competence in developing, designing and building appropriate hardware and software in space technology as an essential tool for its socio-economic development and the enhancement of the quality of life of its people. NASDRA is the implementing Agency for the National Space Programme (NSP)
The scope of the NSP is:
- The study of basic space science in order to lay the foundation for deriving maximum benefits from the nation's participation in the space enterprise;
- The attainment of space capabilities, Nigeria's efforts should focus on research and rigorous education, engineering development, design and manufacture, particularly in the areas of instrumentation, rocketry and

small satellites as well as in satellite data acquisition, processing, analysis and management of related software;

- The establishment of a national earth observation station for remote sensing and satellite meteorology data acquisition. Such an infrastructure will enhance the indigenous ability to adopt, modify and create new techniques for national resources inventories, monitoring, evaluation and management;
- The provision of efficient, reliable and adequate telecommunications services in Nigeria in order to enhance the growth of the industrial, commercial and administrative sectors of the economy.

4. Conclusion

The UN/ESA efforts to improve the state of astronomy research in Africa have not been in vain. Tremendous success has been achieved by many African countries just within four years. Other countries in Africa should also endeavor to adopt in the first instance intercontinental and international collaborations as a starting point. There is a strong indication that astronomy research/education in Africa will accelerate in the forthcoming years. The establishment of a large central observational facilities is the highest priority item for the future development of optical and infrared astronomy in Africa.

References

Abiodun, A.A. and Odingo, R.S., 1983, Adv. in Space Res., **3**, 79.
Elsässer, H., 2003, this volume.
Hassan, S.M.,2003, this volume.
Haubold, H.J., and Wamsteker, W., 2003, this volume.
Haubold, H.J. and Onuora, L.I., 1995, Eds. Basic Space Science, AIP Conference Proceedings, **320**.
Martinez, P., 2000, Bulletin on Teaching of Astronomy in the Asia-Pacific Region, **16**, 6.
Martinez, P., 2003, this volume.
Onuora, L.I., 1999, "The UN/ESA Workshop on Basic Space Science in Nigeria: Looking Back", COSPAR Information Bulletin, **144**, 15.
Stobie, R.S., 1994, "The Future Development of Astronomy in South Africa", in Southern African Large Telescope (SALT) Proposal (South Africa: South African Astronomical Observatory).
Warner, B., 2003, this volume.

Chapter 4.5

TOWARDS INTRODUCING SPACE SCIENCE IN UGANDA

S. Anguma and J. Ayikoru
Mbarara University of Science and Technology
P.O. Box 1410, Mbarara, Uganda
mustmed@infocom.co.ug and anguma_simon@hotmail.com

Abstract This paper discusses the strategies and importance of introducing space science in Uganda. It proposes that Mbarara University, as a new university focusing on science and technology, would be ideally situated to spearhead the introduction of space science in Uganda. It is our expectation that this will have a spin-off effect to other higher institutions of learning and that consequently space science will become fully incorporated into the national teaching curriculum for all schools in Uganda. Based on the fact that the Government has a deliberate policy of popularizing science and technology to accelerate national economic development, the introduction of space science in the school system is to be enhanced by these efforts. We have charted the way forward for space science in Uganda and outlined the conceptual framework illustrating the spin-off effect into the education system.

Introduction

Uganda is a country in the East African region and is located between the longitudes of 30 and 35° east. The equator passes through Uganda, and the country stretches 4° north and 1.5° south of the equator. The country has borders with Sudan in the north, Kenya in the east, the Democratic Republic of Congo in the west and Tanzania and Rwanda in the south. It covers a total area of about 242,554 square kilometres and the entire country lies about 1,000-m above sea level, generally decreasing in altitude from south to north. The shaded area in the map shows the location of Uganda on the African continent. (figure 4.5-1).

W. Wamsteker et al. (eds.), Developing Basic Space Science World-Wide, 211-218.

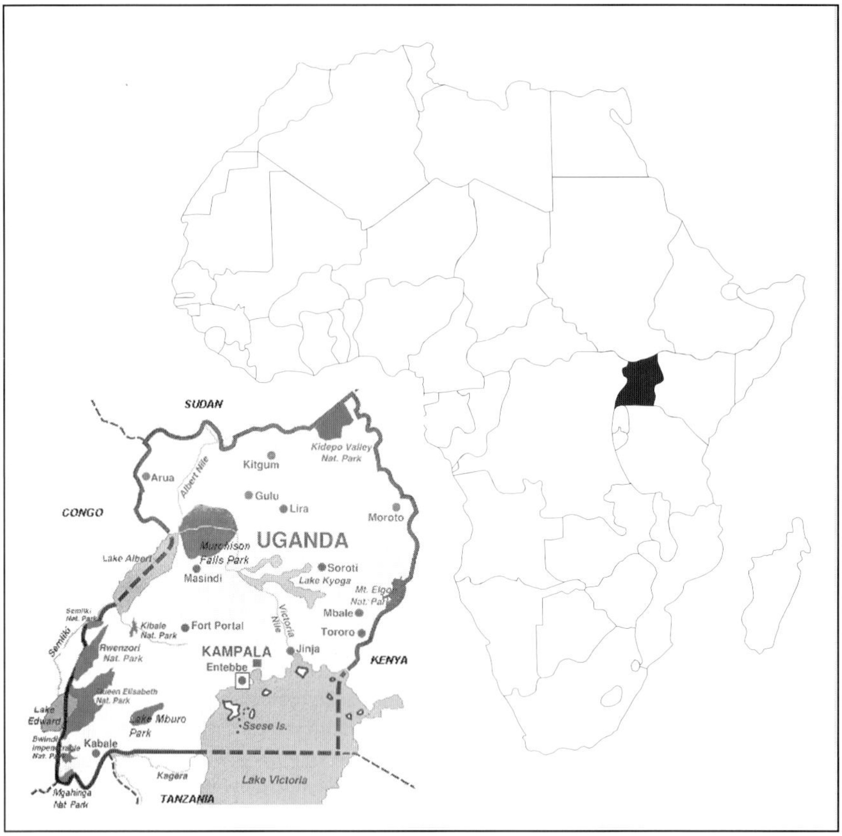

Figure 4.5-1. Map of Africa with inset of enlarged Uganda

The Government of Uganda has, over the years, embarked on an expansion of the country's industrial base, mechanization of agriculture and development of improved and drought-resistant varieties of crops and seeds in order to realize sustainable economic development. However, the availability of a skilled, dedicated and well-motivated human resource base is essential for such a programme to be implemented effectively. It has therefore been the Government's deliberate policy to popularize and develop science and technology, in order to provide the knowledge, skills and professional and technical expertise that the country requires.

Unfortunately, not many students opt for science subject combinations at high school. This is a result of many factors, including a lack of motivation arising from, among other things, the scarcity of qualified and competent science teachers in schools. Moreover, the expertise in science and technology that is required for economic development can be nurtured only by people with an interest in science. As a result, over the past decade, the Government has implemented structural changes in some existing institutions and created new science-oriented schools and universities in an effort to popularize and further develop the culture of science and technology in the country. Listed below are some of the many approaches.

Mbarara University of Science and Technology (MUST) was established by an act of Parliament in 1989. Among other objectives, it was created to address the acute shortage of science teachers and medical personnel in the country and to further develop the culture of science and technology countrywide. Over the years, numerous students have graduated with a "Bachelor of Science degree with Education" (BSc Ed), specializing in either physics or mathematics (see Table 1). Biology and chemistry options were introduced in 1997. It will not be long before Uganda starts witnessing the production of trained science teachers in all the natural sciences. The faculty of science with education is also under an obligation to start a programme to train computer science teachers for secondary schools and to start other new programmes that will support the country's development agenda.

The situation of rural schools, which used to suffer from a scarcity of professionally trained graduate science teachers, is beginning to improve. It is our sincere hope that the next generation of science teachers graduating from Mbarara University will include teachers of astronomy if the current window of opportunity, during which promotion of science and technology is at the top of the Government's national agenda, is used to introduce space science into our university teaching curriculum. This will begin a process to build capacity in human resources, setting the cornerstone for the development of space science in the country.

***Table** 4.5-1.* **Bachelor of Medicine and Bachelor of Surgery (MBChB) and Bachelor of Science with Education (BSc Ed) graduates from Mbarara University of Science and Technology since 1995.**

YEAR	MBChB	BSc Ed
1995	29	–
1996	57	–
1997	41	–
1998	44	–
1999	23	28
2000	34	39
2001	54	29
Total	282	96

(source: Office of The Academic Registrar, Mbararara University of Science and Technology)

Another example of the Government's commitment is the transformation plan for changing Uganda Polytechnic Kyambogo into a university with effect from October 2001. Kyambogo University will be born through a merger between the Institute of Teacher Education Kyambogo (ITEK), the Uganda National Institute for Special Education (UNISE) and Uganda Polytechnic Kyambogo (UPK). Another public university designed to train experts in agriculture will open in northern Uganda next October. This will bring the number of national universities in the country from one to four. In a recent speech, delivered on behalf of President Museveni by Mrs. Betty Akech, the State Minister for Higher Education, the President stated:

"...the Government is committed to establishing the universities to enhance its policy of supplying human resources to all levels of education…"

Earlier, the President had promised to raise government scholarships in Makerere University from the current 2,000 to 4,000 students a year and proposed a district quota system for admission, as opposed to the current system where the 2,000 brightest students benefit from state sponsorship, irrespective of where they come from. This initiative is designed to be implemented by October 2001. The beneficiaries of government sponsorship will be selected for courses with a heavy scientific content (New Vision, 2001).

In another development, Makerere University authorities have decided that most of the $1-million (1.8 billion Uganda Shillings) Carnegie Corporation scholarships it received last year will benefit female science students (New Vision, 2001). The Acting Academic Registrar, Sebastian Ngobi, specifically emphasized that about 70% of the scholarships will be granted to science students (loc.cit).

There is immense support from most stakeholders to popularize and develop science and technology to boost economic development. This is therefore an ideal opportunity to introduce a subject that will further enhance motivation, harness the curiosity and imagination of young people and hence attract them to study science subjects and technology with maximum enthusiasm.

These and many other structural, administrative and organizational transformations in some institutions, as well as the Government's new education policy plans, are all geared towards popularizing and developing science and technology for national economic and human resource development. Whereas the strategies listed above are commendable and strategically functional in the broader sense of development, it is important to recognize the power of space science to attract students to science subjects in general.

Before discussing how the outlined strategies will help introduce space science in schools, it is instructive to appreciate the advantages of space science in the development of a scientific culture in society. In the first place, space science is unique in itself and presents an attractive recipe for motivating young minds to pursue an interest in science subjects. It is therefore important for the following reasons:

1. Astrophysics is a “frontier science”. Unlike other natural sciences, it uses the universe as its laboratory in which physical laws and theories are applied, tested and refined at temperatures, pressures and scales unobtainable in laboratories on Earth (Stobie, 1995). It is quite exciting to deal with such monstrous quantities, which are a physical reality! Astrophysics is not only amazing in terms of the scales and quantities it deals with, but it also helps students appreciate mathematical concepts by applying them to the physical world.
2. Many students merely memorize physics course units, but astrophysics breaks them out of this rote learning, forcing them to think independently. It is an attractive science not only because it stretches the imagination but also because it is highly interdisciplinary. It involves many aspects of physics and other subjects. Due to this interdisciplinary nature, students are trained to solve problems that give them a broader view of science (Wentzel, 2003), something vital for the creation of a scientific culture in society.
3. Space science is concerned not only with the study of celestial bodies, as is the general misconception, but involves interesting disciplines of practical value such as remote sensing, geodesy, satellite communications, satellite meteorology and studies of the global climate, and space and atmospheric sciences. It is quite important to make people appreciate the general and inclusive nature of space science as opposed to the misconception that it is alien, narrow and beneficial exclusively to space-faring nations and superpowers.
4. Astronomy deals with our place in time and space and with our cosmic roots, the origin of our star, our planet, the elements in our bodies, and life, and reveals a universe which is vast, varied and beautiful. It also puts our planet into perspective and makes us realize that if we continue overpopulating it, we put all earthly life forms in danger (Fiero, 2000). Therefore, it is fundamentally important that space science be introduced at all levels of education in order to create awareness about our home in the universe.
5. Space science is the ultimate interdisciplinary subject. It harnesses curiosity, imagination and a sense of shared exploration and discovery. It attracts young people to science and technology and can

increase awareness, understanding and appreciation of these disciplines (Fierro, 2000)

6. Astronomy has influenced our history and culture, through both its practical applications and its philosophical and religious implications. This is reflected in calendars, mythology and a variety of art forms. It has practical applications to navigation, timekeeping, calendars, climate and other external influences on our environment (Fiero,2000).
7. Most interestingly, astronomy can be pursued as a hobby, in comparison to other science subjects, which not many people would wish to pursue and enjoy as a serious hobby. Amateur astronomers who enjoy the subject as a hobby at times contribute scientifically useful observations, something uncommon in other sciences!.

1. They Way Forward

Uganda does not have a rich and well-documented historical legacy of indigenous concepts in astronomy, although the different tribal groups do attach meanings to spatio-temporal positions of celestial objects. Of special interest are the positions of the Sun and Moon and the rising time of some planets and stars. These events, which are mainly useful for activities such as measuring time and predicting the onset and reliability of rainy seasons, have all remained undocumented.

Time is now ripe for us collectively to make the most of the limited human resource capacity available to improve the state of scientific literacy and to document our cultural practices in astronomy before they are totally forgotten or lost. It is important that we should initiate a collaborative network with countries and institutions that have well-developed space science programmes, and become fully involved in space science research.

Using the strategic advantage created by an enabling environment in which the development of science and technology is at the top of the national agenda, the introduction of space science in the teaching curriculum in Mbarara University is expected to kick-start the evolution of astronomy in Uganda. This will go a long way towards conveying scientific knowledge in a way that makes it attractive to basic school-level students, thus motivating the young minds to take an interest in science subjects, something that is consistent with the Government's effort in popularizing science for economic development.

2. Stages of Introducing Astronomy

To address the currently inadequate human resource capacity in Uganda, the initial goal should be to create a pool of "intermediaries" by training science teachers with a strong focus in astronomy. Mbarara University provides an excellent institutional facility for this. The faculty of science with education was established in 1995 to address the acute shortage of science teachers nationwide. Once astronomy is introduced in the University, teachers sufficiently knowledgeable about astronomy will graduate, constituting an initial, intermediate human resource base.

At this level, graduates are trained to teach in national teacher colleges, secondary schools and, in extreme cases of scarcity, teacher training colleges as shown in the conceptual framework in figure 4.5-2.

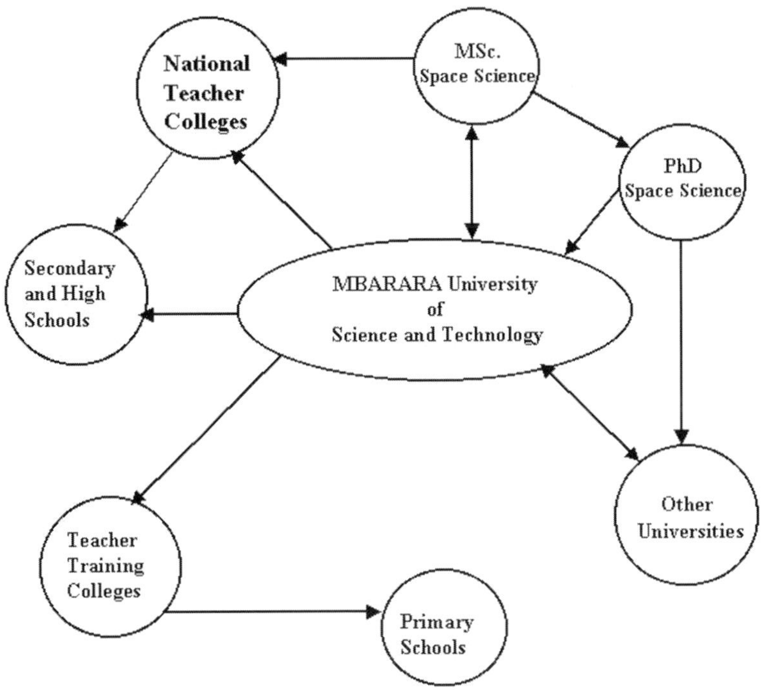

Figure 4.5-2. The conceptual framework for developing astronomy in Uganda

Some of these graduate teachers may continue pursuing M.Sc. and Ph.D. studies in space science, and can eventually become senior staff members in universities and other institutions of higher learning.

With more and more teachers produced and students qualifying with higher degrees in space science, sensitization of the Government and other stakeholders through seminars and workshops, and media coverage about inclusion of space science in the national curriculum, the creation of a national institute for space research will be strongly advocated. The research institute will, among other activities, organize public awareness campaigns on space science, provide advisory services on space issues to the Government and the general public, initiate innovative research projects and develop local research capabilities. It is at this stage that the importance of space science in shaping a scientific culture in society will be appreciated.

Acknowledgements

Firstly, I wish to express my sincere thanks to the administration of Mbarara University for having granted me study leave during which this paper was written, and commend them for their undying enthusiasm for the introduction of space science in the University.

I am indebted to the organizers of the UNESCO Pilot African Academic Exchange (UPAAE) fellowship programme, both for having given me the chance to participate in this programme, and also the for the financial and material support that facilitated this work. I am very grateful!.

This paper was written at the South Africa Astronomical Observatory. I am sincerely grateful to the staff of the Observatory for the support, hospitality, time and encouragement they offered me while I was preparing this work.

Finally, I thank all those who helped me in one way or another – I have not mentioned their names here – for their contributions and good will towards this piece of work.

References

Fierro, J., 2000, Proc. 24th IAU GA, IAU SS, **24**, 11

New Vision: Makerere to admit on District Quotas. Saturday May 19, 2001.

Stobie R.S., 1995, AP&SS, **230**, 9.

Wentzel, D.G., 2003, this volume.

Chapter 4.6

SPACE SCIENCE & TECHNOLOGY IN MAURITIUS: CURRENT STATUS AND FUTURE OPPORTUNITIES

S.D.D.V. Rughooputh, G.K. Beeharry, K. Golap, N.H. Issur, R. Somanah, H.C.S. Rughooputh, N. Udayashankar[1]
University of Mauritius, Reduit, Mauritius

Abstract Space research (with either direct or indirect spin-offs) has been instrumental in leading to accomplishments that are meant to improve our quality of life in its broadest perspective. But are we all acquainted of the now proven-use of these findings and their capabilities? What do these mean to a remote small insular developing state like Mauritius? This paper explores the recent developments in this field in Mauritius and how we intend to optimize the use of the emerging technologies. Initiatives of the University of Mauritius are highlighted.

Introduction

National space strategies define how individual countries can best use and/or adapt relevant technologies. Focus is on food science and technology, meteorology and geo-physics, microelectronics and information technology, biotechnology, energy, marine sciences, science and technology infrastructure, and research & development. Some countries in Africa are already engaged in different applications whilst others are still debating on what science and technology can do for their people. We all need to motivate our youth through a variety of science and technology programmes, promote science and technology education in our institutions, undertake relevant research, build partnerships between government, industry and universities, rebuild of national and regional institutions, and build capacity through regional and international co-operations. Opportunities exist and these need only to be exploited.

W. Wamsteker et al. (eds.), Developing Basic Space Science World-Wide, 219-234.

Mauritius (20°S and 57°E) is situated in the South Indian Ocean with the main island about 1865 km^2 in size and an Exclusive Economic Zone (EEZ) covering approximately 1.7 million km^2. The Mauritian economy relies heavily on agriculture, industry and tourism. The limiting factors of the island's potential developments are the limited land area, vulnerability of agriculture and other sectors to various stresses and extreme weather conditions (in particular, cyclones) and reliance on pollution-free coastline.

This paper explores the recent developments in space science and technology and how we intend to optimize the use of the emerging technologies. Initiatives of the University of Mauritius will also be highlighted in terms of research and teaching efforts in various domains such as remote sensing and geographical information system, oceanographic studies, coastal zone management, wave modeling, radio-astronomy and astrophysics, the introduction of 'Space Science and Technology Applications I & II' as taught modules in our undergraduate curriculum of studies, distance education project resorting to multimedia tools and applications, to name but a few.

1. Education

1.1 Radio-Astronomy: The Largest Activity of Basic Space Science in Mauritius

The Mauritius Radio Telescope (MRT) is an Indo-Mauritian joint venture involving the University of Mauritius (Mauritius), the Raman Research Institute (Bangalore, India) and the Indian Institute of Astrophysics (Bangalore, India). The MRT (operational in 1994) is unique in the world. The MRT is an aperture (Fourier) synthesis radio-telescope (2 Km x 1Km) in which the imaging of the sky is done through the measurements of spatial frequencies using antenna pairs. It is a survey instrument for imaging the southern sky at 151 MHz in the declination range -70 degrees to -10 degrees with a point source sensitivity of 150 mjy and an angular resolution of nearly 4 arcmin by 4 arcmin (to complement the northern 6C Cambridge Survey). Combined with the 6C survey, one of the deepest all-sky Survey at low frequencies will be obtained, which will lead to a lot of interesting astrophysical studies and results. Research activities are focused on hardware design and software development for image processing in radio-astronomy and interpretation of astrophysical data (continuum mode and pulsar modes). Radio-astronomy presents the largest activity of basic space science in Mauritius. The main objectives of the MRT are to make radio maps of the portion of the sky observed by the MRT (continuum low/high resolution surveys), to study pulsars at low frequencies, to catalogue point sources (some $2x10^5$ expected) and to study their spectral behaviour, to study radio emission from our Sun and from extended low-surface brightness features in radio galaxies, to interpret astrophysical radio images, and to enhance local expertise.

Continuum Observations

To date, mainly a low resolution 24-hour image (lri) with a resolution of nearly 1/4 degree by 1/4 degree has been made. More than 2000 sources above 3 jy have been identified with good positional accuracy in the lri. 3 hours of high resolution images (hri) (4 arcmin x 4 arcmin) have already been made and more than 10 extended sources (greater than 8 arcmin) have been identified in the images. Many unidentified radio sources have been found in the MRT radio images and more tests are being made to confirm whether they have not been identified in any other surveys (Udaya Shankar et al., 2002). The techniques for cleaning (deconvolving) the MRT images have already been finalized and the first cleaned MRT images will be published soon. Surveys are "bread and butter" of astronomers as they are the starting point of any astronomical research. Many high redshift radio galaxies have been first seen in low frequency radio surveys. The MRT database will be a powerful tool to look for high redshift radio galaxies, giant radio galaxies, double-double radio galaxies, relics and large scale features like super-galactic bubbles and super-shells within our own galaxy. Thus it is expected that the MRT survey will bring a lot of new and exciting astrophysical results. Figure 4.6-1 shows one of the first high resolution images made at the MRT. At the bottom right of the image is one of the largest (in terms of angular extent) extragalactic radio source in the sky. From the MRT images, one can distinctly see the 3 components of the source. It shows a strong slightly bent bridge or jet between two diffuse lobes. The bridge shows a double structure (inner lobes) near the center, but not a distinct core.

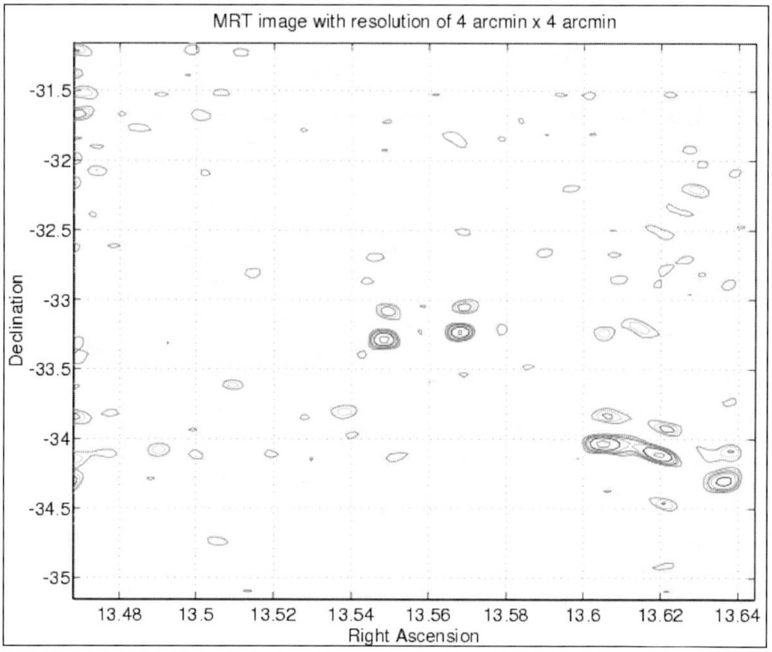

Figure 4.6-1. High Resolution Image of MRT

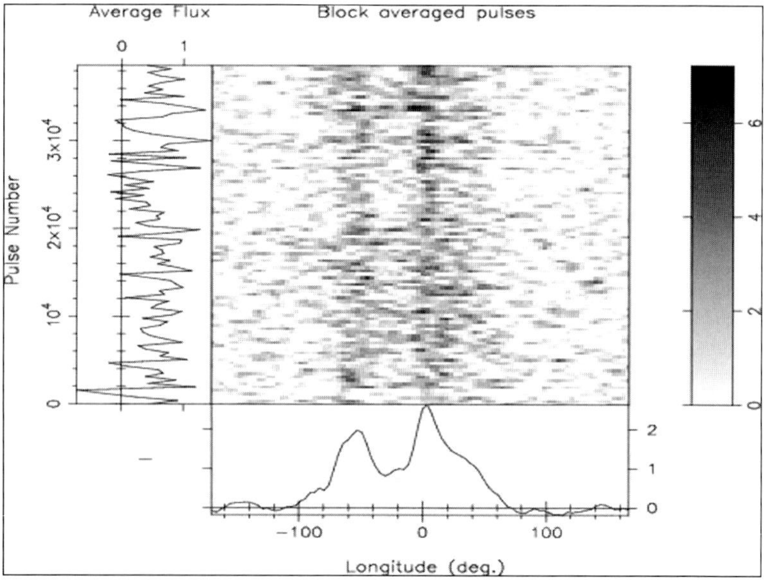

Figure 4.6-2. 4 mins of observation of the millisecond pulsar J0437-4715

Pulsar Observations

The additional features to MRT developed for pulsar observations include a tracking system allowing observations of a source for 2 degrees in the sky, a pulsar receiver of the 'direct-recording' type whereby most of the data processing, including dispersion, is done completely offline. Ten pulsars were detected. The gray-scale in figure 4.6-2 shows the profiles (averaged in blocks of 390 pulses) of J0437-4715 millisecond pulsar revealing 2 prominent peaks. Some additional features are also seen in the overall average profile (39000 pulses) shown in the lower frame. We note that the pulse occupies a very large fraction of the pulse period. The other pulsars that could be studied in detail are J1453-6413 and J1752-2806.

Very Long Baseline Interferometry

The Autralian Long Baseline Array (LBA) is part of the Australia Telescope National Facility (ATNF) and utilizes VLBI (Very Long Baseline Interferometry) techniques with radio telescopes all over Australia, enabling high-resolution imaging. It, however, is seriously limited by a huge hole in UV-coverage due to the absence of any antennas between Australia and South Africa, the only other country in the Southern Hemisphere forming part of the VLBI network for high resolution imaging of Southern sky objects. Mauritius has been identified [Ojha R , private communication] as an ideal place to set up a dish to form part of a VLBI network: apart from its location, Mauritius also has the advantage of an active radio astronomy community. A 25 m class telescope capable of observing between ~1 GHz to 22 GHz would probably be useful to Mauritian astronomers as apart from its vital role in the VLBI network, it can also handle a wide range of projects. The high-resolution images obtainable would allow a deeper study of sources of interest identified in the MRT images. The financial & technical aspects of setting up such a dish in Mauritius is currently being looked into.

1.2 RS/GIS Studies: From Earth Sciences to Astronomy

Looking Down

The RS/GIS Research is a German-Mauritian joint venture involving the University of Mauritius (Mauritius) and the Philipps University of Marburg (Germany) – initiated in 1997. Various studies using Landsat, SPOT and Radarsat images are currently being pursued principally using MapInfo™ and Idrissi™ .Investigations so far carried out include soil studies, agro-suitability studies, modeling the catchment areas and aquifers, land use classification, biodiversity assessment studies and coastal zone mapping/management. figure 4.6-3 shows the sensitivity of Mauritian shorelines to oil spill.

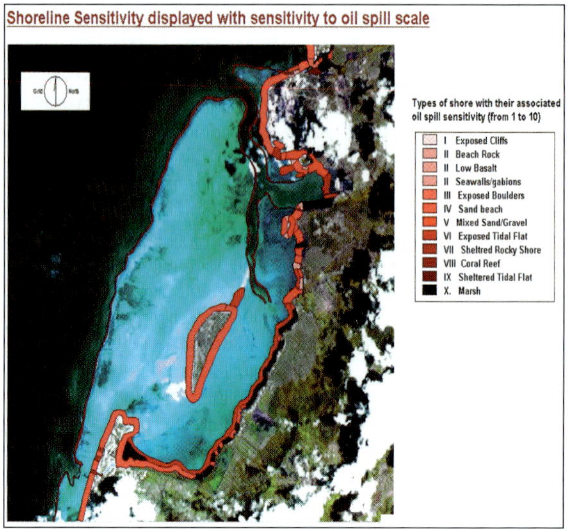

Figure 4.6-3 Shoreline sensitivity to oil spills of the SW coast of Mauritius

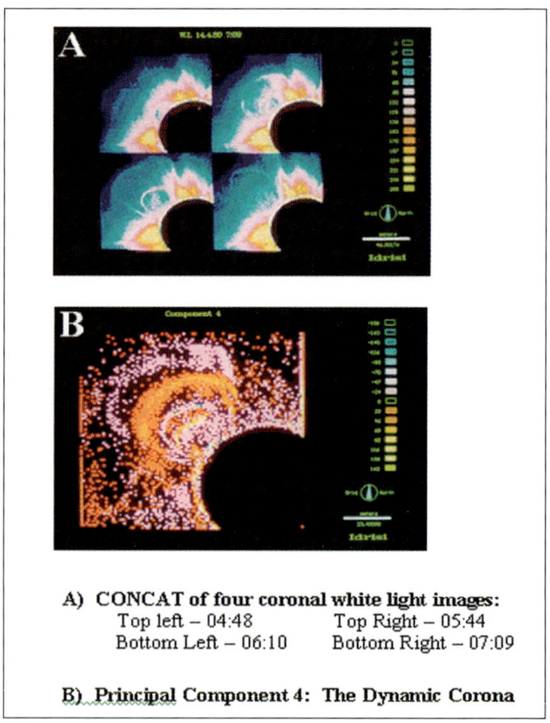

Figure 4.6-4. The Dynamic Corona of our Sun revealed using PCA

Looking Up

Research efforts are also directed towards optimizing on the decision-making advantages of GIS in the field of Astronomy and Astrophysics besides the Earth Sciences. Conventional software routinely used for the manipulation of astrophysical data lacks the decision-making aspect. We have explored the possible use of a Geographic Information System to study the Sun (figure 4.6-4) and the Southern Sky (using data from the MRT Low Resolution Survey and the PARKES Observatory) (Rughooputh *et al.*, 2000).

1.3 Other Areas of Interests

Space Communications

Space Communication receivers operate at very low signal-to-noise power ratios due to limitations such as antenna size and transmitted power and long transmission distances. Since it is costly and difficult to increase the SNR on such channels, channel coding is widely used to increase the reliability of transmitted data. Recently, a novel error correcting scheme (Turbo-codes) has received much enthusiasm amongst the research community. Currently, one of us (HCSR) is investigating the suitability of turbo-codes for space applications. These investigations also deal with the use of artificial intelligence techniques for the decoding stages.

Space Materials

Polymers, by virtue of ease of processing/fabrication, low cost, light weight, structural flexibility and durability, make polymer devices competitive so that they are gradually replacing semiconductors and metals in several areas of applications including space applications. Of added advantage is its potential to combine with many materials, as advanced composites or hybrid materials, for improved properties. Demonstration devices (electronic, photonic. sensors, lasers) have already been tested. Some polymers are available as conducting fabrics, camouflage radar nets, light weight batteries and, more recently, LEDs (and, in the near future, as flat, flexible, displays). Other potential applications include metal deposition on insulated supports, transparent speakers, EM shields, anti-static coatings and screens, anti-corona coatings, environment-friendly soldering material, optical recording media, smart plastic windows, electromechanical actuators (as artificial muscles) and bio-sensors. Although several polymers have now been prepared, it will be extremely difficult for these to compete with metals in traditional electronic applications like, wiring, transmission cables, etc. Research focus is on applications that exploit the existence of extended conjugation in these polymers (Rughooputh, 1998). More importantly, for optical applications, the absorption and emission of conjugated polymers can be tuned over the whole visible spectrum by altering the chemical structure.

1.4 Capacity Building and Consulting Activities

Undergraduate/Postgraduate Courses

Undergraduate teaching in basic space science and technology applications at the University of Mauritius is supplemented by research projects in such domains as communications, oceanography, aerosols, climate, coastal zone management, wave modeling, radio-astronomy and astrophysics, RS/GIS. Two course modules (Space Science and Technology Applications I & II) - aimed respectively at presenting an overview of basic space sciences and space technologies and, at a more advanced level, RS/GIS applications, communications systems, etc. which includes case studies are offered. Courses related to space science (astronomy and astrophysics, electromagnetic theory, solar and terrestrial physics, atmospheric physics, relativity, satellite communications, microwave engineering, opto-electronics, and surveying) are also incorporated in our degree programmes. More than 75 BSc undergraduate research projects relate to basic space science and related areas. MRT continues to play an important role in the development local manpower in the field of Astrophysics, Electronics, Computer Engineering, & Software Engineering and Communications. Engineering students on industrial placement at MRT are exposed to electronics and data-processing techniques.

So far, the MRT group has produced 6 postgraduate theses:

- A PC-based Interferometer for Radioastronomy, N Jayprakash (M.Sc, 1991)
- MRT and a Study of Selected Supernova Remnants Associated with Pulsars, R Dodson (PhD, 1997)
- Study of Low Spatial Frequency Response of MRT, M Katwaroo (MPhil, 1997)
- Synthesis Imaging at 151.5 MHz using the MRT, KGolap (PhD, 1998).
- Wide Field Imaging with the MRT, S Sachdev (PhD, 1999)
- Study of some Southern Pulsars at 151.6 MHz using the Mauritius Radio Telescope, N H Issur (PhD, 2000)

and the RS/GIS Group has produced 5 postgraduate theses:

- Spatial Distribution of precipitation in Mauritius, Martin Seul (Diplome-Geograph, 1999)
- Aspekte der Bodenerosion auf Mauritius. Analyse verschiedener Einflußfaktoren an ausgewählten Standorten mit Hilfe Geographischer Informationssysteme (GIS), Mona Kremer (Diplome-Geograph, 2000)
- Böden und Landnutzung auf Mauritius. Aspekte einer geoökologischen Detailuntersuchung mit Hilfe Geographischer Informationssysteme (GIS), Kranz Olaf (Diplome-Geograph, 2000)
- Die exemplarische Entwicklung von Schutzkonzepten für vom Aussterben bedrohte Pflanzen- und Vogelarten - GIS als funktionales Planungswerkzeug im Bereich Arten- und Naturschutz auf Mauritius, Indischer Ozean, Reinhold Hill (Diplome-Geograph, 2001)
- Environmental Management of Mauritius Coastal Zone through GIS, Yan Gruet (MSc, 2001)

Specialised Training Courses

The National Remote Sensing Center for Mauritius, jointly with the University of Mauritius, will be expected to play an important role in the near future towards consolidating local capacity in support of environmental monitoring and sustainable development. One can expect the running of application-specific courses on renewable resources, regional and town planning, environment and pollution, hydrology, climatology, oceanography, image processing, advanced RS/GIS, satellite communications, space technology, etc..

Consulting Activities

Potential areas in space science where consulting services can also be offered by the university include RS/GIS, communications, mathematical modeling and image processing.

Space Education and Public Awareness

The MRT offers a site of attraction to those members of the public and tourists who are curious to know more about the universe. A dome to house a 10-in optical telescope and a permanent gallery form part of our future plans for construction at the MRT Observatory. It is hoped that the Rajiv Gandhi Science Museum cum Planetarium, under development, will assist in sensitizing the public, especially our younger generations. The 'Space Science and Technology Applications I & II' modules could eventually be further developed into a specialized course for scientists and engineers in the African region.

Meetings

The University of Mauritius was the site for hosting two major international meetings:

1. An International Meeting on Radio Astronomy At Low Frequencies, 23-25 October 1997.
2. Tenth UN/ESA Workshop on Basic Space Science: Exploring the Universe, Sky surveys, Space Exploration and Space Technologies, 25-29 June 2001.

The International Meeting on Radio Astronomy At Low Frequencies, 23-25 October 1997 was the first of its kind involving 9 member states were represented at the meeting with some 28 participants and 32 papers presented. The meeting focused on galactic astronomy., low frequencies radio astronomy, MRT, and extragalactic radio sources. A panel discussion involving the directors of four radio groups (VLA, GMRT, ATNF, and Westerbork) was held on this occasion. The meeting was sponsored by the Raman Research Institute (Bangalore) and the University of Mauritius.

The Tenth UN/ESA Workshop on Basic Space Science is the continuation of the series of UN/ESA workshops on Basic Space Science organized for the benefit of developing countries that was initiated in 1991. The workshop sessions focused on (a) sky surveys, (b) from solar/planetary systems to galactic/extragalactic systems, (c) data manipulation, databases and multi-wavelength analysis, (d) education with and networking of telescopes, with special reference to the southern hemisphere and (e) utilization of space science and technologies and their benefits to society. 28 member states were represented at the workshop with some 65 participants and 52 papers presented. The workshop was co-organized by the Center National d'Etudes Spatiales of France, the German space Agency (DLR), the National Aeronautics and Space Administration (NASA) of USA, the National Astronomical Observatory of Japan and the Planetary Society. The Proceedings of the UN/ESA Workshop on Basic Space Science have been published (Haubold and Rughooputh, 2002).

2. Local Applications of Space Technologies

Application of remote sensing (RS) in Mauritius dates back to the early sixties, with the coverage of the island by aerial photography principally for topographic mapping, insect pest monitoring on sugar cane, and towards the production of land use maps of Mauritius. Whilst aerial photography continues to find useful applications in agriculture, hydrology and other civil engineering studies, satellite-based RS is now becoming a popular tool for periodic monitoring of natural resources pertaining to terrestrial, water, coastal zone and marine resources. Satellite RS should be viewed in the context of its suitability for monitoring dynamic features in different spectral bands; replacing/supplementing/complementing conventional systems for surveys and data collection for planning and monitoring of resources. The inherent advantages of satellite RS are its time and cost-effectiveness over conventional methods with the resolution being comparable to aerial photography. Satellites covering Mauritius include LANDSAT, SPOT, NOAA, INSAT, METEOSAT and RADARSAT.

2.1 National Remote Sensing Center for Mauritius

A National Remote Sensing Centre for Mauritius (NRSCM) has recently been set up next to the Indian Space Research Organisation station for Telemetry, Tracking and Commanding (ISRO-TTC) the Indian Remote Sensing (IRS) satellite (Remote Sensing Centre for Mauritius – A Project Proposal, Dec. 1990). Its main purpose is to co-ordinate and to provide logistic supports to all local RS activities and users in Agricultural Resource Management, Land & Water Resource Management, Coastal Zone & Marine Resources Management, Geology and Meteorology, etc.

2.2 Telecommunications

Converging technologies in the delivery of information offer new services directly into the home, workplace, educational institutions, hospitals, etc.. Such services include on-line business information services, educational applications such as distance-learning and remote teaching, home-shopping and home-banking, television on demand and interactive games; remote health care, tele-medecine, new ways of working such as in-office video-conferencing, e-mail and tele-working. The three main developments that underpin future capabilities are:

a) digital technology developments (producing a significant expansion in the network capacity and an equally significant reduction in the costs of providing such capacity to the end-user)
b) systems developments (giving the customer control of the information being delivered, for example over timing or destination), and
c) technology developments (reducing the delivery costs of broadband services).

One of the objectives of the Government is to increase the telephone density up to 33 by the year 2007. This target corresponds to connecting 85% of the homes and all the businesses and public institutions. The provision of access to this wide community will require considerable investment on the network infrastructure for its expansion, increased penetration and the introduction of broadband technology to support the delivery of multimedia services in the near future. The upgrading of the telecommunications tariff/pricing structure will be needed for the mutual benefit of the operators and the network users. The structure needs to offer value for money and provide enough margin for the operators to constantly improve the quality of service and to invest in the network infrastructure. Mauritius Telecoms (the incumbent operator) has catered for the setting up of 50000 new lines operating on the wireless local loop technology. EUTELSAT provides multimedia platforms in our region to cater for satellite television. RASCOM, a Pan African Communications' satellite with similar capabilities has been launched in June 2001.

Telecommunications Sector Outlook

The Government of Mauritius had presented the broad policy issues that had to be considered in order to formulate the required strategies for the development of the telecommunications sector over the coming decade through a Green paper (Mauritius 2007, 1997). Subsequently, the policies proposed in a White Paper (1997) stresses on Mauritius fulfilling its commitments under the WTO Agreement on Telecommunications and on the expansion and upgrading of the telecommunication infrastructures whilst allowing for open international competition. A new regulatory body, the Mauritius Telecommunications Authority (MTA), was established through 'The Telecommunications Act 1998'. The MTA is responsible for the economic regulation of the telecommunication systems and the services provided over such systems. The Act (1998) allows for the establishment and operation of global mobile personal communication services (GMPCS) whereby the network is satellite-based with the coverage extending over the Mauritian territory. This legislation has been recently reviewed and a new body, the ICT Authority, has been set up to extend the defunct MTA's roles to include other converging technologies.

2.3 Meteorological Applications

Meteorological satellites provide ideal platforms for RS of meteorological phenomena in both the lower and upper atmospheres of the earth. These platforms make regular monitoring of a large part of the earth's atmospheric environment possible. RS-sensed data complement ground-based measurements. The tracking of the paths of cyclones and major frontal systems in the whole of the Indian Ocean region is vital for analysis and forecasting. Mauritius has been using satellite imagery from NOAA 12/14/15, JGMS, Meteosat 5/6, as well as Doppler RADAR microwave imagery and radiosonde data for weather forecasting purposes. Space-based technology has opened new vistas to hitherto unknown weather development patterns, and enhanced the capability to monitor, track, predict and warn with a greater degree of accuracy (hence reductions in the loss of lives). The Indian Ocean basin is currently the site of a meteorological study under the INDOEX programme. The aim of the study is to explain the earth's radiation budget and aerosol source distribution and mobility taking the ITCZ into consideration.

2.4 Agricultural Applications

Rapid and accurate monitoring of the agricultural sector has long been a preoccupation of each country. The island's sugar industry is threatened by increased cost of production and lower prices for sugar on world and preferential markets, its vulnerability to extreme weather conditions (droughts and cyclones), and the decreasing acreage under cane cultivation. For the industry to survive, several strategies such as increasing production per unit area of land cultivated, lowering cost of processing, maximising utilisation of sugar-cane by-products as well as diversifying within sugar are being pursued. As such, RS/GIS is finding useful applications in crop monitoring, yield performance and yield forecasting, analysis of slopes and topography by making use of digital terrain modelling (in the context of mechanisation, soil compaction and irrigation), farm layout and field planning, and land use and management. Several projects are under way for making cane variety recommendations and fertilizer input.

Stock taking and production of fruit trees (mango, litchi, coconut, etc.) and livestock (beef, sheep, deer, etc.) can be better evaluated by using RS/GIS techniques. The latter is suitable for planning island-wise de-rocking scheme programme, irrigation scheduling, advising planters of preferential variety of crops. In some cases, repetitive imaging would be required for crop condition monitoring and yield forecasting. Precision agriculture (precision farming or site-specific management) industry would result through the use of high resolution images when used in conjunction with GPS enabling a more judicious use of inputs, thus providing the basis for improved yields, reduced production costs and decreased environmental impact. Although, these projects can be easily addressed, the main obstacle for an eventual full use is the costs involved in the acquisition of images.

2.5 Coastal & Marine Applications

Critical coastal and marine management issues such as sensitivity analysis, habitat modeling and pollution monitoring require the ability for resource managers to make rapid and appropriate decisions. Satellite imagery has potential applications in coastal/marine research. Noticeable advances have been made in studying coastal erosion, bathymetry and in mapping/classification of coral reefs and other coastal habitats/resources (such as algae, mangroves, sea-weeds, sea-grasses, bright sand and silt, mud, rock, etc.) using Landsat and Spot images. Coral reefs, in particular, need to be effectively managed since they are vulnerable to a wide range of stress conditions (siltation, pollution, coastal development, coral mining, overexploitation of reef species, intensive recreational use, military activities and mineral exploitation). Satellite image analysis revealed corals to be less healthy in the lagoons less exposed to heavy wave action, regions prone to eutrophic conditions, and surface thermal anomalies in waters around Mauritius (Daby, 2000).

Off-shore banking fishing operations contribute approximately 70-80% of the national fish catch. Satellite RS to localize far off-shore shoals of tuna and other fish species is currently under-exploited. Identification, tracking down and monitoring of other living marine resources (dolphins, whales, game fish , sea turtles) are other areas where RS has potentials. Other applications include EEZ maritime surveillance, direct detection of oceanic pollutants, identification of potential sites for OTEC plants, collection of wave data and deep ocean processes, geoid determinations, synoptic measurements of the ocean, locations of oceanic frontal boundaries, current and circulation patterns, and mapping non-living resources such as magnesium nodules. GIS will find useful applications through spatio-temporal studies of the RS data for enhancing coastal and marine resource management besides weather forecasting.

2.6 Hydrological Applications

The Water Resources Unit (Ministry of Public Utilities and Infrastructure, Mauritius) uses aerial photography and satellite RS to address hydrology problems involving their catchment areas and trends to be solved, thus complementing the conventional data. When studied together with geology, geomorphology and climatology, many sites with underground water potential have been identified. Daby (1999) have detected underground seepage into our lagoons using Landsat 4 TM thermal infrared imagery.

2.7 Other Areas and Potential Areas Applications

Many potential areas for applications of space technology exist in Mauritius for flora and fauna conservation; forestry; tourism; emergency management - fire/flood/etc; traffic navigation; land use and planning; health; environment and pollution monitoring; development of distribution information support systems and their applications in water and electric power distribution.

3. Conclusions

We are witnessing the convergence of technologies resulting in our space (information) age – with RS as the *visual imagery* component, GPS as the *location* component, and satellite telecommunications, as *the channel for speech/data transmission* component.

Mauritius, a remote insular and under-developed island, is striving hard to keep in pace with developments elsewhere. Space science culture is very much alive and, thanks to the political will and the continued support of the Government of Mauritius, we find ourselves amongst the leading nations in Africa in this domain. The MRT is a testimony of this fact. The new legislation now enables for more global competition for communications, in particular, satellite-based broadcasting and telephony.

We have highlighted several initiatives of the University of Mauritius. Other institutions are also using space technologies in a very involved way. The future looks bright and promising.

Acknowledgements

We thank the University of Mauritius for providing the necessary facilities for carrying out this work and for hosting the 2001 UN/ESA Workshop on BSS in Mauritius.

References

Daby D., 2000, PhD thesis, University of Mauritius

Golap K., etal., 1998, Astron. and Astrophys., **19**, 35.

Haubold, H.J. and Rughooputh,S.D.D.V. , 2002, AP&SS, **282**, Vol.1.

Mauritius 2007, 1997 – Green Paper on Telecommunications: Strategies and Policies for the Info-Communications sector, 1997-2007, Sept. 1997 - Ministry of Telecomms & Info. Tech. Remote Sensing Centre for Mauritius, 1990 – A Project Proposal, DOS:MRSC:PR:01, Indian Space Reseach Organisation, Department of Space, Bangalore, India.

Rughooputh S.D., Oodit S., Persand S., Golap K., Somanah R., 2000, AP&SS, **273, 245**

Rughooputh S.D. and Rughooputh H.C.S., 1998, in Proc. of International Conference on 'The Engineering Profession Beyond Year 2000' Institution of Engineers (Mauritius)- University of Mauritius, Mauritius.

The Telecommunication Act, 1998, Act No. 25 of 1998, Republic of Mauritius.

White Paper, 1997, White Paper on the Telecommunication Sectors: Fostering the Info-Communications Society, Dec. 1997, Ministry of Telecomms. and Info. Tech.

SECTION:

5. WESTERN ASIA AND NORTHERN AFRICA

Chapter 5.1

KOTTAMIA TELESCOPE UPGRADING

S.M. Hassan
National Research Institute of Astronomy and Geophysics
Helwan Observatory
Cairo, Egypt

Abstract This paper provides a summary of the upgrading of the Kottamia Telescope. This includes a new optical system as well as auxiliary instrumentation.

Introduction

A Historical Overview

In spite of the relatively good climate in Egypt for astronomical observations, no great deal of work has been done during the 19th century. It was not until 1905 when Mr. Reynolds - at that time an English amateur astronomer, and later treasurer of the Royal Astronomical Society in London - presented Helwan Observatory with a 30" reflector telescope. The Kottamia Observing Station is located far away from disturbing influences. With 250 clear nights a year, this site provides favorable conditions for astronomical observations.

Due to the clear sky of Helwan, the astronomers were able to collect a large number of valuable photographic plates on Nebulae, Galaxies, Comets (especially Comet Halley during the 1910 perihelion passage), Satellites of Jupiter, the planet Pluto, the Moon, and many stellar observations. These observational programs continued for nearly half a century. Confronting the desires of astronomers to achieve more, a larger telescope had been proposed, and a 74" telescope equipped with Cassegrain and Coudé spectrographs was recommended to the Egyptian Government. The acquisition of the telescope and instrumentation from Grubb Parsons (Newcastle-U.K.) was approved by the Egyptian Government in 1948 (the same year the 200" Hale – Telescope at Mount Palomar (California) was constructed in the USA. The new Kottamia

W. Wamsteker et al. (eds.), Developing Basic Space Science World-Wide, 237-240.

telescope was among the largest in the world at that time. Considering the favourable seeing conditions in Egypt for astronomical observations, one might have expected important contributions in the general domain of astronomy. The Telescope and spectrographs were to be delivered in 1955. However, due to many difficulties arose in the meantime, the delivery was delayed until 1960 (Samaha, 1962). In May 1964 the telescope with the spectrographs were ready for the first light after the completion off the optics and spectrograph acceptance testing program.

The Site

Kottamia Mountain (figure 5.1-1) is in the NE desert, about 70 Kms NE of Helwan and 476 meters above sea-level. Its geographical coordinates are:

Latitude: + 29°55'35.24"

Longitude: 31°49'45.85" E

Height: 482.7 m

The prevailing seeing conditions at Kotamia are can be summarized as follows:

50% in the range of 1.5 - 2.0 arcsec in autumn and winter (best in winter).

30% in the range of 2.0 - 3.0 arc sec. in spring and summer (better in spring).

The remainder is worse.

The telescope is conventionally designed with:

1. A f/4.9 Newtonian focus with Camera for direct imaging
2. A f/18 Cassegrain focus with a prism-spectrograph provided with two Cameras giving dispersions 48 Å/mm and 100Å/mm at Hγ.Later, a multi-colour photoelectric photometer was designed to be installed in this focus for photoelectric observations.
3. A f/28.9 Coudé focus equipped with a grating spectrograph with two Cameras giving dispersions 5.6 Å /mm and 20 Å/mm for further spectroscopic work.

Main programs undertaken with the 74" telescope:

1. Observations of the Moon and Planets in collaboration with American astronomers, NASA, and Manchester University.
2. Radial velocities measurements in collaboration with Greenwich Observatory.
3. Photometry of Star Clusters for the determination of Galactic Structure in collaboration with Basel Astronomical Institute, Asiago Astrophysical Observatory, and Padua University Observatory.
4. Extragalactic research in collaboration with the Asiago Astrophysical Observatory, Padua University Observatory and Institutes in U.K., Germany, FSU countries.

Figure 5.1-1. Helwan Observatory in the Egyptian desert south of Cairo, housing the 74" Kottamia telescope.

3.1 Kottamia Telescope Upgrading.

In the early 1990's it was decided to make use of the advances in optical manufacturing, instrument design and detector technology. A proposal was prepared to upgrade and modernize the 74¨ telescope. Supported by the conclusions of the 4th UN/ESA Workshop on Basic Space Science (A/AC.105.580), hosted by the Government of Egypt in 1994 (Haubold and Mikhail,1995 and 1995a), the proposal was submitted to the authorities and after an extensive evaluation, the proposal from NRIAG was approved (Hassan, 1998). This would assure to maintain the importance of this telescope facility, which is the major telescope in North and Middle Africa as well as Western Asia. The importance of this facility, which would supply major experimental capabilities for basic space science in the region, is obvious.

The Optical system

In 1995 an agreement was signed with Zeiss Company (Germany) to refit the Kottamia telescope with a new optical system (Primary M1and a secondary Cassegrain M2). The optical system, made of "Zerodur", successfully passed the acceptance test at Schott Glaswork in Mainz. To obtain better optical performance, a more efficient primary mirror support system has been developed. The on-site installation was completed in May 1997. The results of the final tests of the combined optical system (M1+Cell +M2) show that the encircled energy E is:

- Primary mirror: E 80% within 0.247 arcsec using a 20x20 pixel grid;
- Primary mirror in Cell: E 80% within 0.279 arcsec using a 20x20 pixel grid;
- Secondary mirror: E 80% M2 within 0.26 arcsec using a 20x20 pixel grid;
- Optical system: E 80% within 0.35 arcsec using a 20x20 pixel grid;

These results were well within the contractual specifications.
First light for the upgraded telescope is expected in the near future.

The CCD - system

Astromed in U.K. provided the Kottamia Telescope with a CCD – system which comprises:

- An imaging back-illuminated CCD Camera (TK 1024):

  No. of pixels: 1024 x 1024
  Size of pixels: 24 nm x 24 nm
  Typical peakwell capacity: 350,000 electrons
  Typical dark current for liquid
  Nitrogen cooler: 1 e/pixel/hr
  Readout Noise:10 electrons r.m.s.
  Output data rate: 50 pixels/s

- A guiding CCD Camera KODAK (KAF - 1300L)

  No. of pixels: 350 x 520
  Size of pixels: 16 nm x 16 nm
  Typical peakwell capacity: 150,000 electrons
  Typical dark current for the thermoelectric cooler: 0.1 e/pixel/s
  Readout Notice:10 electrons r.m.s.

Refurbishment of the Kottamia 74" Mirror Aluminizing plant:

Negotiations are taking place with the Royal Greenwich Observatory for the refurbishment of the plant to overcome the difficulties taking place previously.

References

A/AC.105/580, 1995, Report on the 4th UN/ESA Workshop on Basic Space Science, Cairo, Egypt, *Earth Space Review*, **4**, 28.

Hassan, S.M. 1998, *African Skies/Cieux Africains*, **2**, 16.

Haubold, H.J. and Mikhail, J.S. (Eds.) 1995, *Earth, Moon ,and Planets*, **70**, 1-233.

Haubold, H.J. and Mikhail, J.S. (Eds.), 1995a, *Astrophysics and Space Science*, **228**, 1-405.

Samaha, A.H., 1962, Helwan Observatory Bulletin, **62**,6.

Chapter 5.2

THE MARSKHOD EGYPTIAN DRILL PROJECT

M.A.M. Shaltout
Space Research Center
Desert Environment Research Institute
Minufiya University
El-Sadat City, Egypt
mamshaltout@frcu.eun.eg

Abstract We describe a possible participation of Egypt in a future Mars rover Mission. It was suggested that Egypt participate through involvement in the design, building and testing of a drill to obtain sub-surface samples. The Space Research Institute of the Russian Academy of Sciences (IKI), formally invited the Egyptian Ministry of Scientific Research to study the concept for potential use on the Russian Mars 2001 Mission. As one of the objectives of the Marskhod mission was the analysis of sub-surface samples, a drilling mechanism in the payload would be essential. The Egyptian expertise in drill development is associated with the archaeological exploration of the Pyramids. A sophisticated drilling system perforated limestone to a depth of 2 m without the use of lubricants or cooling fluids that might have contaminated the Pit's environment. This experience could have been applied to a drill development Mars 2001 mission, which was unfortunately canceled due to economic problems.

Introduction

Exobiology is the study of the origin, evolution, and distribution of life in the universe. On Earth, there is water and there is life. On Mars, there may have been water. Has there been life? It is obvious that this question can only be resolved at or below the Martian surface. The exobiological strategy for Mars exploration has been developed to determine the planet's pre-biotic/biotic status. This has been formulated such that the operational search for possible life, becomes the search for liquid water, past or present. The optimal approach is a series of missions with increasing capabilities that progress from global reconnaissance, to

W. Wamsteker et al. (eds.), Developing Basic Space Science World-Wide, 241-249.

high-resolution imaging of minerals characteristic of aqueous activity and those lithologies capable of preserving biosignatures, using advanced Landers.

The search for biosignatures involves a search in both time and space for the pre-biotic/biotic history of Mars (Mayer and Kerridge, 1996). The Martian climate is essentially the combination of the seasonal cycles of atmospheric temperature, pressure, aerosols, and water vapor, along with the related forcing of (and by) the regolith around the polar caps. The Viking spacecraft provided data for an entire Martian year. However, observations made in other Mars years suggest that the "Viking" years may not be representative of the present epoch. The Mars Global Surveyor (MGS) mission of NASA has, since late 1999 till the present, (http://mars.jpl.nasa.gov/mgs/) made a highly detailed survey of the Martian surface and given important new insights in the climate of Mars. The 2001 Mars Odyssey mission is designed to do more detailed analysis of the Martian surface and subsurface through more advanced remote sensing (thermal emission and γ-Ray spectroscopy) (http://mars.jpl.nasa.gov/missions/present/odyssey.html).

These missions will help to remove uncertainties in our ability to extrapolate the climate and on the nature of the chemical characteristics of the surface and subsurface. The changing orbital elements, on timescales of 10^5 - 10^7 years cause changes in the nature of the seasonal cycles. In fact, it is generally assumed that the changing seasonal cycles are reflected in the layering of the polar deposits and in the relative youth of the deposits themselves. These issues have been put into the context of the current and future missions to Mars (Jakosky, 1996). During the summer of 1976 the two Viking spacecraft arrived at Mars. Their primary objective was to determine if there was life on Mars. The results of the three life detection experiments aboard the two Landers, did not show the presence of extant life. In addition, no evidence was found for any organic matter in the Martian regolith at either landing site. The Viking Landers provided data to help explain why no extant life was found at the Martian surface: "*The surface of Mars is an oxidizing, desiccated environment devoid of liquid water*". However, data from the chemical analyses of the regolith conducted during the Viking mission as well as recent analysis of SNC meteorites (shergottite, nakhlite, and chassigny classes) considered to have originated from a planet-sized "parent body" (i.c. Mars) in the inner solar system, suggests that all of the chemical elements necessary for life, occur on Mars. Also more recent studies on Martian meteorites such as ALH84001, have been suggested to carry evidence of fossilized microbes from Mars (McKay et al., 1996). These claims which still are controversial drive much of the current exobiological discussions on "Life on Mars". Data gathered to date suggests that if liquid water existed on the surface of Mars for sufficient time the probability for life to have arisen is high (Mancinelli, 1996).

While the Viking Landers, and orbiters were based on technologies of the late 1960's, the MGS orbiter and 2001 Mars Odyssey missions are mainly based on 1980's technologies. The next phase of technology started with the Mars Pathfinder Lander. Reapplying commercial, military, and/or Earth-orbiting space technologies, a new generation of instruments is being developed. As numerous examples and comparisons testify, through innovative and often bold approaches, the near-term and future instrumentation could provide a significantly greater scientific return on cost invested, in the context of a systematic, and long-term exploration of Mars (Clark, 1996).

Unfortunately the MARS 96 mission from Russia failed. The economic problems associated with the restructuring of the Russia created a postponement of the Mars 2001 Mission, which would include a Marskhod rover under international corporation. Egypt would have participated in the Mars 2001 mission through the design, building and testing of a drill to obtain sub-surface samples. The associated preparatory activities are the subject of this paper.

1. Egyptian Expertise in Drill Development

In 1985, the Egyptian Antiquities Organization and the National Geographic Society launched a project called the Nondestructive Investigation of the Second Boat Pit of Pharaoh Khufu. The research plan involved archaeology, chemistry, geophysics, imaging technology and remote sensing (Holden, 1987), The idea was to survey the chamber and its contents by remote sensing techniques and to select a proper drilling site (Yoshimura et al., 1987). This involved drilling a vertical hole in a block of chalky limestone up to 2m thick without contaminating the pit's environment through the entry of ambient air or the use of lubricants and cooling liquids. This would allow also to sample the air inside the cavity at different levels, making sure that no chlorofluorocarbons were introduced into the sample and passing the air through filters to separate any pollen grains or other microorganisms for analysis. In addition, the pressure, temperature, and relative humidity inside the chamber were to be measured, and the interior was to be photographed with a video system and a 35-mm still camera without raising the temperature inside. Finally, the drill hole had to be resealed, with sensors left inside the pit to allow periodic measurements of the temperature and humidity.

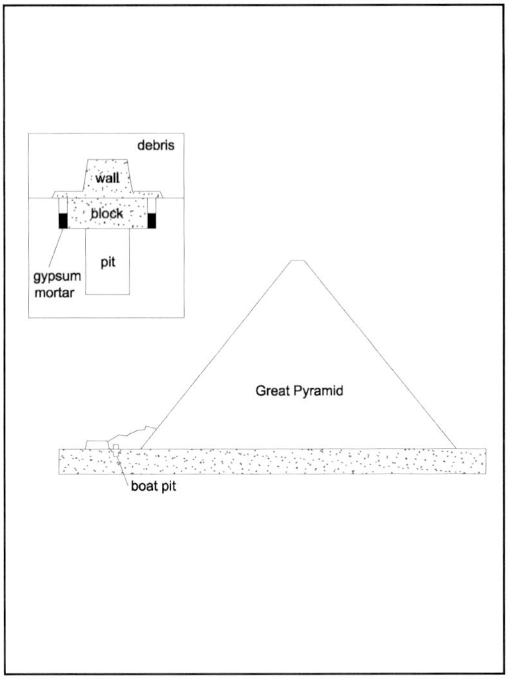

Figure 5.2-1. The dissembled boats of Pharaoh Khufu were hidden in pits beneath blocks of stone and a mound of debris just south of the Great Pyramid of Giza. The first such pit was excavated and its boat reconstructed for permanent exhibit; the second pit, shown schematically here, was not excavated but was explored by special techniques of remote sensing and photography to leave its seal of gypsum mortar intact (after El-Baz, et al., 1989). Courtesy American Scientist.

To accommodate these stringent requirements, a unique rotary air lock was developed allowing to seal the work surface. When the lock is sealed, probes can be safely exchanged by means of a bolted connection at the upper part of a rotating assembly.

Work at the site began on a block of about 4 m long and about 90 cm wide near the eastern end of the pit. Only part of the wall was cleared to expose the block and on top a scaffold was erected to hoist the equipment. An area in the center of the block was made perfectly level, and the surface of the stone was sealed with a thin layer of epoxy resin. A lead gasket formed an inner barrier to the possible emission of freons from the neoprene perimeter seal. As a further precaution against the introduction of contaminants to the air in the pit, the volume between the primary lead seal and its surrounding neoprene secondary seal was evacuated to about 0.5 atmosphere. This "leak-scavenging" partial vacuum was also maintained between other seals in the assembly.

The special design of the drill bit precluded the need to remove a large core segment from the bottom of the hole. The 90-mm-diameter drill bit was operated at a slow speed of 375 rpm to avoid heating of the rock. Only 2.5 cm was drilled at a time, and before each cycle the drill

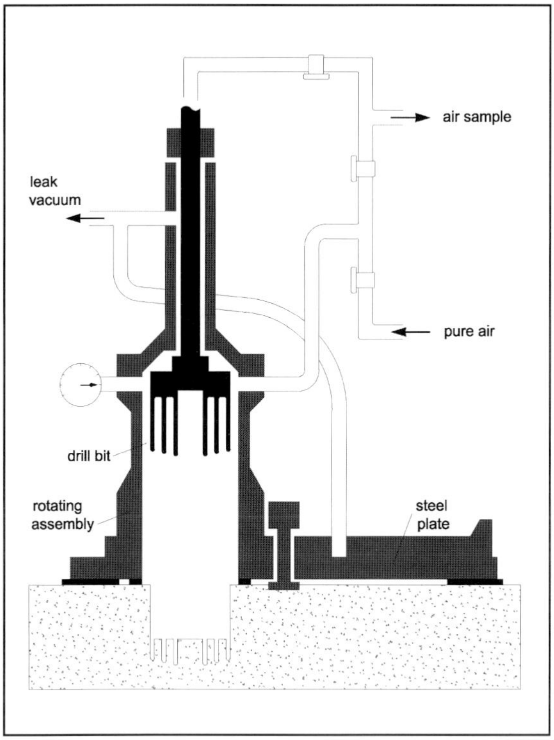

Figure 5.2-2. Because there was a possibility that the second boat pit had been hermetically sealed since its construction, sampling the air inside it offered a chance to gather information about the air of northern Egypt some 4,500 years ago. The air lock used was specially designed to avoid introducing any gases or particles from the air outside into the pit during drilling, air sampling, environmental monitoring, and photography. The leak vacuum, in which the volume between two seals is evacuated to 0.5 atmosphere, acts as an extra precaution against introducing contaminants into the pit (after El-Baz, et al., 1989); Courtesy American Scientist.

chamber was pumped and charged with freon-free air to 10-30 mm of mercury less than the barometric pressure. A special core-breaking tool was used to snap off the two standing rings and central core of material produced by the triple core bit, which was removed with a vacuum cleaner. The drill head was then reassembled to the air lock, the drill chamber was purged of outside air, and another drilling cycle commenced. After two days and drilling through 159 mm, the drill assembly suddenly fell 15 mm, indicating breakthrough. The drill bit was withdrawn into the airlock, which was turned 90 to seal the hole. Now the drill head was replaced with the air-sampling probe. A stainless steel tube 12.7 mm in diameter, and 3.7 mm long was pushed through two O-ring seals to pump 70 liters of air from various depths in the pit, to check for the stratification of gases. The air samples were obtained in three stainless steel and three aluminum to be analyzed in the laboratories of the National Oceanic and Atmospheric Administration in Boulder, Colorado through gas chromatography (El-Baz et al., 1989).

2. Marskhod Drill: Exobiology Science Objectives and Design Considerations

The science foreseen with the drill was to continue the investigations of the geochemistry of the Martian surface (El-Baz, 1995).
The related issues would have been:

- Search for carbonates and nitrates

Important compounds to be sought on Mars are carbonates and nitrates. Both are expected to be deposited from liquid water and their discovery would confirm that Mars did have standing bodies of water at some time in its past. The presence of carbonates is suggested by models of the early thick atmosphere. Furthermore determining the presence and abundance of nitrates (a major reservoir of nitrogen) would be key to understanding the volatile abundance and outgassing history of Mars.

- Organics

Models suggest that there may be organics in the deeper layers of Mars. The search for organics remains a key objective for exobiology exploration. Apart from the effects of the surface oxidants, organics should be well preserved on Mars: the cold, dry climate should be a preservative; tectonics, which on Earth result in the heating and metamorphism of sedimentary material, is absent on Mars, and the lack of water should also help to isolate and preserve organics buried at depth. Preliminary measurements would focus on the detection of organics to be followed by more complex analysis - isotopic studies comparing organic and inorganic carbon as well as isotopic ratios of the organics.

- Water

Exobiology needs to understand the distribution and duration of bodies of liquid water throughout Mars' history. The primary reason Mars holds such interest for exobiology is the possible geo-morphological evidence for liquid water on its surface in the past. This is the only firm evidence we now have for liquid water outside the earth's biosphere. Water, in the liquid state, is the most critical environmental requirement for life. Therefore the search for past and present life on Mars has naturally focused on searching for liquid water or ice. Studies of water on Mars primarily influence the exobiology mission strategy in terms of site selection. The ultimate goal for exobiology on Mars is to find direct evidence of life. ***If*** investigations into its geologic and climatic history show that Mars enjoyed biologically friendly conditions such as liquid water habitats, and ***if*** orbital imaging and surface analysis allow the selection of suitable sites (e.g. ancient lake sediments), ***then*** an intensive search could be supported for the detection of organics in the subsurface sediments.

- Need for subsurface samples

There are several reasons why it will be necessary to obtain samples from below the surface on future Mars missions. Below the mantle of windblown dust. Previous samples were from depths up to 10 cm only. At this depth, only windblown material was collected. The elemental composition of this windblown material was similar at the two Viking sites. It is probable, therefore, that this material represents a globally uniform mantle.

- To reach the organics.

The GCMS instrument on the Viking Lander was used to search for organics in the surface materials. No organics at the level of sensitivity of the instrument (ppb) were detected. This result has lead to the suggestion of the existence of one or more strong oxidants in the Martian soil.

3. The Preliminary Design of the Marskhod Drill

The Marskhod drilling could be done with a power-head similar to B& D's Apollo Lunar Surface Drill (ALSD, with a head mass of 4.04 kg and a drill bit of 1.9 kg). This is the only mechanism that would do the job. Basically it is a motor driven single-tooth cam that lifts a striker mass that is then propelled forward by a helical compression spring, and impacts the end of the bit while the bit is being rotated. Air-spring mechanisms would not work on Mars (Moores, 1995).

If the Mars drill doesn't have to obtain deep cores, as did the ALSD, its mechanism could be of smaller scale. This is a critical point - the mass of an impact-drilling machine, including the drill bit, is closely related to the required hole depth. For example, if the Mars drill needs to drill only 0.5 meter deep, obtain only a few grams of material for analysis, and a degree of mixing between strata is permitted, then an auger bit, such as Friedman envisions, would probably be a good choice. The drill bit would be 10 mm diameter by 0.6 m long and have a mass of about 0.25 kg, but the proposed total mass of 1.0 kg for power-head, bit, and control electronics, is unrealistic.

A more realistic target of 2.0 kg would already be difficult. The bit must be driven in a straight line or it will bind and energy will be wasted. If the drill is attached to the revolving "experiment head" we can assume that a control algorithm can be written so the articulated arm system can drive the drill in a straight line. Otherwise, the drill must be mounted on a slide rail with its own feed motor/mechanism. This would bring an additional mass penalty. A rock dust sample could have been collected with an articulated scoop that is controlled separately from the drill. That way, if the bit became stuck or the drill failed, a surface sampling system would still be available.

With a weekly power allocation of 50 watt-hrs (about 10 minutes drilling time), there should have been time enough to drill at least one, and possibly two holes.

***Table** 5.2-1.* **Some rough estimates of drill parameters:**

Parameter	Condition
Impact energy per blow	2 Joules
Blows per minute	2200
Bit rotation speed (load)	500 rpm
Motor specifications	Based on existing B& D cordless hammer drill motor
Input volts	14.4
Input watts (load)	300
Output watts (load)	200
Mass	0.4kg
Housing	Magnesium, investment cast & machined
Drip bit	Percussion carbide tipped steel shank

The questions which were left open were:

1. How many holes need to be drilled?
   (The bit proposed, would only be good for about five holes in granite, with penetration rate ever-decreasing due to dulling of the carbide edge).
2. Does the system have to allow for bit replacement?
3. What if the bit gets stuck?
   The control system should have the ability to eject the bit (or drill system) if this occurs. Otherwise, the rover would be pinned to that site for the rest of its days.
4. What are the reliability parameters of the drill system?

5. How much redundancy has to be designed in?

These factors will affect development time, and particularly weight of the system. On the Egypt drill you don't have to worry about these things. You could just have made everything three times as strong as you thought was needed.

References

Clark, B.C. (1996). "Overview of the Techniques of Science on Mars - The Past, Present, and Future"; COSPAR '96, Symposium F 3.5, Birmingham, UK, 14-21 July 1996.

El-Baz, F. (1995). "Marskhod Drill : Exobiology Science Objectives and Design Consideration", private communication.

El-Baz, F. (1997) "Space Age Archeology", *Scientific American*, No. **8**, 40-45.

El-Baz, F., Moores, B., and Petrone, C.E. 1989, *American Scientist*, **77**, 61.

Galeev, A.A., Moroz, V.I., Zakharov, A.V., Linkin, V.M., Surkov, Yu.A., and Kremnev, S. (1996). Mars '96 Mission, COSPAR '96, Symposium F 3.5, Birmingham, UK, 14-21 July 1996.

Gorevan,S., Myrick,T., Rafeek,S., Shaltout,M.A.M., 1997, Space Technology, **17**, 231.

Holden, C. A (1987). "Questor Ancient Egyptian Air", *Science*, **236**, 1419.

Jakosky, B.M. (1996). "Seasonal Cycles, Climate and Climate Change on Mars", COSPAR '96, Symposium F 3.5, Birmingham, UK, 14-21 July 1996.

Mancinelli, R.L. (1996). "Prospects for the Evolution of Life on Mars : Viking 20 Years Later", COSPAR '96, Symposium F 3.5, Birmingham, UK, 14-21 July 1996.

McKay, D.S., et al., 1996, Science, **273**, 924.

Meyer, M.A. and Kerridge, J.F. (1996). "An Exobiology Strategy For Mars Exploration", COSPAR '96, Symposium F 3.5, Birmingham, England, 14- 21 July 1996.

Moores, B. (1995). "Marskhod Drill: Statement and Preliminary Design", private communication.

Shaltout, M.A.M. (2000) "Space Research Activities in Egypt", *COSPAR Information Bulletin*, **147**, 38-41.

Yoshimura, S., Nakagawa, T., Tonouichi, S., and Seki, K. (1987). "Studies in Egyptian Culture", Tokyo, Weseda University.

Chapter 5.3

SMALL ASTRONOMICAL TELESCOPES FOR RESEARCH AND EDUCATION AT HELWAN, EGYPT

M.A.M. Shaltout
Space Research Center
Desert Environment Research Institute
Minufiya University
El-Sadat City, Egypt
mamshaltout@frcu.eun.eg

Abstract This paper reviews the current status of small astronomical telescopes in use at Helwan Observatory (NRIAG). It highlights the capabilities of small telescopes with modern instrumentation in the context of astrophysics of the 21st century. An overview of research and education programs is included. An international project with Intercosmos countries and the SAO in the field of artificial satellite observations and tracking is commented. Meteorological and solar data collaborations with ESA and NASA for applied and academic research is done at Helwan Observatory.

Introduction

Among other branches of science, astronomy has been studied in Egypt since its early history. The Ancient Egyptians set rules for and accumulated chronological data from observed solar eclipses and other phenomena. Inscriptions on the ancient monuments and temples, the location of the Pyramids and paintings in tombs prove beyond any doubt the great knowledge in this field of science.

Later during the Alexandrian epoch (150-200 A.D.), these efforts culminated in the production of the ALMAGISTI by Ptolemy, which led in to many discoveries of later eras. Numerous important discoveries were made in the Arabic culture. Theories, which brought about an understanding of the motion of the sun, moon and the planets, were the milestones for more refined theories in the future.

W. Wamsteker et al. (eds.), Developing Basic Space Science World-Wide, 251-259.

Famous Arab Astronomers were: Mohammed El-Batani, Abdel Rahman Al-Soufi and Ibn Younes. Ibn Younes lived in Cairo about 1000 years ago and composed the "Hakimid Zig". He had an observatory on the Mokattam Hills in Cairo East. Another observatory, also in Cairo, was known as the Al-Aftal Hansha Ansha on the limestone plateau opposite Ather El-Nabi in the direction of Maadi. A third observatory was erected in the Fiela Mosque and was transferred to the Guishi Mosque. The last of these observatories was situated on Bab-El-Nasr and was known as the "Maamoun Observatory".

During the last century, an astronomical observatory was founded in Boulac (Cairo) in 1800 by the French at the time of Napoleon Bonaparte's regime. This observatory was closed in 1860 when another one was founded at Abbussia (Cairo) by Ismail Basha. The 10" reflector telescope of the Abbussia Observatory was eventually moved to Helwan in 1903 where the present observatory stands on a limestone Plateau, 116 meters above sea level 5 kilometers east of Nile, and 25 kilometers south of Cairo. At that time, Helwan was a little village of only around 5000 inhabitants. It was characterized by sunny days and clear nights with high atmospheric transparency. Astronomical observations at Helwan started in 1905 using a 30" reflecting telescope, and the observatory has participated in various international projects (see also Hassan,2003). Due to the expansion of the city of Helwan which is now the centre of heavy industry in Egypt, and the resultant effects of light and atmospheric pollution, the quality for astronomical observations has been greatly affected. A new site was chosen at Kottamia in the northern eastern desert, 80 km from Helwan on Cairo-Suez road on a hill of altitude 476 meter. The Kottamia observatory was erected and started functioning in 1962. The observatory houses a 74" reflecting telescope manufactured by Grubb Parsons of England.

Since 1964 the astronomical observations at Helwan observatory concentrated on solar research and the tracking of artificial satellites. All of the small telescopes in Helwan observatory belong to the "Department of Solar Research and Space Sciences" and are used for research programs as well as education. The department contains two laboratories, one for solar research, and one for space sciences.

1. Small Telescopes for Solar Research

A solar station was installed on the observatory grounds in 1957 for observing the various solar phenomena visually, photographically and spectrographically.

The station is equipped with:

- a 10" Horizontal Coelostat;
- an Auto Collimating Hilger interchangeable high dispersion spectrograph
- a special camera for direct photography of the solar disk.
- a photoelectric scanner was developed to work photo-electrically
- and the spectroscopic capabilities of the Station were increased by using a grating in
- addition to the prism;
- the primary image of the solar disk was increased from 3 cm to 10 cm on the slit of the spectrograph to measure the spectrum of the sunspots and other solar phenomena.

Now Helwan is one of the most polluted cities in the world due to the industrial activity, and in particular the cement industry. One of our research programs makes spectroscopic studies for the atmospheric spectral lines to determine the types and abundance of the air pollutants in Helwan atmosphere.

In 1964, a 6" Coude refractor manufactured by Carl Zeiss Jena with a photographic solar camera was installed at the observatory.

The daily observations for sunspots have covered the three solar cycles 20, 21 and 22 and monthly reports on solar photospheric aspects are sent to the World Solar Data centres for more than thirty years. In cooperation with the Debresen Observatory in Hungary photographic solar observations techniques wee implemented to study the proper motions of sunspots.

Also, a Hγ (FOV ~0.6°) and a Ca II K-line (FOV ~ 1.0°) interference filter have been installed on Helwan 6" Coude solar refractor, for studying the chromospheric phenomena. These filters were manufactured at Ondrjov observatory in the Czech Republic. This telescope is also used for observing the new moon to determine the first day of each Arabic month (Lunar months). This is very important for various Islamic Festivals.

The 6" Coude refractor has been in operation for a long time for education for under- and post-graduate students from the Astronomy Departments of astronomy at Cairo and Al-Azaher Universities as well as for training courses for NRIAG researcher assistants and for training courses associated with international summer schools in Cairo in 1981 and 1994 in cooperation of International Astronomical Union (Andersen, 2003).

A complete station for solar radiation measurements (manufactured by Eppley Laboratory, Inc.). was installed in 1986 on the roof of NRIAG building in Helwan,. This station comprises the following instrumentation:

- Precision Spectral Pyranometers;
- Normal Incidence Pyrheliometers with solar tracking mount.
- Ultraviolet Radiometers (Photometers);
- Precision Infrared Radiometers (Pyranometers);
- H-F Cavity Radiometers; and
- Electronic Integrators.

Figure 5.3-1. The normal incidence pyrheliometer with solar tracking mount on the roof of Helwan Observatory.

The station coordinates with the national meteorological network of stations to measure ambient air temperature, relative humidity, wind speed and direction, and atmospheric pressure. Through measurements of the direct, diffuse, and global solar radiation in different spectral bands, the impact of air pollution in Helwan is studied. Various test programs solar cells and photovoltaic panels in outdoor conditions are also made.

The research programme can be summarized to include the following elements:

1. Observation and analysis of solar phenomena;
2. Studies of the Solar-Terrestrial interaction;
3. High energetic solar flare forecasting;
4. Variability of solar irradiance and its relation to climate change;

5. Short and long term periodicities in solar X-ray and EW and its influence
6. on the Ozone Layer;
7. Ground verification of Meteosat data for solar energy and moisture distribution over Egypt;
8. Photometeric and spectroscopic studies of active regions in the solar atmosphere;
9. Solar radiation measurements for air pollution estimation; and
10. Solar cell and PV panel testing in outdoor conditions.

Small Telescopes for Space Sciences

In 1905, Mr. Reynolds, a British amateur astronomer and later president of the Royal Astronomical Society, London, came to Egypt to observe the total solar eclipse. Due to the clear sky in Egypt and its suitability for astronomical observations, he presented the Helwan Observatory with a 30-inch reflecting telescope for observing faint nebula. This telescope was used extensively to photograph nebula covering the zones of the sky allocated to the observatory by the IAU. Also many studies were made of comets, asteroids, and planets.

The work on the ninth planet Pluto in 1930 (early after its discovery) was of worldwide importance. Observations of the planet Mars during its opposition in 1956 were carried out and a great number of photographs were taken of the planet on this unique occasion, in accordance with the IAU recommendations. A Macrovitch Camera was fixed on the mounting of the 30-inch reflector as a separate 5-inch telescope for the occasion of the IGY Program in 1957. This camera has been heavily used.

An extensive observing program Nova Herculis in 1961 was carried out in the Newtonian focus with the objective prism and camera.. Extensive observations were made of Comet Arend-Roland (C/1956 R1), Comet Mrkos (C/1956 E1) and others.

A photographic photometer was build at Helwan to be used on the 30-inch telescope for the observation of fainter stars. Artificial satellites observing and tracking started at Helwan in 1961. In 1966 a bilateral agreement for a joint program in satellite tracking was signed with the Astronomical Council of the USSR Academy of Science.

In this context a satellite tracking station was placed at Helwan both for visual and photographic observations by means of a NAFA 25, two AFU 75 Cameras. In 1974 laser ranging equipment from the " INTERCOSMOS" Countries was installed at Helwan for high accuracy satellite ranging. Geostationary satellite tracking was is done in cooperation with the Institute of Theoretical Astrophysics in Moscow with a 30-inch telescope equipped with a CCD Camera. This program was extended to include the SAO as a third party in 1997.

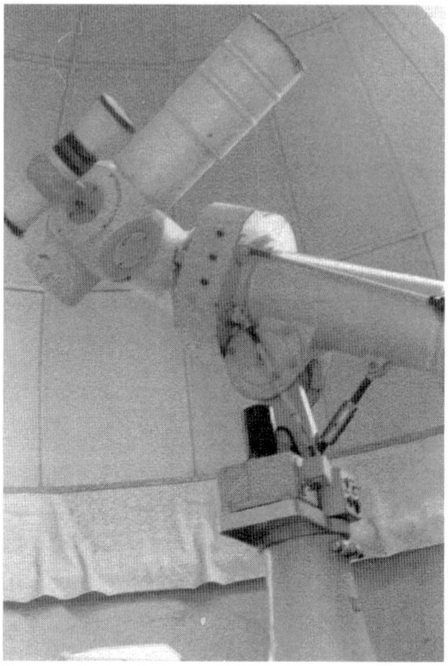

Figure 5.3-2. The 6-inch Coude Refractor Telescope at Helwan Observatory

Recently a new automatic laser ranging station has been installed. This is accurate to a few centimeters at about ten thousand kilometers and has a timing accuracy of 1 second. A small photoelectric telescope with wide band glass filters for photometric studies of night sky, zodiacal light and twilight is also present at the Observatory. This is used for site testing and visibility studies of the new Moon at different sites in Egypt such as Abu Simbul and Daraw in Upper Egypt. This is to determine the first day of each Arabic month (Lunar months), important for Islamic Festivals. In summary the research programme includes the following elements:

- Photographic and laser tracking of artificial Earth's satellites;
- Determination of artificial satellites and the perturbed forces to improve the theory of satellites dynamics;
- Studying the upper atmosphere;
- Satellite geodesy and geodynamics; and
- Observing the moon and the planets, photometric studies of night sky, zodiacal light and twilight.

2. Helwan Observatory from Small Telescopes to Space Missions

Remote Sensing Research

Every day, the Meteorological Authority in Cairo receives three photographs of cloudiness over Egypt from the Meteosat satellite, one in the visible, and two in the infrared bands (10.5- 12.5 μm) and (5.7-7.1 μm). The monthly average cloudiness for 24 sites over Egypt have been measured and calculated based on these data during the 1980s. Ground truth correlation analysis between the cloudiness and global radiation measured from ground stations was carried out. Models and empirical relations for estimating the global and diffuse solar radiation cloudiness data over Egypt are deduced and tested and seasonal maps for the global and diffuse radiation over Egypt are made (Shaltout and Hassan, 1990).

Also, the same (ground truth) technique was used to estimate the moisture over Egypt, where correlation analysis between the cloudiness observed with Meteosat and water vapor content, relative humidity, and evaporation measured on the surface was carried out. Models and empirical relations for estimating the moisture over Egypt were also deduced and tested (Shaltout et al., 1996). The water loss from Nasser Lake in the south of Egypt (Shaltout and El-Hosary, 1996)is one of the national problems of Egypt, as this is the water bank of Egypt. Evaporated water quantities range from 10 to 16 billion cubic meter every year (20-30 % of the Nile water).

Solar Research by Data from Artificial Satellites of NASA

The Space Environment Monitor (SEM) aboard the Synchronous Meteorological Satellites (SMS) and Geostationary Operational Environmental Satellites (GOES) included 0.5 -4 Å and1 -8 Å X-ray ion chambers. Currently, GOES-7 is the primary source of solar X-ray, electron and proton flux data to estimate the short term periodicity of solar X-ray bursts and its relation to electron density of the lower ionosphere (Shaltout, 1980a,b, and c; 1990d). Also, solar X-ray data is used to predict and estimate the strength of solar proton and electron events (Shaltout, 1984), and find a relation between solar X-ray bursts with interplanetary shock waves and geomagnetic storms (Hady and Shaltout, 1984). Various periodicities of solar X-ray emissions are also studied at Helwan (Hady and Shaltout, 1988) and (Shaltout and Hady, 1987).

The proton and electron flux measured by GOES are used to study the influence on radio communication from solar proton flares (Shaltout, 1989; 1990b and c). It is also used for estimating the method for predicting the high energetic solar flares (Shaltout, 1991; 1995a; 1995b; Shaltout et al., 1995).

The first active Cavity Radiometer Irradiance Monitor (ACRIM I) experiment was lunched on the NASA Solar Maximum Mission (SMM) spacecraft in February 1980 to make regular observations of the total solar irradiance. The second active Cavity Radiometer Irradiance Monitor Satellite Solar Monitoring Experiment (ACRIM II) has been providing total solar irradiance observations since its launch as part to the Upper Atmospheric Research Satellite (UARS) in late 1991. Also, the Nimbus-7 EAB began taking useful data on solar irradiance on 16 November 1978 and seven years of data have been received.

This data was analyzed and established the variability of the solar constant measured at Earth's surface and by artificial satellites and its relation to Climatic change and the flooding of the river Nile (Shaltout and Tadros, 1990a; 1990b; Shaltout, 1993).

Also, the relation between solar activity and earthquakes on a global scale and in the region of North Africa was discussed by using worldwide data (Shaltout et al., 1991; 1999).

Egypt Participation In Future Mars Rover Mission

During the Fourth UN/ESA Workshop on Basic Space Science, held in Cairo, in July 1994, the possible participation of Egypt in the future Mars rover Mission was discussed. One concept suggested was that Egypt participates in this mission through involvement in the design, building and testing of a drill for obtaining subsurface samples. This is further discussed in Shaltout (2003). Unfortunately, in the end this project was discontinued.

References

Andersen, J., 2003, this volume.

Hady, A.A. and Shaltout, M.A.M. (1988). "Seconds Periodicities in the Solar Soft X-ray Emission", Proceedings of the Thirteen International Congress for Statistics, Computer, Science, Social and Demographic Research, Cairo, Egypt, 26-31 March 1988.

Hassan A.H., 2003, this volume.

Shaltout, M.A.M., 2003, this volume.

Shaltout, M.A.M., (1980a), "Short-term periodicity of the Solar Soft and Hard X-ray Burst and its relation with the Electron Density of Lower Ionosphere, Helwan Observatory Bulletin, No. **243**.

Shaltout, M.A.M. (1980b), "High-Resolution Spectra of the Time Series of the Solar X-ray Bursts", Helwan Observatory Bulletin, No. **244**.

Shaltout, M.A.M. (1980c). "Solar Seismology and X-ray Bursts Activity", Helwan Observatory Bulletin, No. **245**.

Shaltout, M.A.M. (1984a). "Predicting and Estimating the Strength of Solar Proton and Electron Events from X-ray Bursts Observations", NRIAG Bulletin Vol. **IV**, Ser. A, 19-29.

Shaltout, M.A.M. (1984b). "Solar X-ray Bursts and Their Relation with Interplanetary Shock Waves and Geomagnetic Storms", NRIAG Bulletin Vol. **IV**, Ser. A, 1-18.

Shaltout, M.A.M. (1989). "High Energetic Solar Flares of 24 June 1988 and Their Influence on Radio Communication", Proceedings of the Sixth National Radio Science Conference, Paper E3, Military Technical College, Cairo, Egypt, 19-21 February 1989.

Shaltout, M.A.M. (1990a). "Solar Radiation and Heiloseismology Periodicities from Power Spectra Analysis of Solar X-ray Emission Measured be Satellites", IAU Colloquium No. 130 *Sun and Cool Stars*, Helsinki, Finland, Helsinki University, July 17-21, 1990.

Shaltout, M.A.M. (1990b). "Solar Proton Flares and Their Influence on Aircraft Radio Communication", Proceedings of the Seven National Radio Science Conference, Paper R2, Military Technical College, Cairo, Egypt, 20-22 February 1990.

Shaltout, M.A.M. (1990c). "High Energetic Solar Proton Flares of 19 October 1989 and its Terrestrial Effects", Proceedings of the Conference on Chromospheric and Coronal Heating Mechanisms, Heidelberg, Germany, University of Heidelberg, 5-8 June 1990.

Shaltout, M.A.M. (1991). "High Energetic Solar Proton Flares of 19 to 20 October 1989", Proceedings of IAU Coll. 133, 1-5 August 1991, Iguauso, Argentina.

Shaltout, M.A.M. (1993). "Variability of the Solar Constant Measured at Earth's Surface and by Artificial Satellite Nimbus-7", Proceedings of the International Conference on Application of Time Series Analysis in Astronomy and Meteorology, University of Padua, Italy, 6-10 September 1993, 383-386.

Shaltout, M.A.M. (1995a). "High Energetic Solar Proton Flares on the Declining Phase of Solar Cycle 22", COSPAR'94, Hamburg, Germany, 11-21 July 1994, Symposium Dl.l, *Advances in Space Research*, **16**, No. 9, 47-50.

Shaltout, M.A.M. (1995b). "High Energetic Solar Proton Flares of 22 March 1991", IAU Colloquium 153, Magnetic and Dynamic Phenomena in the Solar Atmosphere, Makuhari, Japan, 22-26 May 1995.

Shaltout, M.A.M. and El-Hosary, T.N. (1996). "The Estimation of Evaporation Over Naser Lake in Egypt from Meteosat Satellite Observations", COSPAR'96, Birmingham, 14-21 July 1996, England, *Advances in Space Research*, **19**, No. 3, 515-518.

Shaltout, M.A.M. and Hady, A.A. (1987). "Hours and Minutes Periodicities in the Solar Soft X-ray Emission", Proceedings of the 5th annual meeting of the Egyptian Geophysical Society, pp. 487-499.

Shaltout, M.A.M. and Hassan, A.H. (1990). "Solar Energy Distribution Over Egypt Using Cloudiness from Meteosat Photos", *Solar Energy*, **45**, 345-351.

Shaltout, M.A.M. and Tadros, M.T.Y. (1990a). "Variation of the Solar Activity and Irradiance, and Their Influence on the Flooding of the River Nile", Mausam, **41**, No. 3, 393-402.

Shaltout, M.A.M. and Tadros, M.T.Y. (1990b). "Correlation Analysis Between the Sunspots and the Nile Flood", Mausam, **41**, No.4, 565-568.

Shaltout, M.A.M., Hassan, A.H., and El-Hosary, T.N. (1996)., "The Estimation of Moisture Over Egypt from Meteosat Satellite Observations", COSPAR'94, Hamburg, Germany, 11-21 July 1994, Symposium A.3, *Advances in Space Research*, **18**.

Shaltout, M.A.M., Tadros, M.T.Y., and Mesiha, S.L. (1991). "Power Spectra Analysis for WorldWide and North Africa Historical Earthquakes Data in Relation with Sunspot Periodicities", General Assembly, IUGG, Vienna, Austria, 11-24 August 1991, Renewable Energy, **17**, 499-507, 1999.

Shaltout, M.A.M., Yousef, S., and El-Said, O.M. (1995). "Statistical Analysis for Solar Proton Events Measured by GOE Spacecraft During the Period 1976-1990, COSPAR'94, Hamburg, Germany, 11-21 July 1994, Symposium E2.2, *Advances in Space Research*, **17**, No. 4/5, 171-175.

Chapter 5.4

ASTRONOMY IN SAUDI ARABIA: THE CHALLENGES

M.B. Al-Malki
Teachers College of Riyadh
P.O. Box 4341, Riyadh 1149
Saudi Arabia
malmalki@yahoo.com

Abstract In the last thirty years, more than hundred astronomers graduated from the astronomical departments of King Saud University and King Abdalaziz University. About 15-20% of them are working in basic space sciences or education, related to astronomy. In this paper, the problems and challenges that astronomers in Saudi Arabia face are presented, such as the limited number of astronomers as compared to the astronomical facilities available in S.A. e.g. more than eight reflectors (range from 14" to 21"), two solar labs. and some others small instruments. The lack of under or post-graduate students, and the limited support from higher authorities in the universities. The paper gives some recommendations to overcome some of these problems.

Introduction

Astronomy, as a science, is important in Saudi Arabia, because it is closely related to their Islamic religious practice. Astronomical knowledge is needed to calculate prayer times, to find the Kebla (Makkeh direction), and to define the date of Ramadan (Fasting month) and Thoalhijah (Pilgrimage month) by new moon sightings (Helaal). Also the holy book "Al-Quraan" directs the attention to the universe around them, and asks to try to know "how the creation started". Such stimulus encouraged, in past times, Moslem scientists to study astronomy very deeply, and make great efforts over many generations.

One would expect this tradition to have continued in the Moslem religion into the current times. Unfortunately that is not the case. This is mainly due to a weak back-ground in astronomical knowledge among the

W. Wamsteker et al. (eds.), Developing Basic Space Science World-Wide, 261-266.

people, the lack of under-graduate students, and the lack astronomical research and researchers.

1. Professional Astronomy in Saudi Arabia

Modern astronomy has now been taught in Saudi Arabia's (S.A.) universities for more than thirty years. There were two full departments, in the King Abdulaziz University (KAU) in Jeddah, which has an Astronomy degree program. The second is at the in King Saud University (KSU) in Riyadh. Although the Physics Dept. has the name “Physics and Astronomy Dept.”, it does have neither a graduate nor an undergraduate Astronomy or Astrophysics degree program.

In addition to these two departments there is the Institute of Astronomical and Geophysical researches (IAGR) in the King Abdulaziz City for Science and Technology (KACST).

More than 100 astronomers have graduated from both KAU and KSU, less than ten continue through the Ph.D., about five graduate with an M.Sc. and continue to work for Ph.D., and less than five are studying for a M.Sc. Also some physics graduates work in astronomy.

In KACST some astronomers with B.Sc. degrees work in IAGR. A few astronomers are employed in the Science museums and planetariums. Less than five astronomers work for the military, and most of the remaining astronomers work in education, teaching mathematics and/or physics. So approximately 15%-20% of the astronomy graduates are employed now in basic space science related to astronomy, the rest are not, and that illustrates the number of interested students. Most these astronomers are Saudis, the first Ph.D. was granted in 1985, and the last was in 1999. Some non-Saudis work as university staff members in KAU and KAU.

Both KSU and KAU have Celestron 14" (C14) telescopes with photometers, a 15cm Coudé refractor, a Solar laboratory, a small planetarium and other astronomical instruments. In addition KAU has a CCD camera, and KSU has a double telescope with a 45cm Richy-Chrétien and a 24cm Schmidt telescope. The observatories at KSU and KAU are located near the universities, where the light pollution is very high, limiting the observational possibilities.

Through a special Royal Decree, KACST is in charge of large-scale astronomical projects, such as the National Observatory Project (NOP). The KACST site selection program for NOP started towards the end of the 1970’s with the help of a Canadian team and an American consultant. The oil fires in the region and various conflicts have strongly affected the Saudi atmosphere, and delayed the project. Also, the earthquakes in

Egypt, which directs most of the efforts in Geophysics projects at IAGR and KACST, generated delays for many astronomy projects.

KACST have four 15cm Coudé installed throughout S.A. for NOP site selection, and three C14" for Islamic crescent visibility. KACST has a large database of ten years of solar observations made at the Solar Village in Oiaina (near Riyadh). Moreover, KACST has acquired a Laser/Lunar ranging telescope (75 cm). This telescope will be used for Geodesy and Geodynamic studies and research, mainly for earth rotation, polar motion and Time-Service.

A 3m Radio Telescope has been designed by Electrical Engineering Dept. and Astronomy Department of KSU to observe sun and may be stars. Figure 5.4-1 shows the location of all these astronomical facilities.

For the Time-Service, still elementary in S.A., KSU has three atomic clocks in the Astronomy Department with a GPS receiver in the seismological observatory. KACST has a GPS in the Remote Sensing Institute. The responsible organization for time keeping and dissemination is the Standardization and Metrology Organization for G.C.C. countries (SMO). SMO has two atomic clocks. A closer communication and cooperation started between KSU and SMO some time ago, but has stopped now.

2. Amateur Astronomy in Saudi Arabia

During the last few years the General Presidency of Youth Affairs has supported free courses for amateur astronomers in Riyadh, and plan exist to establish a association. A L200 Meade 10" telescope has been acquired for use by amateur and professional astronomers. In Jeddah, the private sector supports a Science Club, which develops most of its activities in the field of Astronomy through Jeddah Center for Science and Technology.

There are three large planetariums (each seating more than 100 visitors) all of them in Riyadh. None of them are open for the public visits till now. The oldest is used for military studied in the Department of military survey, the second is used to teach students of the Islamic University of Umam Mohammed bin Saud under the supervision of the Geography Department, and the third is in King Fahd Cultural Center which has not yet officially opened. Most of the shows are related to astronomy. In the Science Oasis a small plastic planetarium is open for visitors, twice a day.

There are two science museums in S.A.. The largest is the Jeddah Center of Science and Technology in Jeddah, which has no planetarium. The second one is at the Science Oasis in Riyadh.

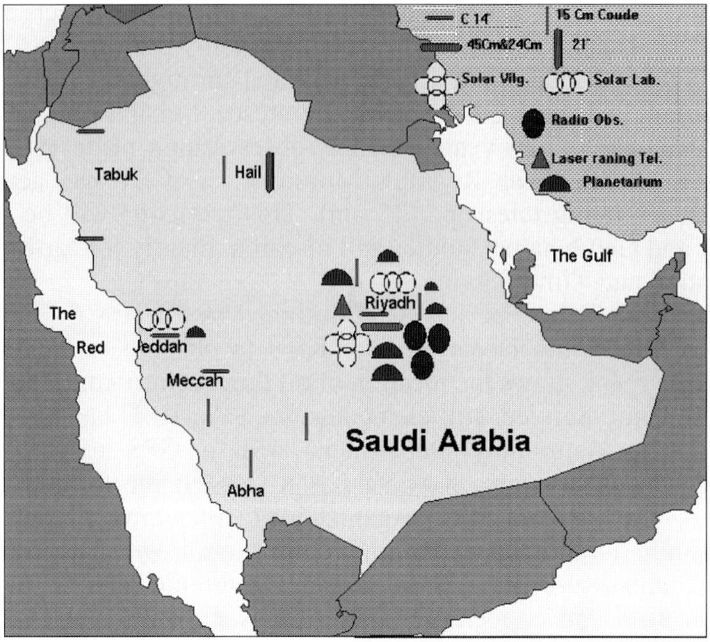

Figure 5.4-1. The Distribution of astronomical instrumentation within Saudi Arabia.

Both Departments in KSU and KAU are open for schools to visit and learn about astronomy. In KSU, open nights for the public are regularly scheduled.

It is expected that all the activities, described above, will improve the astronomical and general basic space science awareness and increase the interest for professional and amateur astronomy. It is estimated that among the two million people in Riyadh, about 200 people have a private telescope, a ratio of (0.001%) not that good, but the hope of a fast improving is there.

Some amateur astronomers have very good astronomical equipment, for example H.M. prince Moqren Al-Saud has a fully computerized 21" reflector telescope (in Hail- in the north of S.A.) with a CCD camera and spectrograph. H.M. is offering his observatory for all astronomers to use, but the location of the observatory is not easy to access from the work places of most of the astronomers.

Table 1 and 2 summarize the professional astronomical manpower and astronomical instruments in Saudi Arabia.

***Table** 5.4-1.* **Astronomical manpower in S.A.**[1)]

	Ph.D.	Graduate Students	M.Sc.	Pre-Grad. Students	B.Sc.	Under Graduate Students	Technical personnel
KSU	1s	1s	0s	-	2s	-	1s
	2n	-	1n	-	-	-	-
KAU	5s	0s	0s	-	-	-	-
	1n	-	2n	-	-	-	1n
KACST	2s	0s	2s	2s	7s	-	4s
	0n						
SMO	-	-	-	-	-	-	-
Others	3	-	-	-	5s	-	3n
Total	14	1	5	2	14	-	9

[1)] In the table (s) is for Saudis and (n) for non-Saudis

***Table** 5.4-2.* **Astronomical instruments in S.A.**

	C14"	15 cm Coude	45 cm & 24 cm	C8" or Meade	21"refr .	Solar lab.	Atomic clocks	Plane-tarium	Photo-meter	CCD	Radio obs.	Laser ranging
KSU	1	1	1	-	-	1	3	1	1	2	1not in use	-
KAU	1	1	-	-	-	1	1	1	2	2	-	
KACST	3	4	-	6	-	Solar village	2	1	2	3	2can be used	1
SMO	-	-	-	-	-	-	2	-	-	-	-	-
Amateurs & others	2	3	-	14	1	-	-	6	1	2	-	-
Total	7	7	1	20	1	3	8	9	6	9	3	1

3. Problems and challenges' face Astronomy in S.A.

The following are the most important problems and challenges that face basic space science and astronomy in S.A. specially the researches:

1. As can be seen in Table 2 and comparing with table 1 (and figure 5.4-1) the number of active astronomers in S.A. is relatively low compared to the astronomical facilities available.
2. The large distances and limited transport infrastructure between the observatories make travel hard.
3. The observatories associated with the Universities suffer from light pollution.
4. The limited career possibilities for graduate astronomers, limit the number of undergraduate students.

5. The small number of undergraduate students that finish the career now, together with the essential absence of postgraduate students, create a heavy load for the academic staff, limiting the time available for research. Of course the absence of students in the basic space sciences also affects the departmental budget negatively and it's image in the University, resulting in a decrease in resources, such as professional journal subscriptions.
6. The differences in specialization among Ph.D. astronomers, make it difficult to do team research.
7. The language of publication in the basic space science is very different from Arabic.
8. Astronomical outreach is very important for local astronomers for the departmental survival, but it does take time away from research activities.
9. Astronomy and basic space science education is not available in the curriculums in the public schools. Also a clear shortage of teachers at college level is present.

All the above reasons affect basic space science research in Saudi Arabia. Urgent action to remedy this situation is needed. The first step should be taken by the scientists themselves. The active participation in the programs of international organizations such as IAU support offers many opportunities. I think here e.g. of free periodicals and observatories preprints, reduced cost subscriptions. Participation of Saudi astronomers in the IAU astronomers exchange programs will help to upgrade the level locally. Saudi Arabia could request the IAU, COSPAR or UN support in NOP through advise and consultancy for KACST. This is important since the foundation of a big national telescope in Saudi Arabia will clearly enhance the support and improve the image of basic space science in the country.

4. Conclusion

The number of astronomers in Saudi Arabia small compared to the available astronomical facilities. A good and organized cooperation between African and Western Asian astronomers can produce an important stimulus for basic space sciences in Saudi Arabia.

Even though astronomical outreach publicity and stimulation of students′ interest the task of the local astronomers, they can benefit very much from experiences of others, and learn how to overcome the national problems. International cooperation is essential to improve the ability and the quality of basic space science research in Saudi Arabia. The local astronomers hope that efforts in this can be stimulated and will contribute to productive collaborations.

Chapter 5.5

THE STATUS OF THE NETWORK OF ORIENTAL ROBOTIC TELESCOPES PROJECT

F.R. Querci and M. Querci
Observatoire Midi-Pyrénées
Toulouse, France
fquerci@ast.obs-mip.fr

Abstract The NORT project is an effort to establish a network of telescopes, for the dual purpose of training and to supply a series of telescopes with which coordinated observations can be made essentially round-the-clock. The astrophysical problems to which such a facility could be applied are identified, and the current state of affairs is presented.

Introduction

The Network of Oriental Robotic Telescopes (NORT) plans for research-grade telescopes to be installed on high mountains around the Tropic of Cancer, from Morocco to India and Malaysia, and in east African countries such as Ethiopia, Kenya and Madagascar. Its objectives, in terms of preliminary and future cooperative research projects, are:

- to create and/or support lectures and training in astrophysics and space sciences at national universities by national astrophysicists returning to their home countries or by invited professors specializing in various fields of astrophysics;
- to organize summer schools for students; basic courses in theoretical astrophysics, technological equipment and data processing on variable-star research have already been attended by students from Mauritania, Morocco, Tunisia, Libya, Jordan, Palestine and Syria in Observatoire de Haute-Provence (OHP), France;

W. Wamsteker et al. (eds.), Developing Basic Space Science World-Wide, 267-271.

- to train astrophysicists on conception of observational equipment, observational techniques and data reduction techniques in OHP; astrophysicists from Bahrain, Morocco and Libya have already attended long training periods.

The final objective is to enhance research in astronomy and astrophysics in developing countries through their participation in the NORT by setting up robotic 2-metre-class telescopes on high mountains, telescopes which would be involved in stellar variability studies and in other scientific projects. This will permit these countries to develop international-level expertise in astrophysics and space sciences.

NORT is based on non-stop observations (24 hours per day) of variable objects, thanks to international collaboration with other robotic telescopes, both individual and networked. A database of variable stars/objects will be achieved and will help to prepare variable-star observations with large telescopes and interferometers such as VLTI, and Keck I and II. NORT fosters the development of new technologies in developing countries and participation of these countries in large projects with institutions in Europe and the United States of America.

Today, the availability of robotic telescopes on the market (see, e.g., Querci and Querci 2000) gives many countries the opportunity to participate in astrophysics and space sciences (see, e.g., Hajjar et al. 2001).

1. Site Selection

We have explained in previous papers (e.g. Hajjar, Querci, Querci, 2001) why Arab and African countries are interesting locations for astronomical sites. In fact, the climate in some of these countries, such as Sudan, Ethiopia and Kenya, is currently becoming drier (see figure 5.5-1).

As we have discussed in Querci and Querci (2001), a preliminary selection of sites is first carried out based on data archives from meteorological satellites, see also (http://eosweb.larc.nasa.gov/sse/). This method permits identification of very good meteorological sites and selection of sites not subject to the same windstreams. In-situ studies of selected sites are then carried out, analyzing the quality of the atmosphere with respect to opacity, seeing and sky brightness, using the Generalized Seeing Monitor (GSM) or Differential Image Motion Monitor (DIMM) techniques, as already practised at the European Southern Observatory (ESO)'s laboratories in Munich, Nice and Marrakech. Light pollution from villages or oil-fields as well as electromagnetic pollution have to be considered. Accessibility (roads) and the availability of services such as electricity and water are also determining factors.

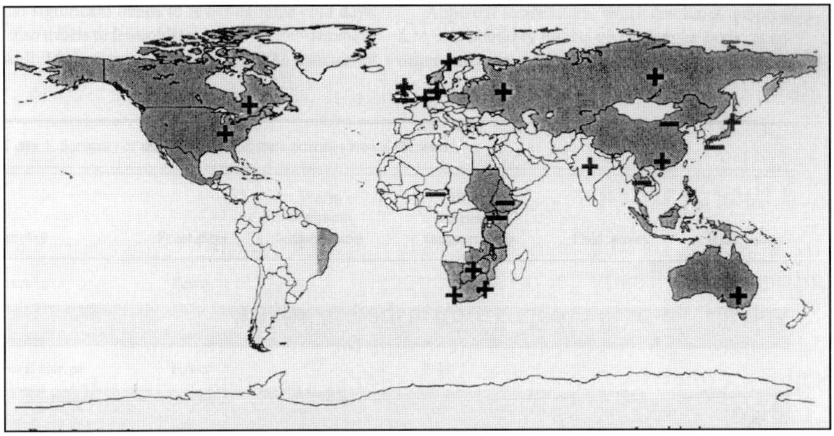

Figure 5.5-1. Regions where analyses of heavy participation have been completed using large sets of daily precipitation time-series. Signs (pluses and minuses) indicate regions where significant changes in heavy precipitation have occured over the past few decades (from Easterling et al. 2000).

2. Scientific Objectives

Many scientific objectives can be achieved with 1-metre class and medium 2-metre class robotic telescopes equipped for photometry, polarimetry and spectroscopy. Let us recall some fields of research:

- eruptive variables such as Be stars and RCB stars;
- binary stars, mainly with matter exchanges;
- searches for meteors and comets, the behaviour of comet coma (condensation, tail, flares, jets, etc.);
- survey-type studies, including searches for Be stars, clustering in star forming regions, and polarimetric surveys;
- discovery and tracking of near-Earth objects with a high angular speed (around 2 or 3 degrees per day);
- searches for planets around nearby stars; and
- X-ray source visible counterparts observed in cooperation with X-ray satellites.

These fields of research can serve as an introduction to contemporary astrophysics for many countries. Some examples of existing studies, together with the countries involved, include:

- binary systems and X-ray binaries (Jordan/France);

- circumstellar envelopes of Ae/Be and young star imaging polarimetry (Lebanon);
- stellar evolution (Lebanon);
- Be and B[e] stars by photometry, spectroscopy and interferometry (Lebanon/France); and
- circumstellar envelopes and polarization of AGB and post-AGB stars (Saudi Arabia/France).

3. Observatories

In support of the astrophysical research and/or education, some countries have already set up telescopes:

- Egypt, a 1.93-m telescope (refurbished in 2000);
- Morocco, a 55-cm telescope at Rabat for public education;

Others countries have planned national telescopes and have started site testing:

- Libya, a 2.0-m national telescope;
- Iran, a 1.0 to 2.0-m national telescope;
- Lebanon, a 1-m national telescope, under a project developed by the universities and the Lebanese National Council for Scientific Research (CNRSL);
- Pakistan, a 1.5-m national telescope;

Some countries have started to consider the development of astrophysics, and are planning the following telescopes:

- Jordan, a 1.5-m telescope;
- Kuwait, a 1- or 2-m national telescope;
- Tunisia, a 1-m telescope by OTEAS.

In Syria, a 15-year plan for developing astronomy, astrophysics and other space sciences is in progress.

4. Conclusion

The main objectives of the NORT are:

- To promote education and training in universities towards a rapid participation of Arab and African countries in international astrophysics;
- To contribute to a complete time coverage and follow-up of variable astronomical objects with 2-m class robotic telescopes, objectives that cannot be achieved using large specialized telescopes (8- to 10-m diameter);
- To promote a new type of cooperation with larger facilities such as optical and infra-red long-baseline interferometers (GI2T, ISI, VLTI, Keck I and II), which are mainly required at critical phases of

variability of the studied objects, phases that are detected through permanent monitoring by networks like NORT.

NORT is an outcome of the United Nations/European Space Agency workshops on basic space science. It has also been adopted by Arab Union for Astronomy and Space Sciences (AUASS). Translating the strong interest in NORT shown by Arab and African countries into reality requires:

- Coordination of efforts between AUASS and the Working Group for Space Sciences in Africa (WGSSA);
- An active collaboration with international organizations such as the United Nations and the Organization of African Unity (OAU), to implement the project on a short timescale in some African countries.

References

Easterling, D.R.,et al., 2000, Bull. American Meteorological Soc., **81**, 417.

Hajjar, R., et al., 2001, Seminars of the United Nations Programme on Space Applications, ST/SPACE/**5**, 43.

Hajjar, R., Querci, F.R., Querci, M., 2001, in Proc. IAU Colloquium 183 "Small Telescope Astronomy on Global Scales", eds. B. Paczynski, W-P. Chen and W. Ip, ASP Conf. Series, **246**, 33.

Querci, F.R., Querci, M., 2001, in Proc. on "Preserving the Astronomical Sky", eds. R.J. Cohen and W.T. Sullivan, ASP Conf. Series, IAU Symposium **196,** 170.

Querci, F.R., Querci, M., 2000, AP&SS, **273**, 257.

Chapter 5.6

BASIC SPACE SCIENCES IN JORDAN

H.M.K. Al-Naimy and Khalil Konsul
Physics Department, UAE University
P.O. Box 130302
Mafraq, Jordan
alnaimiy@yahoo.com

Arab Union for Astronomy and Space Sciences (AUASS)
Jordanian Astronomical Society (JAS)
Amman, Jordan

Abstract The aim of this paper is to summarize the activities and research projects of Basic Space Sciences (Astronomy and Space Sciences (AASS)) in the following Jordanian organizations and Institutions:

1. Jordanian Astronomical Society (JAS).
2. Universities {Mainly Al al-Bayt University}.
Institute of Astronomy and Space Sciences (IAASS).
Maragha Astronomical Observatory (MAO).
3. Arab Union for Astronomy and Space Sciences (AUASS).
4. ICOP Activities: Islamic Crescent Observational Program.
The paper summarizes also other activities in some Jordanian organizations and the future expectation, for AASS in Jordan.

Introduction

Most of us know that astronomy and space sciences (AASS) are important field of research, knowledge and culture. They have been the parents of both eastern and western sciences. Today we might convince our self that astronomy is Human's attempt to study and understand celestial phenomena, part of his never-ending urge to discover order of nature. The science of astronomy was well advanced in Mesopotamia, Egypt, Greece, India and China. Over the past 4000 years it has established in many aspects the roots of our culture, and has been the

W. Wamsteker et al. (eds.), Developing Basic Space Science World-Wide, 273-289.

stimulus and the bases for the birth and the growth of science and technology, in a continuous interplay with religions substantially influencing history.

In all Muslim countries astronomy has been used as a Basic science and applied successfully for the Islamic Mawaqeets (i.e. determinations of the Islamic praying times, and for fixing the beginning of the Hijra months in general and Ramadan, Shawal & the Al-Hija in particular and to calculate the Qibla direction). That's why astronomy and space sciences established in different Arab countries, especially in Jordan, were started in a good manner.

The Hashemite Kingdom of Jordan, is situated in the northwest corner of the Arabian Peninsula, with a population of around 5 million people and about 57×10^4 km^2 in area. Jordan features highly diverse topography, from the desert areas in the east and southeast, to the hilly areas in the north and south, down to the Jordan Rift Valley, which includes the Dead Sea (The lowest point on the Earth's surface, averaging 400m below sea level). Amman, the Capital of Jordan lies in Jordan's mountainous north-central region and enjoys a moderate, Mediterranean - like climate. Some parts of Jordan are ideal for astronomical observations, with about 200 photometric nights per year.

The first official major step of astronomy in Jordan was started in September 1987, with the foundation of the Jordanian Astronomical Society (JAS) in Amman. Later on and as an initiative of JAS, the Institute for AASS in Al al-Bayt University (IAASS) was founded in September 1994. In September 1998 the Arab Union for Astronomy & Space Sciences (AUASS) was also established. Nowadays, there is a plan by UN to establish in Jordan, a Regional Center for Basic Space Sciences & Educational Technology in Western Asia (Al-Naimiy, 2000).

The aim of this paper is to describe the development of AASS in Jordan in last 15-years and it's future outlook.

1. Jordanian Astronomical Society (JAS)

JAS is the only organization in Jordan that popularizes amateur astronomy, formerly known as the Jordanian Amateur Astronomers Society. JAS was found in Amman in September 1987, it is located in Haya Cultural Center. This society has been promoting astronomy & space sciences not only throughout Jordan but also to the rest of the Arab world. Currently JAS have about 300 members, 100 of who are active in club programs. (http://www.jas.org.jo) (Odeh, 2000).

Figure 5.6-1. Haya Cultural Center

Objectives

- Assembling amateur astronomers in Jordan and in the Arab World, in order to develop their astronomical hobby through the exchange of astronomical data, information, articles, and observational expertise, through the provision of all what an amateur astronomer may need (e.g. astronomical equipment, periodicals, books, references, Internet, planetarium ...etc).
- Popularization of Astronomy, that is, spreading astronomical culture and awareness among all classes of the Community, and ascertaining the importance of astronomy and allied sciences in every-day life, and their inter-relationship with all fields of knowledge in Basic Sciences and Technology.
- Transferring astronomy in Jordan and the Arab World into a professional activity, and from convert individual initiatives into institutional frames.
- Contributing in astronomy education in primary and high -schools, and sometimes in universities (at the undergraduate level).

Ways and Means of Fulfilling JAS's Objectives

- Weekly scientific lectures at Haya Cultural Center.
- Publications, such as "Pleiades", and "Al-Debran" the Monthly Internal Publication of JAS.

- Publishing astronomical articles, essays and astronomical news in local newspapers and mass media.
- Provision of astronomical periodicals (Such as "Astronomy", "Sky & Telescope", "Astronomy Now", "Popular Astronomy", "The Planetary Report"...), as well as astronomical books, videotapes, posters, astro-software...etc.
- Weekly and monthly observation of the sky and of the major celestial events, using the naked eyes and astronomical equipment (binoculars and telescopes).
- Organizing "Astronomical Camps" in the Jordanian deserts, which are usually devoted to certain astronomical events, such as meteor showers, comets...etc. (http://www.jas.org.jo/camp.html)
- Cooperation with other institutions (Jordanian, Arab or International), possessing similar interests, for example JAS is a member of the International Meteor Organization (IMO), the International Occultation Timing Association (IOTA), and the American Meteor Society (AMS). And the Royal Astronomical Society of Canada (RASC). (http://www.jas.org/out.html).
- Delivering lectures and seminars, and organizing " Star Nights", at various universities, schools, clubs, societies, youth campus...etc.
- Establishing " Astronomy Clubs" at various schools and universities with continued support from the Society (http://www.jas.org.jo/clubs.html).
- Organizing "Astrofests", activities dedicated only to schools, which may include an opening ceremony, astronomical lectures, an astronomical exposition, and a star night.
- Organizing "Scientific Astronomical Days", in cooperation with universities and/or institutions (such as the professional unions).
- Organizing conferences about astronomy and space sciences, usually in cooperation with universities or other agencies. JAS paid special attention for organizing the 1st (1992), 2nd (1995), 3rd (1998), 4th (2002) and the 5th Arab Conferences in AASS. These are organized in cooperation with other universities, such as, the Al al-Bayt University, the Jordanian University and the Princess Sumaya University College for Technology. The outcome of 3rd conference was the establishment of AUASS. Last activity was the 5th ACASS organized by AUASS & JAS at Princess Sumaya University College for Technology in August 2002. Beside AUASS & JAS have organized two Islamic conferences on the "***Astronomical Applications on Islamic Sharia'a***", in cooperation with the Ministry of Awqaf & Islamic Affairs; the 2nd Islamic conference held in Amman in October 2001, the conference attended by more than 50

Scientists and Astronomers from 18 countries. The 3rd will be held in Amman from 20th-22nd of October 2003 (http://www.jas.org.jo/confe.html).

- Organizing special training courses on topics in astronomy, both for students or teachers. Some of these are held in cooperation with the ministry of Education. Until now more than 25 courses have been organized for 2500 teachers from 250 schools.

Other JAS Activities

Since its inception, JAS has paid special attention to the visual observation of meteors, comets, planets and the crescent for fixing the beginning Hijra months. More than 50 astronomical camps were organized by the society so far. More than 20 international camps have been dedicated to observe meteor showers.

In May 1997, the President of the International Meteor organization (IMO), Jurgen Rendtel, and two other members visited the JAS during the Eta Aquarids shower. The IMO delegation gave valuable lectures about observing techniques, including how to monitor meteors with an FM radio receiver and Yagi antenna, which can detect smaller meteors than visual observation could, (http://www.jas.org.jo/radio.html), in 2002 Jas started to observe meteors by Radio using VHF. The most important activities for meteor shower was the one of November 1999, JAS organized the 1999 Leonied conference in cooperation with (IAASS) during the period of (12th –21st) November 1999. More the 40 distinguish astronomers attend the conference, among them, Professor Jack Baggaley, Robert McNaught, David Asher and Daniel Fisher. The first 2-day was given more than 15 papers at Al al-Bayt University on the Meteor Showers. Then the participants moved to Al-Azraq camp (belongs to Jas), which locate at a desert 170km east of Amman, they observed more than 4000 Leonid Meteors.

2. Institute of Astronomy and Space Sciences (IAASS) at Al al-Bayt University (AABU)

The establishment of IAASS http://www.aabu.edu.jo/space/mainl.htm) at AABU (http://www.aabu.edu.jo) in 1994, was a concrete step towards linking the University with developed countries worldwide and to keep in touch with scientific innovations and technological know-how in these countries. IAASS represents the nucleus for basic space science in the Arab world. It is hoped that it will continue to develop with a view to fulfilling the needs and aspirations of the University community, the local community, and the region as a whole. The campus sky is very good for astronomical observations with more than 200 photometric nights per year, international astronomers who visited the site, assured that the Hamza Camp at Al-Azraq area is an astronomical paradise.

Objectives

- Preparing and qualifying national human resources to work in astronomy and space sciences and progress in technology, to render services to the Arab and Islamic societies, in general, and to the Jordanian society, in particular.
- Research and observations, in various wavelengths, on celestial objects.
- Research and studies on the atmosphere and its layers to improve the efficiency of wireless networks and natural earth resources using space photographs and images.
- Mathematical models of the ionosphere and magnetosphere to describe the motion of space plasma in those regions.
- Preparing highly qualified graduate studies to supply the local and Arab societies with human resources capable to assume the professional responsibilities required in the future.

Ways and Means of Fulfilling the Institute's Objectives

- Establishing scientific and technological links with foundations and centers in the Arab and Islamic word for exchange of results, studies, and research.
- Organizing international conferences, seminars, scientific meetings, workshops, and local training courses. To form specialized teams to promote understanding of space sciences and increased participation in analogous activities inside and outside Jordan.
- Prepare international and regional scientific agreements for the exchange of experiences and experts, to exchange updated information and to consult in matters of scientific expertise
- Collecting information, data sources, and various documents that are necessary to follow-up relevant technological developments in areas of specialization.
- Secure the availability of specialists in the fields of astronomy and space sciences.
- Acquire and install laboratories and workshops to boost research activities at the Institute.
- Attracting human resources to assure the creation of a selection of scientists, researchers, staff members and technicians capable of working with astronomy and space sciences and their related technologies.
- Deepening the formulae and relations between the setup of theoretical, and applied and basic research projects conducted by Al al-Bayt University and other universities, on the one hand, and the

requirements of applied research conducted by Research and Development (R&D) Departments in the public sector, on the other, for the sake of translating research projects and adapting them for the service of this region.

Department of IAASS

- Department of Astronomy:

This Department (Education, Astronomical Observatory and Astronomical Heritage of Arabs and Muslims Unit) will cooperate with international observatories and organizations in these domains. In addition, the Institute carries out practical studies on designing observational devices and accessories, analytic and computer control and its future developments, and applications used in fixing the beginning of lunar months and prayers time.

- Department of Space Sciences:

The Department (Space Physics, Remote Sensing, and Space Communications) is in charge of teaching at the graduate level and carries out atmospheric research on the ionic layers in relevance to wave propagation inside the atmosphere for commercial links and radio frequencies.

Research fields include space technologies for remote sensing, as well as space physics, like, for instance, cosmic rays, geomagnetism, physics of the upper atmosphere, and geophysics. The Institute will build up the basic infrastructure of image analysis and its applications.

- Strategic Environment & Water Resources Research Unit:

This unit (http://www.aabu.edu.jo/water/home.htm) in charge of graduate and post-graduate teaching and research for the development of resources.

- Information and Computer Unit:

This unit is responsible for the computer networks with other Units and Departments for the analysis of scientific data and space images. The Unit will also store information concerning astronomical and space research projects, and keep open communication lines with international institutions.

Curriculum

The study plan at the Institute is divided as follows:

- Nine credit hours for compulsory courses.
- Nine credit hours for the thesis.
- Six credit hours for elective courses.

In addition, the Institute offers six credit hours as university requirements and three credit hours as remedial requirements. The total number of credit hours needed for fulfilling the requirements of the M.Sc degree is 33.

The contents of compulsory and elective courses in the fields of astronomy and space sciences, are coordinated with those offered by similar departments in international universities. The compulsory courses concentrate on the fundamentals of astronomy and space sciences, mathematical physics and astronomical and space techniques. Elective courses for astronomy students, on the other hand, concentrate on astrophysics, radio astronomy, cosmology, celestial mechanics, stellar structure and galactic and extragalactic structure, while elective courses for the space sciences students include space physics, remote sensing and technologies, plasma physics, electromagnetic theory, computational techniques, and radio astronomy.

To be eligible for admission to the M. Sc. Program offered by the Institute, candidates should be holders of the B.Sc. degree in one of the following fields: astronomy-physics-mathematics-space, engineering-electric, engineering, communication and control system engineering.

Academic Facilities (Equipment and Observatory)

The Institute's laboratories, equipment, and academic facilities include the following:

- The Computational Laboratory: This consists of a network equipped with terminals that provide the academic departments and units at the Institute with data significant to research in space and astronomy and to studies relevant to the concerns of the Institute.
- Remote Sensing Laboratory: This is formed of a number of devices particularly linked with computer scanning and remote sensing relevant to space images and their analysis (it has various related packages, software such as, Remote Sensing ERDAS Software and DRISI Geographic Information System Software).

Astronomical Observatory (Maragha Observatory)

The observatory is located at the campus of AABU.

Longitude: 36° 14' 19"

Latitude: 32° 20' 30"

The site seeing is 2-3 arcsec. The number of clear nights is around 200 days/year.

The observatory is named after an old observatory, which was installed by the Muslim astronomer Nasir Al-Dean Al-Tousi in 627 H/1264 A.H. in the city of the Maragha (The capital of the old Azerbaijan). This old observatory contained most of the books, which has been brought from Iraq, Syria and Al-Jezzera during Baghdad collapse in 656 H/ 1258 A.H. Many Islamic astronomers used the observatory in that time.

- The Meade 16" LX200 Telescope:

The Meade Schmidt-Cassegrain Optical System: In the Schmidt-Cassegrain design of the Meade 16" model, light enters from the right, passes through a thin lens with 2-sided aspheric correction ("correcting plate"), proceeds to a spherical primary mirror, and then to a convex aspheric secondary mirror. The convex secondary mirror multiplies the effective focal length of the primary mirror and results in a focus at the focal plane, with light passing through a central perforation in the primary mirror.

The 16" mode includes an oversize 16.3759" (about 41.59 cm) primary mirror, yielding a fully illuminated field-of-view significantly wider than is possible with standard-size primary mirrors. It is this phenomenon which results in Meade 16" Schmidt-Cassegrain having off-axis field illumination 10% greater, aperture-for-aperture, than other Schmidt-Cassegrain utilizing standard-size primary mirrors. The telescope is equipped with a CCD ((Pictor 1616), controlled by Pentium II 440 BX computer, which can be used for data acquisition too. The astronomy software is: *The sky levels IV* and *Epoch 2000* image processing with interface cable.

***Table** 5.6-1.* **Main Specifications and Features: 16" LX200.**

Optical Design	Schmidt-Cassegrain
Clear Aperture	406.4 mm (16")
Primary Mirror Diameter	415.9 mm (16.375")
Focal Length	4064 mm
Resolving Power (arc sec.)	0.28
Limited Photographic Magnitude	18.0
Image Scale (arc sec/mm)	50.75
Maximum Practical Visual Power	800
Telescope Mounting	One-piece fork, double-tine
Materials: Tube body	Aluminum
Mirrors	Pyrex glass, Grade-A
Correcting Plate	BK7 optical glass
Telescope Dimensions, swung down	18"x26"x51"

Computers for image transfer and display, contrast control, smoothing, sharpening filters, image alignment. On-line programs: automatic exposure level, dark current exposure and flat-field sequence, pixel binning and accepts color filter system.

Main purpose of the observatory:

- Training of students from Universities and schools to develop astronomical education

- Research projects (variable stars) for staff and M.Sc. and Ph.D. students, also incorporating global network of small telescopes.
- crescent observations.

The Observatory is open for international for scientific cooperation (Al-Naimiy & Kandalyan 2000a).

Figure 5.6-2 The permanent Open air Planetary system model mounted for visitors at the grounds of of Al al-Bayt University.

Figures 5.6-3 The telescope at the Maragha Observatory during a sky viewing evening for students and interested general public. & The dome and workshop building of the Maragha Observatory close to the Institute of Astronomy and Space Sciences (IAASS) at the Campus of Al al-Bayt University (AABU).

Aspirations

The future plans for the Institute include following goals and objectives:

Establish in the next 5 years a Astronomy Department and Al al-Bayt Observatory, a Space Physics Department, a Remote Sensing and Environmental Department, and a Communication and Meteorological Department. These will include the necessary laboratories, workshops and telescopes (a 1.5m optical and 20m Radio Telescope, which gives Jordan a good opportunity for cooperation with astronomical institutions from other countries).

1. International Cooperation.

Generate more efficient frameworks of International collaboration, through scientific agreements, personal contacts and participation in conferences and related scientific activities. To establish links with scientific international organizations such as the International Astronomical Union (IAU), the International Geophysical Union (IGU), and others. The hosting of the 8th UN/ESA workshops on Basic Space Sciences is a first step on this road.

2. Teaching, Scientific Research, and Technical Staff.

Given the fact that research techniques and technologies in space sciences are developing at high pace, the Institute has plans to achieve the following:

- Employing scientists of a high standard to remain up-to-date with the developments in AASS at the international level.
- Use available training avenues such as scholarships, sabbatical leaves, and training courses to secure a sufficient number of staff in all specializations necessary.
- Offering intensive training programs for technicians who work on ready-made projects, on contract basis. The programs should include design and scientific activities for different technological aspects included in these projects. This is needed in order to increase the abilities of these technicians and to help them keep up with technological developments, and to satisfy research and teaching requirements.

3. Plans for the coming five years.

The Institute foresees a staffing plan with some 20 Ph.D. scientists in space physics, astrophysics, cosmic ray physics, atmospheric physics, electronic engineering, communication engineering, remote sensing and computer science. This will be supported by 30 scientific and technical assistants as well as the necessary administrative staff.

4. Selection of projects.

In order to maintain the quality of teaching and research, the Institute plans to move in two directions:

1. Ready – made Projects

This includes signing contracts with international agencies for designing, manufacturing and establishing large and developed scientific research facilities that can serve two goals:

- Providing staff members with recent technologies needed for conducting excellent research programs in accordance with up-to-the-minute advancement worldwide.
- Proving to the international community the academic seriousness of AABU in pursuing scientific research and seeking knowledge, besides encouraging international organizations, committees and pronounced scientists cooperate with the Institute.

2. Local Projects

Initiating the design of certain projects and executing them completely or partially by staff members of the Institute. Given the limited experience and facilities available at the Institute, projects at the very beginning will be modest. However, they will, hopefully, develop and expand in the future, as the Institute will, gradually, is able to depend on its own.

5. Future Scientific Projects
   1. Mathematical Models for ionosphere physics, magnetosphere and upper atmospheric layers.
   2. Establishing a database for x-ray stars, incorporating all currently available information of these objects. .
   3. Establishing an astronomical map for the Hashemite Kingdome of Jordan including correction of al-Qibla direction for the Kingdom mosques, tables of prayer times and calculations of the beginning of the lunar months for the next hundred years, especially the religious months (i.e. Ramadan, Shawwal and Thu al-Hija).
   4. Natural resources through remote sensing methods for the Arab and Islamic world, and share the results with neighboring countries.
   5. Building a large Astronomical Observatory (AO), (Optical and Radio) with a diameter for the optical mirror not less than 1.5 meters, and a diameter for the radio dish of (20-30) meter.
   6. To remain involved in international scientific projects and taking an active part in some of them. Especially relating to multi-wavelength studies of stars. For this purpose the Institute has obtained a small optical telescope (40cm). In addition, the Institute is now looking forward to bolstering its cooperation in this field with Arab countries.

7. Initiate a Ph.D. degree program in AASS.
8. Expansion of its activities at the local and regional levels.
   IAASS Activities
9. Teaching activities of IAASS during the period, September 1994 up to September 2001.

***Table** 5.6-2.* **Student population and results 1994 through 2001**

M.Sc Students	No. of Students
Attended M.Sc Program	60
Graduated	20
Continuing there program in different levels	25
Continuing there Ph.D. Program	4

The students graduated in Astronomy & Astrophysics; Remote Sensing and GIS; Environment; Space Communications; and Space physics.

10. Conferences and Meetings

Many meetings have been organized over the last three years in Space communications, Remote Sensing, Astrophysics and Space Physics. Many astronomical observations were made at the Maragha Observatory, e.g. to observing the comet Hale-Bopp as well as many planetary and variable star observations. Some of these events were attended by a large number of professors and engineers and other members from Jordanian Universities and Scientific centers. The regular seminars are attended by interested staff members and graduate students.

The most important activities are:

- The First International Conference was held at the IAASS of the AABU in Mafraq, in May 1998. The proceedings for this conference, which was attended by more than 50 space scientists from 18 countries have been published by AL al-Bayt University press (Al-Naimiy & Kandalyan 2000a).
- IAASS Hosted the 8th UN/ESA workshop: Scientific Exploration from Space, in March 1999. attend the workshop. (http://www.seas.columbia.edu/~ah297/un-esa/activities-html#1999)The proceedings for the workshop (attended by more than 120 scientists and students of basic space sciences from 35 countries have also been published (Haubold and Al-Naimiy, 2000).

3. Arab Union for Astronomy and Space Sciences (AUASS) (http://www.jas.org.jo/union.html) or (http://www.auass.org/srd)

This union was established in August 1998 with its headquarters in Amman as an outcome of the 3rd Arab Conference on (AASS), held in Amman Jordan in September 1998. More than 100 participants from 14 Arab countries and observers from France, Italy and USA. The aim of the Union is to develop (AASS) in the Arab Countries through conferences, meetings, publications and, joint research projects through the cooperation of the International (AASS) institutions.

Objectives

- Through the introduction of astronomy and basic space science in the Arab countries as scientific disciplines on their own support their role in the scientific and technical progress of Arab societies.
- To stimulate the exchange of scientific data and information, as well as expertise, in the fields of astronomy and basic space science. Through this improve the scientific and technical competence of the scientists working in Arab countries.
- Unifying the scientific and technical terms of astronomy and basic space science that are in use in the Arabic language.
- Historical review of the contribution of the Arab Islamic civilization and other past civilizations in the region in the realm of astronomy and related sciences.

Tools to fulfill the objectives

- Promoting astronomy education; organizing conferences, seminars, and workshops, in basic space science; stimulate the popularization of astronomy and basic space sciences, through an active Outreach program; collection of information about specialists, professional institutions, and societies dealing with astronomy and basic space science in Arab countries; device some long-term (large-scale) astronomical projects with the participation of s many Arab specialists and their institutions; establish the collaborations at international level, necessary to realize AUASS's objectives.

Some of the AUASS activities

AUASS established many scientific committees for to achieve of its goals.

- The Crescents observation and Mawaqeet Committee of the AUASS, to follow up the problems of the young crescent moon visibility prediction (for the use of fixing the beginning of the Islamic Holly Lunar months). (http://www.jas.org.jo/icop.html). Initiate an Arabic refereed Journal for Astronomy and Space Sciences (AJAASS) starting from September 2003.

- The scientific research division (SRD). The SDR/AUASS aims at supporting scientific research in the various fields of AASS through interdisciplinary working groups of common interest. Also the emergence of working groups in social fields, such as the history of AASS, or even AASS in culture and education, seem inevitable.

4. Islamic Crescents Observation Project (ICOP)

This is a global project, organized by the AUASS and JAS. It aims to gather a largest possible number of lunar observers worldwide for the prediction of the visibility of the young crescent (http://www.jas.org.jo). ICOP is supervised by committee from AUASS, IAASS & JAS. Their members are from different part of the Islamic countries. These committee's organizes the Islamic conference regularly and twice a year.

ICOP's objectives are to develop the crescent's observation criteria in collaboration with the relevant authorities and to supply the international community of the important Islamic calendar months.

5. Teaching of (AASS) in Jordanian Schools and Universities

This consist mainly of general astronomy courses in secondary schools and astronomy and astrophysical courses in physics and mathematics departments of most of Jordanian Universities at the undergraduate level.

6. Future Expectations

The Hashemite Kingdom of Jordan was chosen to host the Regional Center for Basic Space Sciences and Technology Education in Western Asia (affiliated to the United Nations. The Jordanian astronomical organizations are planning to build a large astronomical observatory (Optical & Radio) with a 1.5m optical and a 20 m Radio telescope for use by local and international scientific community.

Following the European example of ESA the AUASS proposed and plans to create an Arab Research Space Agency (ARSA). This proposal has been discussed in the 5th Arab Astronomy and Space Science Conference in August 2002. And a committee from most of the Arab Countries has been created to follow up the creation of the Agency.

References

Al-Naimiy, H. M. K., 2000, Bull. of Teaching of Astronomy in Asian-Pacific Region, **17**,13.
Al-Naimiy H. M. K. and Kandalyan, R., 2000a, Maragha Astronomical Observatory, Al-Al-BAYT University Publications 1999.
Al-Naimiy, H. M. K. and Kandalyan, R., 2000b, Proceeding of the 1st International conference in AASS, 4-6/May/1999.
Huabold,H.J., Al-Naimiy, H. M. K., 2000 , AP&SS, **273**, part 1-2.
Odeh, M. S., 2002, Sky & Telescope, **99**, 85.
Jas Publications (1987-2002); AUASS Publications (1998-).

SECTION:

6. SMALL ASTRONOMICAL TELESCOPE FACILITIES

Chapter 6.1

THE ROLE OF PUBLIC OBSERVATORIES IN ASTRONOMICAL OBSERVATIONS

T. Kogure
1-10 Togano-o Hashimoto,
Yawata, Kyoto 614-8322
Japan
tkogure@pa2.so-net.ne.jp

Abstract Recently in Japan, the number of public observatories equipped with moderate to large telescopes (60 - 100 cm), having high observational capabilities, is rapidly increasing. These observatories, which were mostly established in the 1990s, have started their astronomical observations with the collaboration of amateurs and research institutes. Whereas most of them are working in CCD imaging and/or CCD photometry, some observatories are carrying out spectroscopic observations of stars and galaxies. This is a new trend in public observatories in Japan, and they are working both for astronomy popularization and astronomical observations as local centers of their own districts. Although they are still faced with many problems that also affect public observatories such as a limited number of staff and insufficient budgetary background, their efforts for observations will provide a new epoch of domestic as well as international network observations in the near future. Some examples of public observatories along with their expected roles will be discussed briefly.

Introduction

Astronomy in the 18th and 19th centuries in Europe was often promoted by advanced amateur astronomers, in the construction of large telescopes, in observations of non-stellar objects, and others. In the 20th century, particularly in its latter half, with the beginning of the era of big science and of space science, to which astronomy was assigned, activities of amateurs and professionals are separated greatly from each other. In Japan, activities of advanced amateurs have long been concentrated on some specific fields such as comet hunting, sky patrols for finding

W. Wamsteker et al. (eds.), Developing Basic Space Science World-Wide, 293-298.

variable objects, like novae, variable stars, and so on. They have had no close connection with research institutes in general. Recently, however, the situation has changed towards the stimulation of amateur activities and the promotion of amateur-professional collaboration in Japan. This new trend, which become clear in the late 1980s and accelerated in 1990's, can be summarized as follows:

1. Public observatories equipped with large telescopes have been constructed by local governments (prefecture, city, town, village) and private persons/groups. Not only has the number of telescopes been increased, but also the size of the telescope has increased remarkably. Between 1970 and 1990, the size of public telescopes has nearly doubled in size, from 60-cm to 1.1-m (Kuroda, 1993). It should also be noted that professional astronomers including graduate students are finding jobs in public observatories. This is a stimulating factor for carrying out astronomical observations in such observatories.

***Table** 6.1-1.* **Main public observatories**

Name of PAO	Year	Telescope	Instrument
Nishi-Harima A.O.	1990	60cm refl.	CCD
		50cm Solar tel.	Lyot Filter
Bisei A.O.	1993	101cm refl.	CCD, Pe, Sp
Saji A.O.	1994	103cm refl.	CCD, Planetarium
		15cm Solar refr.	
		4 tel.mounted.	
Ayabe A.O.	1995	95cm refl.	CCD, Sp.
Misato A.O.	1995	105cm refl.	CCD, Planetarium
Toyama A.O.	1997	100cm refl.	CCD, Sp, Video.
Gunma A.O.	1998	150cm refl.	CCD, Sp.IR

Abbrev.: CCD = cooled CCD camera, Pe = photoelectric photometer, Sp = spectrograph, IR = infrared detector.

2. CCD cameras as the new high-sensitivity detectors in imaging and spectroscopy are getting more and more familiar in astronomical observations, even for amateurs in recent years. Thus the observing power of telescopes has increased remarkably. Now the telescopes of 1-m aperture have the capabilities of astronomical observations comparable with those of 2-m aperture and even more than those of decades ago. Reflecting this new trend, many public observatories are providing observational instruments such as CCD cameras, photoelectric photometers, and spectrographs.
3. The observation of variable objects suitable is for amateur-professional collaboration. Variable objects, such as novae,

supernovas, variable stars, comets, etc., mostly attract amateur astronomers, and these are good subjects for promoting amateur-professional collaboration with large telescopes in public observatories.

4. New waves of astronomy education and popularization are emerging using large telescopes. Public observatories with large telescopes offer the opportunity of close star gazing for children, families, and citizens. These experiences give valuable effects particularly for young people. Large telescopes are also effective in the training of young astronomers and even for graduate and under-graduate students. This situation enables the participation of amateur astronomers in astronomical observations of stars, nebulae, and galaxies, with their own interest. They are also able to participate in some domestic as well as international, network observations and/or observational campaigns of the future.

1. Activities in Public Observatories

Table 1 gives a list of the main public observatories which were established in recent years and are active in popularization, education, and astronomical observations. One should mention that there are many other public or private observatories active in these areas. The activities in these public observatories include the followings:

Popularization

Needless to say, the proper task of any public observatory is the popularization and education of astronomy, particularly for local peoples. There are many forms of different activities undertaken by different observatories such as:

a) Nightly sky-watching with large telescopes Close-ups of planets, nebulae, star clusters, galaxies;
b) Astronomical events; star-watching, festivals, photographic contest, among others;
c) Public lectures, seminars, for the public or for school pupils;
d) Support for star-watching clubs/groups; and
e) Facilities associated with an observatory (See Table 3) such as a planetarium, a science center (exhibition), a lodge or dormitory for stay (school pupils, families, other individuals) parks (field athletics, camping sites, cultural halls, and others).

The public observatories are undertaking these popularization activities under the auspices their respective programs.

Figure 6.1-1. Bisei Astronomical Observatory which is located in one of the best sites for astronomical observations in Japan and equipped with a telescope of 101-cm aperture, constructed with many new technical ideas for observing stars. The dome is separated from the main building to avoid every possible heat source which may distort airflows in and around the dome. The open space under the dome was left for the same reason to keep the steady flow of winds.

Astronomical Observations

New-type public observatories with large telescopes are ordinarily equipped with some observational instruments, such as cooled CCD cameras, photoelectric photometers, video cameras, and spectrographs, according to their observational programs.

These public observatories were established in the 1990s, so the observational instruments are still mostly in the stage of test observations. Among these, Nishi-Harima A.O. which is the first new-type public observatory, are active in CCD imaging of nebulae and galaxies and CCD photometry of variable stars. They have started regular publications on their research activities. Bisei A.O. has a 101 cm telescope equipped with cooled CCD camera, photoelectric photometer, and a spectrograph. The spectrograph can be used in two modes of high dispersion (13 A/mm) and low-dispersion (100 - 160 A/mm) by changing the grating and the CCD camera. Recently this observatory has carried out some astronomical observations with the collaboration of amateurs and/or University staff, among which remarkable results were obtained in: (a) Multi-color imaging observations of Jupiter-SL9 comet impact sites; and (b) Collaborative observations of Nova Cas 1995, which was discovered by an amateur astronomer and follow-up photometric observations also made by a group of amateurs astronomers. Spectroscopic observations were carried out at Bisei A.O. Saji A.O. is now preparing CCD observations with its 103cm re ector and solar observations with its 15 cm Solar refractor which has 4 telescopes mounted simultaneously. At Ayabe A.O. the 95 cm telescope is equipped with a CCD camera and a low-dispersion spectrograph. Occultation of stars by asteroids, search for novae, supernovas, etc are now under consideration for future activities. Misato A.O. has a 105 cm telescope equipped with a cooled CCD camera, and CCD imaging and photometry are expected. This observatory is active in the promotion of an imaging network system through the Internet.

2. International Cooperation

Among public observatories, Bisei A.O. is active in international cooperation, mostly with Asian countries. Their activities include :

a) Organization of a "Mini-workshop on astronomy popularization in Asian countries", which was held on 23 July 1995, and its Proceedings were published by the Observatory.
b) Invitation of a foreign astronomer. Recently, a distinguished astronomer was invited from Yunnam Observatory, China, for a short stay. A public lecture and a scientific colloquium were also successfully organized at the Observatory.
c) Proposal for the training program for astronomers from Sri Lanka. It is expected that international cooperation activities will be promoted in other public observatories in the near future.

3. Conclusions

Activities of the new-type of public observatories in Japan can be summarized as follows:

a) Local center of astronomy popularization: Most public observatories are associated with some planetariums or science centers or parks, all of which are attracting families, and individuals, and assist the activities of public observatories in their respective ways. They are all serving as the local centers of astronomy popularization in prefectures, cities, and towns.
b) Promotion of astronomy education and popularization: The basic purpose of a public observatory lies in the popularization of astronomy. Some Observatories (e.g. Nishi-Harima A.O.) have regular contact with surrounding primary schools, accepting pupils for

several days for the purpose of science education. Public lectures are regularly organized in most of the observatories.

***Table** 6.1-2.* **Situation of the public observatories**

Name of Obs.	Year Establ.	Popul.	Staff			Assoc.facil.
			Perm.	Part	Admin.	
Nishi-Harima A.O	1990 Pref	millions	5	0	1	Ast.Park
Bisei A.O.	1993 Town	6000	2	2	2	Hist.Park
Saji A.O.	1994 Village	3300	5	6	4	Ast.Park
Ayabe A.O.	1995 City	40000	3	2	2	
Misato A.O.	1995 Town	4700	3	1	2+4	Ast.Park
Toyama A.O.	1997 City	320000	4	2	0	Sci.Center

Note on staff: Staffs is counted by permanent (astronomer), part-time (astronomer) and administrative staff.

c) Promotion of astronomical observations with the collaboration of amateur and professional astronomers in research institutes. Some observatories (e.g. Bisei A.O.) are effectively promoting amateur-professional collaborations.
d) Participation to campaign and/or network observations in the future: The development of domestic as well as international network observations can be expected in the future. At present, a campaign for the search of supernovas in early-type galaxies is proposed connecting some public observatories and universities. However, for effective observations it will take some time, since most public observatories were established very recently and their instruments are still in the test observation stage, and additionally, they are affected with a shortage of manpower in their observatories (Table 2). Strong support from surrounding advanced amateur astronomers and/or from research institutes is expected.

Chapter 6.2

SMALL TELESCOPES IN RESEARCH AND EDUCATION

R.K. Kochhar
National Institute of Science, Technology and Development Studies
New Delhi, India
rkochhar2000@yahoo.com

Historical Perspective

About 200 years ago, a 9.5 cm (3 inch aperture telescope was called a large telescope. London was then the centre of the world optical trade with John Dollond (1706-61), his son-in-law Jesse Ramsden (d.1800) and son Peter Dollond (d.1820) as the leading telescope maker in England. At that time it was not possible to make good quality flint glass blanks of larger size, and the lenses were ground by trial and error. Credit for providing a scientific basis for optical engineering as well as for making large-sized lens telescopes goes to Munich-based Joseph Fraunhofer (1787-1826) who started his career as an 11 year old orphan in virtual bondage at an optics shop and rose to such eminence that on his death he received a state funeral.

Originally, reflecting telescopes were made from the unwieldy speculum metal, an alloy of copper and tin. The invention, in 1856, of the silver-on-glass technique by the chemist Justus von Liebig (1803-73) in a different context made it possible to make large, low-weight astronomical mirrors. The problem of casting strong mechanical mountings suitable for large telescopes was solved by the use of the steam hammer invented by James Nasmyth (1808-90).

With the advent of large telescopes, the centre of astronomical activity shifted from Europe to the United States of America (US) which had the requisite clear skies, funds and a desire for intellectual leadership. In recent times, advances in technology, telecommunications and international cooperation have freed observatories from the limitations of

W. Wamsteker et al. (eds.), Developing Basic Space Science World-Wide, 299-305.

national boundaries and made it possible to collectively install large telescopes at globally-selected sites.

In one important aspect, astronomy is unlike experimental sciences. Setting up of a modern sophisticated laboratory often means the eclipse of earlier ones. The advent of large telescopes however does not mean redundancy of small telescopes (aperture less than 1m). This is so because from any place on the terrestrial globe only half the celestial sphere is visible. Therefore, no observatory, no matter how well equipped, can cover the whole sky. Secondly, since astronomical events cannot be manipulated or repeated, each independent observation is a valuable addition to the databank. Also, for continuous monitoring of certain class of objects, receipt of round-the-clock data is needed.

These data can be obtained by observers located at different longitudes. Thus while astronomy today depends on large (and space-borne) telescopes for breakthroughs, it solicits contribution from as many telescopes of various sizes as possible.

Astronomy is more than a branch of modern science. It is a symbol of the collectivity and continuity of humankind's cultural heritage. That is why many developing countries, particular those with memories of past contribution to science, are keen to renew their acquaintance with astronomy and contribute to its growth. Paradoxically the factors which favoured astronomy in ancient times now work against it. Astronomy was the preferred activity of ancient cultures which owed their prosperity to agriculture which in turn depended on rain.

Thus there was a direct correlation between astronomy and rain. In contrast, modern astronomy today prefers sites that are dry throughout the year and away from civilization. In other words, inheritors of ancient astronomical tradition are unlikely to find suitable sites in their midst worthy of huge investments that large-scale modern astronomical facilities now require. This handicap notwithstanding, developing countries can still contribute to the enrichment of astronomy through small telescopes, because as we have seen astronomical facilities are not competitive or exclusive but additive.

1. Small Telescopes

There is another reason why developing countries should develop facilities centered around small telescopes: Although astronomy has been and still is the most Brahmanical of all sciences, today it requires artisan support of a high order. Astronomical ambition should therefore be pegged at the level of the available infrastructural support. Many developing countries, in their impatience to make up for lost time, have tended to begin with big science -at the top, hoping that the benefits would slowly trickle down. This has not happened, because the law of gravity does not apply in the intellectual domain.

Very often, the facility is acquired from abroad on a turnkey basis on the initiative of a single individual. Countries with a pre-industrial mindset tend to rank an individual above an institution. The facility is often seen as an extension of the founder's personality, so that its golden age becomes coterminous with the workspan of the founder. Even though the observatory does not actually die, it becomes a victim of AIDS, astronomical instrumentation deficiency syndrome.

If the facility is to emerge as an institution, the acquisition of equipment should be accompanied by the provision of (i) training of manpower, (ii) maintenance of equipment and possible upgrading, and (iii) sustenance of interest.

To be productive, a telescope should be supplied with users. No doubt, in the beginning, trained manpower will have to be acquired from outside. But it is essential to have an in-house training programme so that generation after generation of users are available. The most effective way of solving the manpower problem and ensuring the success of long-range research programmes would be to involve the universities. The universities have a regular intake of students, a well defined curriculum, and a rigorous examination system. If a telescope facility is made part of a teaching programme, it will be assured of manpower at a reasonably competent level. Of the large number of students that enter a university, even if a small fraction can be motivated beyond the compulsions of a degree, long-range needs will be met. Furthermore, to meet the university requirements, the observatory will have to be maintained at a certain level, independent of the idiosyncrasies of individuals.

There is need to establish a symbiotic relationship between the telescope and the workshop. An instrument should be a tool in the hands of a user. It should not overwhelm him or her. If a snag develops, the user should feel confident of setting it right and affecting improvements.

2. Research and Education

Small telescopes illustrate the veracity of Ernst Rutherford's aphorism, "We have not got the money, so we have got to think." Given a well-maintained, well-equipped small telescope a persevering astronomer with a well thought-out programme can obtain data of genuine research value.

Two fields are particularly suited for small telescopes: variable stars, and the sun and the solar system objects.

2.1 Variable Stars

Variable stars are stars whose light output varies. The variation can be due to a number of reasons and may involve timescales ranging from a few minutes to many years. More than 50,000 variables are known, divided into a number of groups. Their study is important because it tells us about the properties and evolution of the stars as well as about the physical processes occurring in them.

Variable stars can be observed with small telescopes using the technique of photoelectric photometry. Photometry of bright variables is not very sensitive to background light; observations can be made even when the Moon is full.

It however does require observing time, so that the location of a (small) telescope near the observers is of great help. Actual observations are not difficult to make. Objects should however be chosen with care keeping in mind the characteristics of the site and the instruments and as part of an international programme. Instrumentation should be standardized and observations carried out on a long-term basis.

In addition to the variable stars, solar system objects can be studied using small telescopes. One can begin with the Sun itself. To image the Sun, the light-gathering power of the telescope may be reduced by using neutral density filters. Solar patrols can be maintained at various longitudes, and while-light flares can be searched for.

2.2 Comets

There are a number of reasons why comets should be studied. First, they are a spectacular sight. More fundamentally, their chemical composition tells us about the conditions at the time of formation of the solar system. They contain condensates from the solar nebula and possibly even interstellar grains. Finally, they are the natural tools for probing the solar wind, especially away from the ecliptic.

Comets are not easy objects of study. They are often large in angular size, low in surface brightness, close to the sun when bright, and fast-moving. Their observation requires international cooperation of a high order; often their highly inclined orbit making it necessary to have data from the northern as well as southern hemisphere.

In particular, there is need for wide-field photographs of the comet's tail. Within a matter of hours, the ionic tail "breaks off" and is carried away by the solar wind while a new tail emerges from the coma. The process is too slow for a single observatory to photograph in a single night, but too fast for the observatory to observe the next night. To understand the theoretical processes involved, it is essential to have, courtesy a number of observatories, a good sequence of photographic images. In addition to coordinated photography, coordinated photometry will also be useful.

2.3 Asteroids

Asteroids are sufficiently numerous so that new ones can be discovered photographically and old ones studied. Unlike the comets they retain their original orbits and masses. Unlike the larger planets, they have not chemically evolved.

Consequently, they provide extremely valuable information on the origin and evolution of the solar system. Towards this end, it is important to learn about the asteroids' masses and volumes as well as geometry, rotation and surface properties.

An important project can be the observation of the occultation of background stars by minor planets. Given the large number of asteroids, such events are quite common. These events can tell us about asteroids what we cannot learn from the earth in any other way. A typical occultation lasts 10 to 20 seconds.

If immersion and emersion are timed at a number of selected sites to within 0.1 second, a level of accuracy easily achieved with simple photoelectric equipment, then the data can be combined to obtain a direct estimate of the dimensions of the face of the asteroid.

In a similar manner, photoelectric observations of occultations of stars by the Moon can be made. It is instructive to recall that rings around Uranus were discovered during a stellar occultation by the planet.

3. Teaching

The research programmes carried out with small telescopes can easily be integrated into formal teaching programmes. The observation of variable stars as well as analysis and interpretation of data can form a project at BSc and MSc levels. Similarly, solar spectroscopic studies can easily be integrated with laboratory spectroscopy.

3.1 Initial training

Training of manpower to initiate astronomical studies requires close attention. In this important area, it would be advisable to enlist support of these countries which have had a head start and which are culturally and economically close to countries with manpower requirements.

It has been seen that projects involving an offer of training by industrialized countries do not produce the desired results. Because of the glamour and the pecuniary benefits associated with a "foreign" visit, the opportunities are "grabbed" by persons high up in the hierarchy who are inclined to treat the visit as a personal honour rather than as an opportunity for obtaining hands-on experience. Also, in the case of younger elements, cultural discontinuity can have a demoralizing effect. Additionally, in the case of capable candidates, an opportunity not to return to their own countries can arise.

It would be far more profitable to make arrangements by which experts from neighboring countries can spend a few months helping the hosts in setting up facilities and training young astronomers. It will also be a help if regional refresher courses can be arranged at regular intervals, and handbooks prepared.

4. The Indian Experience

India was the first country outside Europe to acquire, at Madras in 1786, a modern astronomical observatory, thanks to the colonial use of astronomy as a navigational and geographical aid. During its existence in the 19th century, the Madras Observatory remained rather poorly equipped, its chief instruments being two refractors: a 15 cm aperture English-mount telescope by Lerebours & Secretan, acquired in 1850, and a 20 cm equatorial by Troughton & Simms, with optics by Fraunhofer's successor George Merz, acquired in 1861. The Observatory was kept in working order with the help of the mechanics at the workshop maintained by the government's public works department for its own use.

In contrast to the Madras Observatory stood the short-lived Lucknow Observatory established in 1831 by the king of Oudh (correctly Avadh) who had a European wife and a magpie-like fascination for novelties. The Observatory was placed in charge of a British astronomer and equipped, as befit a king's observatory, with the best instruments money could buy. The Observatory was commissioned in 1841 but closed eight years later by the new king. In the early years of the present century Kodaikanal Observatory emerged as a state-of-art solar research facility under the scientific leadership of George Evershed, who built a number of spectrographs for his own use. In addition, in 1911 he made an auxiliary spectroheliograph and bolted it to the existing instrument to make it more versatile. Not surprisingly, Evershed's stay at Kodaikanal coincided with the golden age of the Observatory.

In recent years, a number of small telescopes have been purchased by various agencies in India, some of which are being profitably used. Significantly, research-wise the most successful has been the one that is the most primitive. This telescope is a 34 cm Cassegrain reflector made in the Indian Institute of Astrophysics workshop and attached to a preexisting mounting vacated by an old 15 cm Cooke refractor. Thanks to the pooled efforts of the astronomers and the workshop staff, this rugged telescope has been successfully used for creating a database on the photometry of rotating variable stars, especially of the RS CVn and T Tauri type. The success of this telescope proves that for an astronomical facility to be successful, engineers and astronomers should work together.

5. Conclusions

1. When a new astronomical facility is set up, it should be at a level consistent with the workshop facilities and infrastructure support available. The equipment should not overwhelm the user.
2. For the initial training of manpower, cooperation should preferably be sought from countries which are culturally akin to the host country.
3. Attempts should be made to integrate astronomical facilities with the teaching programme
4. For best results, observational programmes should be chosen so as to form a part of international campaigns.

References

Genet, D.R., Genet, R.M., and Genet, K.A., 1987, The Photoelectric Photometry Handbook (Mesa, AZ: Fairborn Press).

Hearnshaw, J.B. and Cottrell, P.L., 1986, in Instrumentation and Research Programmes for Small Telescopes (Dordrecht: D.Reidel Publishing Co.) , IAU Symposium No.**118**.

Kochhar, R.K. 1991,Vistas in Astronomy, **34**, 69-105.

Kochhar, R.K. and Narlikar, J., 1995, Astronomy in India : A Perspective (New Delhi: Indian National Science Academy).

Percy, J.R. 1980,, Journal of the R.A.S.C., **74**, 334.

Chapter 6.3

NETWORKING OF SMALL ASTRONOMICAL TELESCOPE FACILITIES IN EDUCATION AND RESEARCH PROGRAMMES ON SUBJECTS SUCH AS VARIABLE STARS AND NEAR-EARTH OBJECTS

S. Frandsen
Institute for Physics and Astronomy
Århus University
Bygning 520
DK 8000 Århus C, Denmark
srf@phys.au.dk

Abstract The trend towards larger and larger telescopes has led to discussion about the future of the many smaller telescopes. Not all science can be carried out in a few nights on a giant telescope. At the same time, advances in automation and communication open up far more efficient and time-saving methods of observation, enabling very large datasets to be obtained and handled. Small telescopes allow a much larger community to participate in and carry out real science. Amateurs and students can become part of the scientific community and can contribute in a significant way in some areas of astrophysics. In this paper, I will discuss modern studies of variable stars and briefly mention monitoring of the sky for Near Earth Objects and other special events.

Introduction - Projects involving Networks of Small Telescopes

Time is often an important factor when studying astronomical targets. Observations must be planned and a strategy implemented in order to obtain data at the right time and during the right periods. When observing oscillating starts with a rich spectrum of modes, daily side-lobes in the power spectrum of modes are very disturbing. Continuous

W. Wamsteker et al. (eds.), Developing Basic Space Science World-Wide, 307-318.

viewing, either by networks or from space can remove or at least reduce these parasitic peaks. In other situations, one is looking for events and many hours are spent waiting for the right event to show up. It is a waste of a large telescope to have it run idle for long periods. This paper presents a number of examples of observing programmes involving networks of small telescopes.

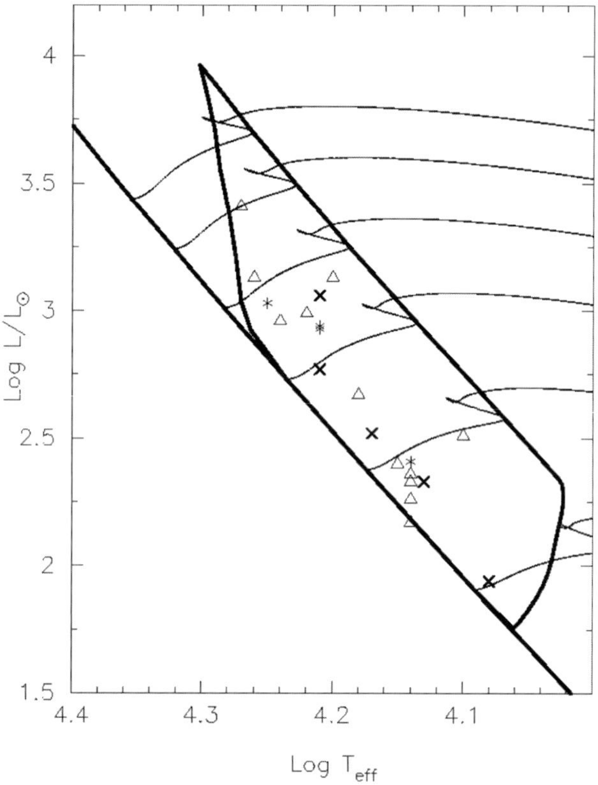

Figure 6.3-1. The location of the Slowly Pulsating B stars (SPB) instability region, with some real SPB's plotted in (from Aerts et al. 2000).

1. Oscillating Stars

Variable stars present a rich source of information on stellar properties. Stars especially, with a rich set of non-radial oscillations excited, offer possibilities for using seismic techniques. Several classes of stars have been studied extensively. Large campaigns have been organized to disentangle the often complicated pulsation spectra. Among targets for such campaigns are white dwarfs, rapidly oscillating Ap stars (roAp's), δ Scuti stars and pulsating B subdwarfs. The stars have relatively short periods, up to a few hours, and mainly present pressure driven modes, with gravity modes appearing in the case of white dwarfs. Campaigns run for periods from one week to several months. Very long campaigns are needed even for the short period variables mentioned, because oscillations show up with very small frequency separation.

1.1 Slowly Pulsating B Stars (SPB's)

An interesting class of stars, SPB's, is located slightly above the main sequence on the hot side of the instability strip (figure 6.3-1). The stars are slowly pulsating with periods in the range of 1 to 3 days, indicating that we are seeing gravity modes, non-radial pulsations driven by gravity. The instability is driven by the same opacity mechanism responsible for driving the β Cephei stars' oscillations with shorter periods (hours). About 30 SPB's are brighter than the 6th magnitude (Aerts et al., 1999a, 1999b). The amplitude variations are quite small, only a few percent. For professional astronomers, these stars present a problem as very long observing runs are needed. It is hard to find so much telescope time at normal observatories.

1.2 Subgiants

The Sun is known to oscillate in millions of different modes. The excitation comes from the random 'kicks' by convective elements, a process that also operates in other stars with convective envelopes. In solar-like stars, the oscillations are unobservable from the ground. The amplitudes are a few parts per million (ppm), which is totally out-of-range for small telescopes, and possibly also for large telescopes. The amplitude increases considerably with luminosity and in the subgiant range the amplitudes are large enough and the periods short enough to change the picture. Oscillations have been seen by Buzasi et al. (1999) in α Ursae Majoris and by Edmonds and Gilliland (1996) in stars in the cluster 47 Tuc. For stars in the range G7 III to K0 III, amplitudes are expected in the range 100-200 ppm, and observing sessions should last 20 to 30 days. With 10 small telescopes and a duty cycle of 30%, the noise level obtained could be in the range 10-20 ppm. The signal-to-noise obtained would be S/N~10, good enough to permit detection of a small set of modes. A simulation of network observations for a δ Scuti star is presented by Kjeldsen (2000).

2. Near Earth Objects

A different strategy is used for the search and study of bodies very close to us. It is obvious that asteroids and comets, which pass close to the Earth, are of great interest. This is true, first, because of the chance, however small, that they would collide with the Earth, but also because the small distances involved permit detailed studies to be carried out. With small telescopes and charge coupled device (CCD) cameras, large fractions of the sky can be monitored and fast-moving objects located. The largest effort at the moment in terms of surveying the sky is the Digital Sloan Sky Survey, which generates 200 Gbyte of data each night, although the Survey is directed towards extragalactic research. The Amateur Sky Survey (TASS) collaboration (Gombert and Droge, 1998, (http://www.tass-survey.org) is aimed at chasing possible Earth-crossing asteroids, but is also useful for many other purposes.

Light curves for variable stars with large amplitudes can be monitored (e.g. semiregular variables). Gamma-ray burst afterglows can be monitored as well as other transient sources such as cataclysmic variables. 'Hot Jupiters' cause eclipses of a few percent, which would not be too difficult to detect were it not for the very small fraction of stars where this can be observed. The major problem in this type of work is to organize the extremely large amount of data generated by CCD cameras. More ideas for targets for networks of small telescopes can be found by looking at the observing programmes of Automatic Photoelectric Telescopes (APT's) described later in the section on *Robotic Telescopes.*

3. Time Versus Size

In science experiments, one constantly attempts to carry out more accurate measurements. A simple but costly way is to increase telescope size. Less costly, but difficult, is to improve instrumentation to increase efficiency and stability. A third option is to use more telescopes, which is straightforward in principle but has its own set of advantages and disadvantages.

Let us begin with one of the positive aspects. The scintillation decreases only slowly with telescope size.

The noise level is given by

$$\sigma_{scint} = 0.09D^{-2/3}\eta^{3/2}\exp(-h/8000m)\Delta t^{-1/2} \qquad (1)$$

where D is the telescope diameter in cm, η the air mass, h the telescope elevation in m and Δt the integration time in sec. For a small telescope (D = 50cm) at sea level (airmass 1.5) and integration of t = 60s, the noise is σ_{scint} = 1.56 mmag. For bright stars using broad band filters, this is the dominating source of noise. Now take 10 widely separated 50cm-telescopes. In this case, the combined noise from a one-minute integration is $\sigma_{10} = \sigma_{scint}/\sqrt{(10)} = 0.49$ mmag. Compare this with the result obtained with a 2.5m-telescope, which is σ_{large} = 0.53 mmag. The 10 small telescopes do slightly better even though they have a collecting area that is 2.5 times smaller. For bright stars and broad filters, the scintillation noise dominates down to $m_v \sim 11$, depending on the exact parameters of the instrument and target star.

The second advantage one can obtain is linked to the possibility of placing the 10 telescopes at different longitudes and obtaining measurements with less interruption. This is illustrated in the following simulation (figure 6.3-2) of the measurement of the light-curve of a δ Scuti star with 5 modes.

4. Existing Networks

4.1 Photoelectric Networks

4.1.1 Whole Earth Telescope (WET)

The best known network is probably the Whole Earth Telescope (WET), which was created first to monitor light-curves of white dwarfs but later expanded to other targets as well. The web page is found a (http://bullwinkle.as.texas.edu). Around 15 sites participate, but only around ten sites are sufficient to provide good time coverage. Campaigns run for a couple of weeks.

4.1.2 Delta Scuti Network (DSN)

The Delta Scuti Network (DSN) mainly observes Delta Scuti stars, as indicated by the name of the network. The campaigns tend to become longer and longer. The activities are described on the Web page located at (http://www.deltascuti.net). The network has existed, as has WET, for some time, and extensive campaigns are carried out. About 15 sites are involved and a dozen δ Scuti stars have been observed.

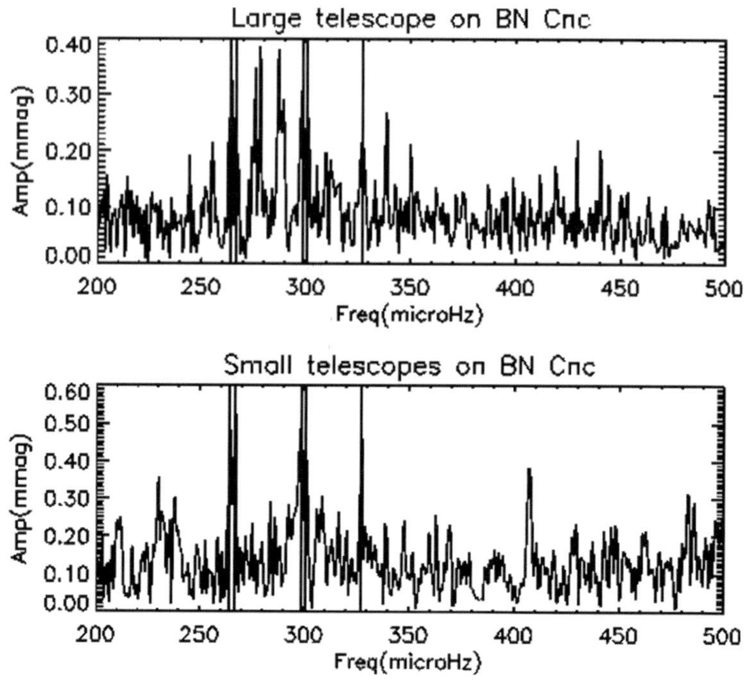

Figure 6.3-2. Amplitude spectrum of two weeks of measurements. A 50% duty cycle and some random weather statistics with a mean 60% of clear nights were assumed. The noise level was also assumed to be twice the scintillation noise. The first panel shows the results using a single large telescope, and the second shows the results using ten smaller telescopes. The oscillations are much easier to identify correctly in the second case.

4.1.3 Step PHI

A much smaller network uses identical photometers. The name is derived from Stellar Photometry International and is controlled from the Paris Observatory with only three observatories involved, in China, Tenerife and Mexico. The network is devoted to δ Scuti stars and has been active for some years. The Web page is located at (http://dasgal.obspm.fr/~stephi).

4.2 Differential CCD Photometry

With the progress in CCD detectors, small telescopes with a large CCD detector are no longer a remote dream. Such systems will probably replace photoelectric photometers at most sites due to their greater flexibility and higher efficiency. Some of the networks mentioned above are using CCD-based instruments already.

4.2.1 Small Telescope Array with CCD Cameras (STACC)

This network is more loosely organized and is directed to observations of multiple targets, such as stars in open clusters. Again, the main target is δ Scuti stars. A Web page describing its activities is located at (http://astro.ifa.au.dk/~srf/STACC).

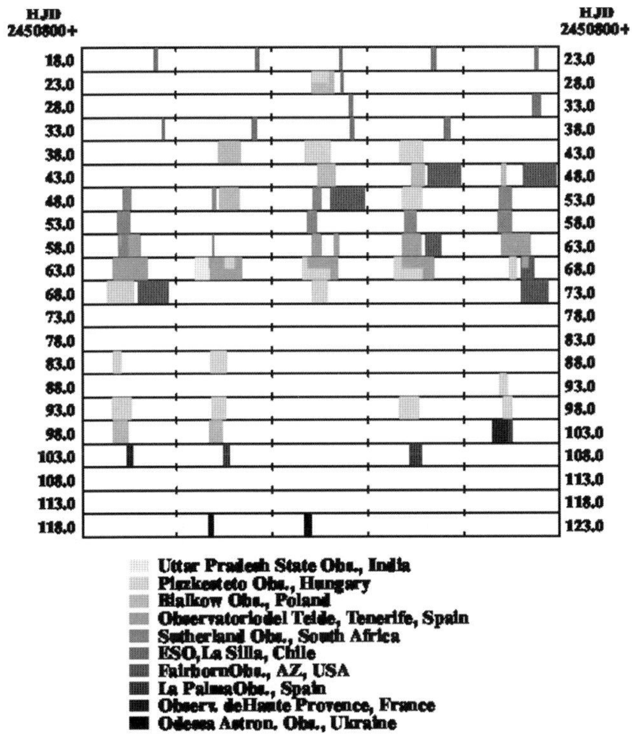

Figure 6.3-3. Distribution of photometric observations during the 1998 STACC campaign. Five consecutive nights are shown in each row (produced by A. Pigulski, Poland).

As this is the area in which my institute has participated mostly, I will discuss this in detail. I will report on our experience with a large campaign on two δ Scuti stars in the Praesepe cluster (BN and BV Cnc), which are close enough together to be observed simultaneously. The campaign involved both a photometric aspect and a spectroscopic aspect. Figure 6.3-3 indicates the nights on which photometry was obtained.

The telescopes had diameters between 40cm and 1m. The results of the campaign were very different from site to site. This is indicated in the table, which gives r.m.s. values for the residuals after the light-curve has been subtracted. The final column gives the 'weight' each site has in the fit of the final light-curve for the star.An illustration of the results can be seen in a light-curve from two of the nights, where data from three sites are present (figure 6.3-4). The point to be illustrated here is that when one relies on different groups at different sites participation on different levels, then a few 'best' sites will show up. Some sites have better instruments than others and some get more nights allocated and benefit from a better climate. The result is a quite inhomogeneous dataset. The results from the STACC 98 campaign are nevertheless among the most precise for a δ Scuti star. And these are observations for two stars in one campaign!.

A second result is the somewhat disappointing residuals that show up, which are more than a factor of two times the theoretical best. We obtained the best residuals around 3 mmag, and would expect the best around 1 mmag. This is partly explained by the nature of the star itself, which certainly is not matched by the light-curve used. It is also due to the difficulty of keeping the instrument response stable for periods of hours. At higher frequencies the noise decreases. A few sites suffered from severe drift problems. Finally, one site was testing a new camera.

***Table** 6.3-1.* **Residuals after fit of light-curve for data obtained at different sites**

Observatory	Residuals (mmag)	%
Tenerife	3.34	40.1
Piszkestetö	4.07	24.1
Fairborn		19.4
(v)*	3.99	13.7
(y)*	5.90	5.7
Bialków	3.96	6.5
Sutherland		5.1
(v)*	4.20	1.8
(b)*	3.43	2.1
(y)*	4.53	1.2
ESO (Dutch)	5.22	2.7
La Palma	3.40	1.8
UPSO	8.44	0.2
Odessa	11.14	0.1
OHP	-	0.0

Some sites were observing in more than one colour and the contribution is shown for each of the passbands as well as the total after converting to the V system.

4.2.2 ISTeC

Another network is the International Small Telescope Cooperative (ISTeC), for which the Web page is located at (*http://www.astro.fit.edu/istec*).

It mainly provides a forum for organizing campaigns and does not have a scientific programme of its own. The list of members of the cooperative is quite long and gives a good indication of the location and capabilities of telescopes in the world. It is a good starting point for contacts to sites if a bright idea pops up and a network operation is required.

4.3 Time Series Spectroscopy

Light-curves can tell us about periodicities and, using different colours, something about mode identification. Time series spectroscopy provides access to a range of other techniques for measuring properties of oscillations. 50cm telescopes may be too small, but on 1.5m telescopes excellent results have been obtained, showing a richness of techniques for probing into details of stellar physics and the oscillations themselves. An illustrative example is the detailed study of the pulsation modes in the roAp star α cir by Baldry et al. (1998).

4.4 Political Aspects

Networks are operated in many different ways. The most streamlined type consists of identical instruments under the control of a center. The solar network called the Global Oscillation Network Group (GONG) is an example of such a network. Everything was constructed and tested by the GONG team in Arizona and all data are shipped back from the stations. Data go through a pipeline reduction and are then made available for teams specializing in various different topics.

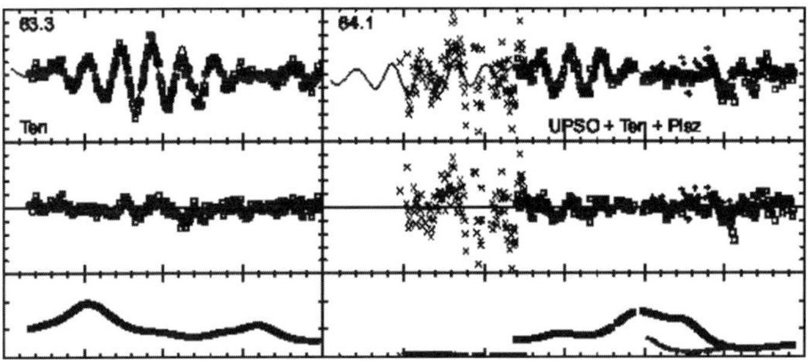

Figure 6.3-4. Observations from two nights in the middle of the 1998 STACC campaign. Upper panels show the light-curve, including a model light-curve. The middle panels show the residuals after subtracting this light-curve. The bottom panel gives the weights applied to the data points during the fit of the light-curve.

Another solar network is called International Research on the Interior of the Sun (IRIS), where the instrument was also constructed at a central site in Nice, France, but where the operations and data reduction is decentralized. The more active role of the people in the institutes connected with the observing sites has a large educational impact in small institutes and may open up opportunities for young people to join the international scientific community.

5. Practical Considerations

The success of a network depends on a number of issues – technical, economic, organizational and political – all in a complicated mix. The best networks operate according to a well-defined strategy.

5.1 Instrumentation

The possibilities with a network depend very much on the nature and location of equipment available. A modern telescope at a good site is a lot more productive than an old telescope in, for instance, Europe, where often only 10-20% of nights allocated are useful. Modern detectors also make a large difference. If one can afford a large field-of-view CCD camera with special software for readout and data reduction, giving light-curves in real-time, then observations can be carried out and good data obtained even in poor weather and close to city lights. This means that campus observatories can contribute.

Spectroscopy is more 'expensive' in all respects. But as high resolution is not necessarily needed, a simple, cheap instrument can be constructed. The most expensive item is still the detector (the CCD camera). Data reduction is probably the most critical issue. Dedicated software must be written that extracts the spectroscopic quantities from the images.

5.2 Storage and Communication

The progress in computing and networking makes the storage and communication easy in case of photoelectric photometry, where the volume is very small. In the case of differential CCD photometry, one should strive to develop real-time reduction software. Then, the amount of storage and data transfer is reduced and becomes similar to the output of a classical photometer. If images have to be stored or transferred to remote sites, then the volume is a problem. For monitoring programmes requiring a large number of CCD images to be compared and checked together, the organization of this part of the programme is the key to success.

5.3 Data Reduction and Manpower

To run a network efficiently, a fair amount of funding and manpower must be available. Managing the operation of the network and the data collected during campaigns is time-consuming and expensive. There is often a considerable delay in the output of the science part after data has been collected. The data quality often varies between different sites and a lot of experience is needed before the data can be combined and results extracted that are as accurate as possible.

5.4 Quality Control

The ideal situation is to have a network where there is good feedback to the participating teams. Few teams obtain perfect data the first or second time around.

6. Robotic Telescopes

Long-term observations of long period variables or the countless hours spent waiting for some spectacular event are often extremely tedious. Only the hope that it will lead to an exciting discovery makes you carry on. In addition, the instrument is often located far from home or the office, so that time spent observing is often not used very efficiently. A very attractive solution is to employ robotic telescopes and observe from home in an armchair or from the office, if the telescope is placed conveniently in longitude on another continent. Such facilities are becoming increasingly popular, also driven by the option of using them for public access to the sky. The role of such telescopes in education will probably explode soon. An example is the Iowa Robotic Telescope Facilities (IRTF). The IRTF Web page is located at (http://denali.physics.uiowa.edu), where you can also find links to some other facilities.

The Institute for Astronomy at the University of Vienna in Vienna, Austria, employs two remote telescopes, with the nicknames 'Wolfgang' and 'Amadeus', placed at the Fairborn Observatory in Arizona, which are used to monitor stellar activity. The Web page (http://www.astro.univie.ac.at/~kgs/APT) also has links to other sites.

7. Conclusion

Even in these days of very large telescopes, there is still a lot of exciting science to be carried out with small telescopes. With better detectors and more automation, these telescopes can participate in many important projects even in the future. Countries that do not have the possibility to join the league of nations with access to large telescopes can join networks and obtain contacts that would give them access to the larger world of astrophysics. They also serve as an excellent platform for students to start learning experimental astronomy, and for giving the general public the opportunity to conduct astronomical research.

References

Aerts, C., De Cat, P., Peeters, E. et al., 1999a, A&A, **343**, 872.

Aerts, C., Uytterhoeven, K., De Ridder, J. et al. 1999b, The Third MONS Workshop: Science Preparation and Target Selection, ed. T.C. Teixeira and T.R. Bedding, 59.

Aerts, C., De Cat, P., De Ridder, J. et al. 2000, in The Impact of Large-Scale Surveys on pulsating Star Research, ed. L. Szabados and D.W. Kurtz, ASP Conf. Ser., **203**, 395.

Baldry, I.K., Bedding, T.R., Viskum, M., Kjeldsen, H. & Frandsen, S., 1998, MNRAS, **295**, 33.

Buzasi, D.L., 2000, Astroph.J., **532**, L133.

Edmonds, P.D. and Gilliland, R.L., 1996, Astroph.J., **464**, L157.

Gombert, G. and Droege, T. 1998, Sky & Telescope Feb. 1998, 43.

Kjeldsen, H. 2000, in The Impact of Large-Scale Surveys on pulsating Star Research, Eds. L. Szabados and D.W. Kurtz, ASP Conf. Ser., **203**, 415.

Chapter 6.4

SINGLE-SITE AND MULTI-SITE PHOTOMETRIC RESEARCH PROGRAMMES FOR SMALL TELESCOPES

P. Martínez
South African Astronomical Observatory
P.O. Box 9, Observatory 7935, South Africa
peter@saao.ac.za

Abstract Small telescopes, though capable of producing significant research, are often neglected by astronomers. This paper examines the reasons for and results of such neglect and argues that to use small telescopes effectively requires one to recognize and exploit their special advantages. These advantages are outlined and the role of small telescopes in multi-site observing projects is discussed. Several examples are given of the photometric programmes running on the 0.5-m telescope of the SAAO at Sutherland.

Introduction

The South African Astronomical Observatory (SAAO) operates four telescopes of aperture 0.5-m, 0.75-m, 1.0-m and 1.9-m at its observing station in Sutherland, about 340 km northeast of Cape Town. By modern standards, the telescopes at Sutherland are all small telescopes. Yet they are used productively and imaginatively every clear night all year round. The situation at Sutherland is atypical, however, and is becoming increasingly more so every day. At some of the world's best photometric sites small telescopes are being shut down to divert resources to larger telescopes. Ironically, these are exactly the sites at which small telescopes should be most exploited - and most productive. The general neglect of small telescopes by observers leads to a downward spiral from poor use, to poor updating, to poor maintenance, to further under-subscription, ending eventually in closure of these telescopes. The present author believes that the underutilization of small telescopes arises from a general lack of appreciation of (i) the unique advantages that

W. Wamsteker et al. (eds.), Developing Basic Space Science World-Wide, 319-330.

small telescopes offer their users and (ii) the quality of science that can be accomplished using such telescopes. This paper discusses the advantages of small telescopes, and the common problems caused by neglect of these instruments. This is followed by a discussion of measures that may be taken to optimise the efficiency and productivity of small telescopes. The latter half of the paper presents a case study of the projects undertaken on the SAAO 0.5-m telescope in recent years.

1. Advantages of Small Telescopes

Many small institutes consider themselves to be incapable of contributing anything of significance to observational astronomy because they do not have access to large telescopes. Nothing could be further from the truth, for provided that a small telescope is properly instrumented and used, it should produce data of the highest quality. In fact, the intrinsic quality of observations acquired on small telescopes is often higher than those acquired using larger instruments because the observations can be obtained more carefully and with more stable equipment. Incidentally, it is important to appreciate that such high-quality observations are no longer restricted to the very brightest objects. Modern CCD detectors have such high quantum efficiency that small telescopes are not so "small" anymore in the sense that they can now explore areas of astronomy that were the exclusive preserve of much larger instruments a decade ago. Moreover small telescopes can be highly cost effective because they are easier to operate and require fewer support staff. In arguing the case for small telescopes we may as well begin by acknowledging that for observations requiring high time resolution and/or high spectral resolution, there is simply no substitute for detecting an adequate number of photons. However, small telescopes are very competitive whenever the integration times are short and the time required for an observation is dominated by overhead. This is normally true for bright objects, especially when they are scattered all over the sky. Thus small telescopes are very efficient instruments to use in large survey programmes. Such programmes often disclose unusual behaviour in familiar objects, rare events or unusual objects, or they reveal subtle trends hidden in the noise of previous statistically incomplete samples.

A major attraction of small telescopes is their ready availability. At major observatories it is quite common to find that large telescopes are oversubscribed by large factors while the small telescopes are underutilized. On small telescopes, time is often allotted in units of weeks rather than nights. Thus poor weather or instrument breakdowns are generally not disastrous. It is also much easier to arrange multi-site campaigns on small telescopes than on large ones.

In addition, the great flexibility of scheduling also allows unanticipated target-of-opportunity observations of rare or unusual astrophysical phenomena. For observers spending more time writing project proposals than observing, there is the added advantage that often only minimal justification is required for small-telescope projects.

Another advantage of the low cost and low pressure of time on small telescopes is that it allows speculative observations to be made more readily than on larger instruments. Observers are often reluctant to propose speculative projects on large telescopes, and time allocation committees are equally reluctant to award time for such projects.

Because large telescopes serve a broad user community they are usually multipurpose instruments. Owing to frequent instrument changes, the accuracy and stability of instrumental calibrations is often rather poorly determined and much time is lost to failures associated with instrumentation changes. On the other hand, small telescopes can be optimized for a single task thus ensuring stability of performance and reduced instrument failure.

There are also ways that astute users of small telescopes surreptitiously make up for their lack of aperture. In studies of variable stars where only the period of variation is of interest, it may be possible to monitor the program object in white light thereby gaining a factor of several times the light measured through filtered telescopes. Moreover, small telescopes usually have better seeing and it is possible to exploit this to get better signal-to-noise.

2. Common Problems Caused by the Neglect of Small Telescopes

Because they are under subscribed, small telescopes at major observatories are sometimes poorly maintained, or equipped with antiquated instrumentation. This creates a bad impression among students, who feel that it is not possible to do cutting-edge astronomy with these small, antiquated instruments. Of course, it is precisely these students who have the most to benefit from using small telescopes. Using a small telescope instils a certain amount of experimental dexterity which is not gained when a student sits in a brightly lit, air-conditioned control room and all telescope operations are handled by a night assistant. Often, astronomers are awarded time on small telescopes for projects deemed to be meritorious by peer review only to have their travel grant application rejected because they are asking for travel money to use a small telescope. When they do arrive at the observing site they may find that the telescope is not well maintained, and several nights may be lost until problems arising from inadequate maintenance are solved.

3. Optimizing the Efficiency of Small Telescopes

Telescope and Instrumentation: Experience shows that when advanced instrumentation is made available on small telescopes the demand for time on them increases significantly. For example, about 15 years ago KPNO placed a direct-imaging CCD camera on their #1 0.9-m and shortly thereafter the subscription rate on that telescope rose to exceed that of the 4.0-m. The efficiency of any telescope, especially a small one, is increased by implementing a vigorous programme of instrument development. Often great improvements in observing efficiency can be gained by appropriate automation, such as placing instruments and data acquisition under computer control. The addition of off-axis guiding and/or autoguiding facilities improves the precision of the observations and facilitates field acquisition. For this purpose one can use inexpensive commercially available CCD cameras developed for the amateur astronomy market.

Where new instrumentation is to be acquired or developed, serious consideration should be given to using CCDs as the detector of choice. As an astronomical detector, CCDs have excellent properties of high quantum efficiency and broad dynamic range with good linearity and stability. For photometry, the CCD represents a major advance over the photomultiplier tube and as such CCD photometry is superseding conventional photoelectric photometry. Until fairly recently CCD-based instruments placed heavy demands on workshop facilities and technical expertise, and this effectively excluded CCD-based systems from many small telescopes. However, there are now commercially available and essentially maintenance-free instruments such as Peltier-cooled CCD camera systems (with impressively low readout noise) that are designed to be operated by end users rather than by trained engineers. With the advent of frame-transfer CCDs even high-speed CCD photometry is now possible since the readout time for these chips no longer limits the sampling process in most cases.

Communications: The more "connected" a small telescope is to the rest of the astronomical world, the more efficiently it can be used. To network telescopes effectively requires good communications, ideally voice, fax and email. Having Internet access (ideally also in the dome) opens a vast realm of possibilities to small telescope users. Nowadays it is possible to obtain off the Internet anything from finding charts to complete papers out of the latest issues of the mainstream journals. In this way poorly funded small observatories can obtain selected items from the recent literature without being able to afford expensive journal subscriptions. Moreover, having email capability provides cheap, fast access to distributed astronomical and technical expertise. For the small observatory having Internet access is almost as important as having access to an astronomical library or having access to workshop facilities.

Time: The importance of having accurate time in the dome cannot be overemphasized, especially if the small telescope is to be networked (see below). Since the oscillators in computers are generally not sufficiently stable to be used for precise timing of astronomical observations an external source of time is required to calibrate the computer clock and to drive hardware interrupts at the correct instants. Fortunately, nowadays one can attain time of high accuracy and precision for about the cost of a personal computer. This is possible by purchasing a commercially available GPS receiving system which can provide time with a precision to 1 μs, depending on the level of sophistication required.

Networking of Small Telescopes: The daily gap in observations from a single site introduces strong side-bands (called "aliases") with a spacing of 1 cycle day^{-1} (11.57 μHz) in Fourier transforms of data combined from several nights. It is a perversity of nature that the periods of many variable stars are either close to one day, or that multiperiodic stars have frequency spacings close to an integral multiple of 1 day^{-1}, making it impossible to study such stars from a single site. To reduce these problems, multisite observations are required from sites well separated in longitude. This increases the duty cycle and lowers the amplitudes of the daily aliases to eliminate confusion. Small telescope users often band together to form consortia of distributed light gathering power and distributed technical and astronomical expertise. This has synergistic effects, which result in outstanding science being produced. The involvement of a small institution in prestigious international collaborations can also lead to enhanced local recognition and improved possibilities for local funding. Another form of multisite observation in which small telescopes are playing an increasing role is in obtaining ground-based support observations for orbiting telescopes.

The Whole Earth Telescope (or "WET") is a network of astronomers spread out over a wide range of longitude and latitude to ensure a high probability of continuous coverage of objects of interest (Nather et al. 1990). The network uses existing small telescopes, usually ranging in size from 0.5 m to 1.0 m. Since its inception in 1988, WET has had over a dozen observing campaigns with contributions from telescopes in Australia, Brazil, Chile, China, France, India, Israel, Norway, South Africa, New Zealand, the United Kingdom and the United States. Each WET campaign typically has two or three targets, and anyone in the

WET consortium can propose targets. Each campaign is coordinated by a Principal Investigator ("PI") who is usually the person who proposed the target(s) for that campaign. Although the PI is normally expected to lead the analysis and write the resulting papers, anyone who contributed to that particular campaign is able to study the entire multisite data set.

4. Examples of Small-Telescope Projects at SAAO

A model of a highly productive small telescope is the SAAO's 0.5-m, located in Sutherland. The Sutherland site lies on a plateau at 1758 m above sea level and about 18 km to the east of the Sutherland village at longitude 20 49'E and latitude 32 23'S. About 50% of the nights at Sutherland are photometric. The site is nonseasonal and has exceptionally stable transparency, largely because of its location and the local weather patterns. The Atlantic Ocean lies 250 km to the west and the Indian Ocean 230 km to the south. There is almost always a mild westerly or southeasterly surface wind of 20 - 40 km/hr blowing, and the prevailing upper atmosphere winds are usually westerly. Only rarely does the wind blow from the northeast, that is, from the interior. The other winds bring in dust-free, pollution-free oceanic air.

The 0.5-m telescope is equipped with a single-channel, computer-controlled pulse counting photometer. The detector is a Hamamatsu 943-02 GaAs photomultiplier tube housed in a thermoelectrically cooled coldbox. The photometer and its associated electronics were designed and manufactured in the Cape Town workshops of the SAAO. All of the examples shown in this paper are based on data acquired using this instrument. By restricting myself to this telescope I hope to suggest ways in which similar sized small telescopes might be put to good use. In giving these examples I claim to be neither comprehensive nor unbiased. My choice of topics was dictated either by my direct involvement in some of the projects described or by my association with colleagues involved in other projects on this telescope.

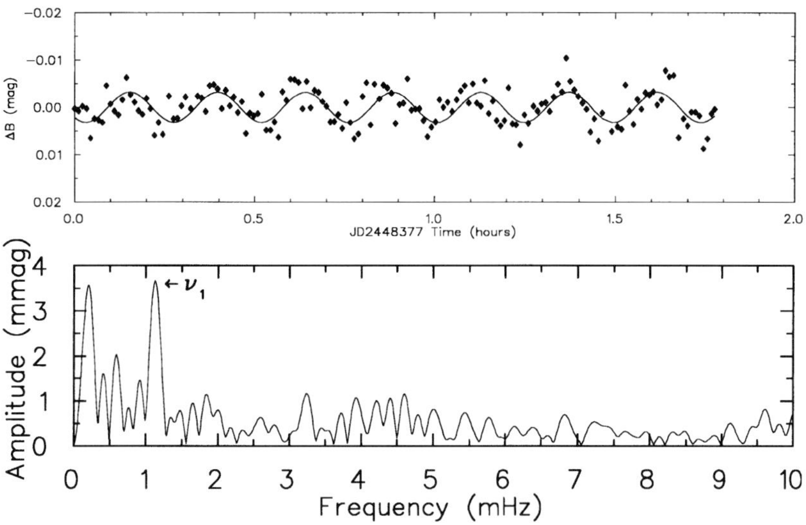

Figure 6.4-1. The Johnson B light curve (top) and amplitude spectrum (bottom) of the rapid oscillations in the Ap star HD84041, discovered using the 0.5-m telescope at Sutherland. The period of pulsation in this star is 15.0 minutes and the pulsation amplitude in this light curve is about 3.6 mmag. The peak labelled "v1" in the Fourier spectrum is the pulsation frequency of the star. The other prominent peak at very low frequency is a gradual sky transparency variation

Rapidly Oscillating Ap Stars

The Ap stars are magnetic, chemically peculiar stars with surface enhancements of the rare earths, lanthanides and Fe-peak elements. The Ap phenomenon ranges in temperature from about spectral type B8 to spectral type F0. Towards the cool end of their temperature range, the Ap stars overlap the instability strip, where both radial and non-radial low-overtone pulsations occur in the delta Scuti variables. It is here that one finds the rapidly oscillating Ap (roAp) stars. These are cool Ap stars that exhibit broadband photometric oscillations with periods in the range 6 - 15 min and Johnson B amplitudes ranging from 0.3 - 8.0 mmag about the mean light level. The roAp stars range in brightness from V = 3.2 to V = 10.3, making them ideal objects for study with small telescopes. The oscillations in the roAp stars are caused by low-degree, high-overtone non-radial p-mode pulsations similar to those of the Sun, but with amplitudes three to four orders of magnitude greater. Another difference from the Sun is that the pulsations in the roAp stars take place in the presence of global magnetic fields with strengths of several hundred to several thousand gauss. The roAp stars allow us to study the interaction of pulsation with magnetic fields and provide an excellent opportunity to study the seismology of stars other than the Sun. Of the 32 known roAp stars, 25 were discovered in Sutherland, and many of those with the 0.5-

m telescope. Figure 6.4-1 shows one of the discovery light curves of the roAp star HD 840401.

Most roAp stars are multiperiodic. They have Fourier oscillation spectra with evenly spaced peaks, and often the p-mode spacing is irritatingly close to an integral multiple of 1 cycle/day (11.57 μHz). To discriminate the real pulsation modes from daily aliases requires multi-site data to suppress the aliases. A number of multi-site studies have been performed using the SAAO 0.5-m and telescopes of similar size in Chile and New Zealand. One such study revealed the presence of up to 17 independent frequencies in the roAp star HD 60435 (Matthews et al. 1987). These frequencies span the range 709 - 1457 μHz and fall into an equally spaced pattern with a separation of 25.8 μHz.

The discovery of frequency variability in roAp stars

Several years ago we discovered that the pulsation frequencies of the roAp stars are not constant. Where possible, on every clear night for the past seven years we have monitored the oscillations of HR 3831 for one hour to establish the pulsation phase for that night. Figure 6.4-2 shows the (O-C) diagram, which suggests a cyclic variation with a peak-to-peak amplitude of 0.12 μHz on a time scale of 1.6 y (Kurtz et al. 1994). It is well established that the frequency of the solar p modes is correlated with the solar cycle (e.g. Libbrecht & Woodard 1990; Bachman & Brown 1993); the range of the frequency variation in the Sun is about 0.5 μHz. This has led Kurtz et al. to suggest that the frequency variations in the Ap stars are caused by some kind of magnetic cycle in HR 3831 which changes the shape of the acoustic cavity over time scales of several years. Regardless of whether this is the correct explanation or not, it is clear that these variations have to be characterized and that further monitoring with small telescopes is required. Similar, albeit less extensive, results have been obtained for several other roAp stars and monitoring of these stars with the SAAO 0.5-m continues.

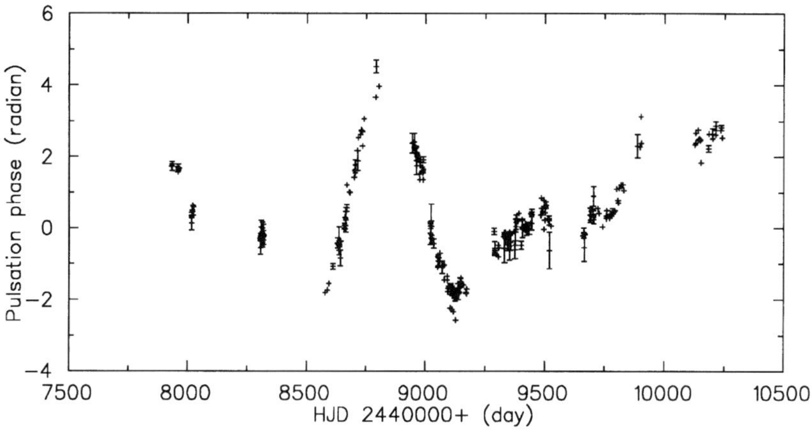

Figure 6.4-2. The long-term behaviour of the pulsation phase of HR 3831. Each point is the pulsation phase determined from a one-hour observation of the rapid oscillations in the star. The ordinates are the differences between observed times of pulsation maxima and calculated times of pulsation maxima for a constant frequency, expressed as a phase difference in radians. In this diagram a constant pulsation frequency should be a horizontal straight line. Sloping straight lines indicate changing frequency and abrupt vertical jumps indicate abrupt phase changes. The smooth, continuous variation in the data suggests a cyclic (but not periodic) frequency variation.

Rotation periods of Ap stars: The cylindrical symmetry of the magnetic fields in Ap stars tends to concentrate the surface abundance anomalies in dark photospheric spots or rings centred on the magnetic poles. These features give rise to photometric variations as the star rotates. The fundamental period of these photometric variations is, of course, equal to the rotation period. We have used the SAAO 0.5-m to determine the rotation period of a number of Ap stars by obtaining Johnson UBVRI photometry of these stars and suitable comparison stars. Figure 6.4-3 shows the rotation light curves obtained in this way for the roAp star HD 84041. The rotation period is of this star is 3.69 d.

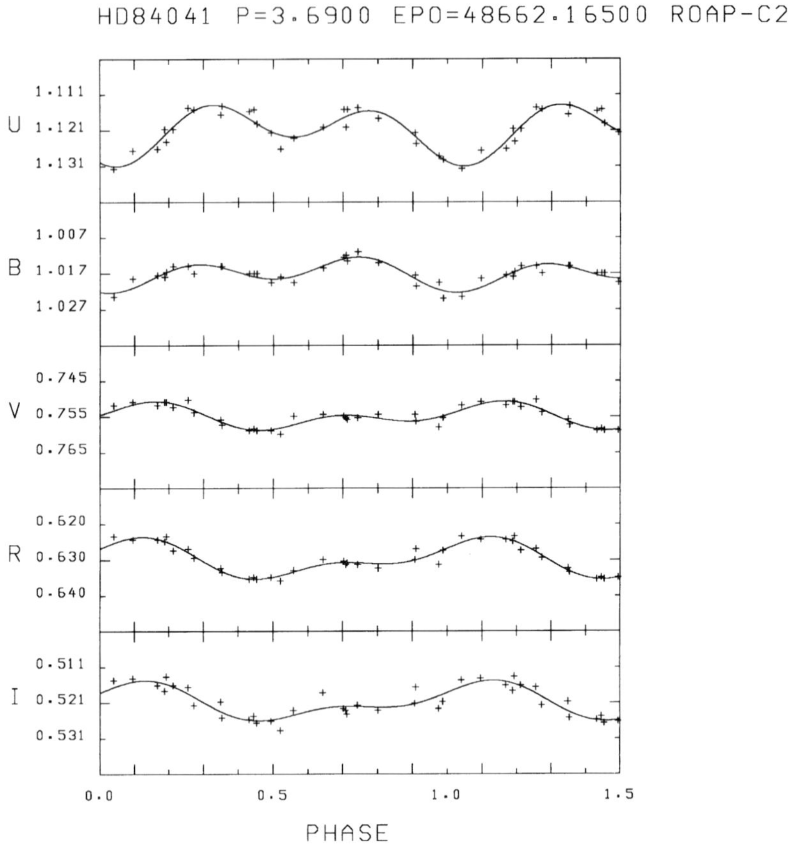

Figure 6.4-3. The rotation light curve of the Ap star HD 84041; the rotation period is 3.69 d. The double wave suggests that one of the magnetic poles is seen at more favourable aspect than the other. Notice the changing amplitudes and the phase shift with filter from U to I.

Simultaneous ground-space observations of the southern K dwarf AB Doradus: AB Doradus is a very active southern K dwarf star with a rotation period of only 12.4 hr. During the night 7/8 November 1993 the SAAO 0.5-m was used to provide ground support observations for space-based X-ray and extreme UV observations using the Japanese X-ray satellite ASCA and the Extreme Ultraviolet Explorer (EUVE) satellite, respectively. Figure 6.4-4 shows flare activity detected during this run (Kilkenny - private communication). The interpretation of these events in the different passbands is not yet clear. This is just one example of the many ground support observations done with this telescope.

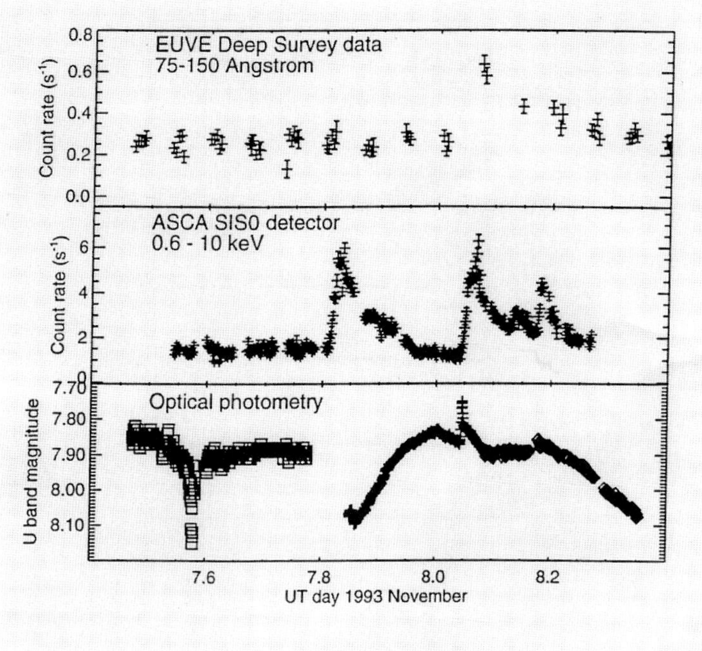

Figure 6.4-4. The extreme ultraviolet (top), X-ray (middle) and optical (bottom) light curves of the K star AB Dor. The optical light curve was acquired at Mount Stromlo, Sutherland and Las Campanas. The Sutherland portion of the light curve runs from UT 7.85 to UT 8.08 November 1993. Taken from Smith, S.M. et al. (1995 preprint).

The zeta Aurigae-type eclipsing binary AL Velorum: Eclipsing binary stars of the zeta Aurigae type comprise a cool giant or supergiant and a hot main sequence star. The light and velocity curves yield data on the mass and luminosity of the evolved cool star, and the hot companion can be used to probe the atmosphere of the extended primary during ingress and egress of primary eclipse. Photometric observations of this star with the SAAO 0.5-m have revealed interesting variations in its light curve (figure 6.4-5). In particular, the 1991-92 and 1992-93 data show that as egress from primary eclipse is almost finished, the brightness decreases by 0.02 mag and then slowly recovers to the level of the previous season's data at around phase 0.2. By using an interactive light curve analysis programme on a PC to test-fit starspot geometries to the AL Vel data, Kilkenny et al. (1995) deduced that the cool K giant has two spots (or spot groups), and that these have probably moved on the star's surface and also possibly changed size/temperature during the three seasons of their observations.

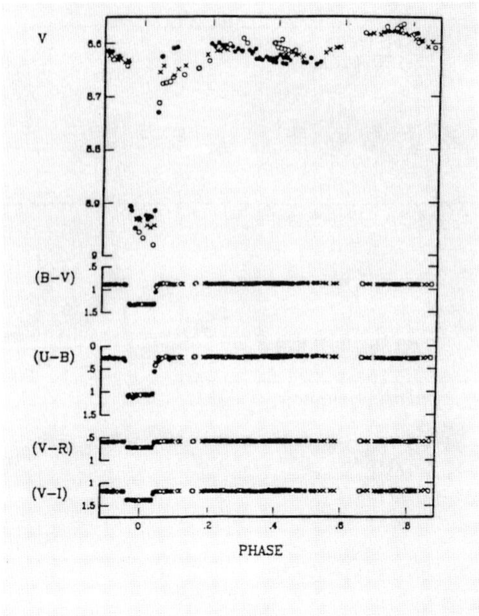

Figure 6.4-5. . Light and colour curves of AL Vel for 1991 (filled circles), 1991-1992 (crosses), and 1992-1993 (open circles) (Kilkenny et al. (1995)).

Further examples: In selecting the above examples, I have restricted myself to the small telescope with which I am the most familiar. A wide variety of other small-telescope projects are described in the Proceedings of I.A.U. Symposium 118 (Hearnshaw & Cottrell 1986) and by Warner (1995).

Acknowledgments

The author thanks Dr Dave Kilkenny of the SAAO and Michael Snowden for helpful comments on an early draft of this paper.

References

Bachman, T.K., Brown, T.M., 1993, Astrophys. J., **411**, L45.

Hearnshaw, J.B., Cottrell, P.L., 1986, IAU Symposium 118, Instrumentation and Research Programmes for Small Telescopes", Reidel, Dordrecht.

Kilkenny, D., et al., 1995, The Observatory, **115**, 31.

Kurtz, D.W., et al., 1994, Mon. Not. R. Astr. Soc., **268**, 641..

Libbrecht, K.G., Woodard, M.F., 1990, Nature, **345**, 779.

Matthews, J.M., Kurtz, D., Wehlau, W.H., 1987, Astrophys. J., **313**, 782.

Nather, R.E., et al., 1990, Astrophys. J., **361**, 309.

Warner, B., 1995, AP&SS, **228**, 283.

Chapter 6.5

"HANDS-ON ASTROPHYSICS"AND BEYOND

J.R. Percy and J. A. Mattei
Erindale Campus, University of Toronto
Mississauga ON Canada
jpercy@erin.utoronto.ca

America Association of Variable Star Observers
Cambridge, MA, USA
aavso@aavso.org

Abstract *Hands-on Astrophysics* is a project which develops and integrates a wide range of science and math skills, through the measurement and analysis of variable stars. It can be used in a variety of settings, from junior high school to university. In this paper, we describe how it can be used in astronomically-developing countries or regions, as a stepping stone to more advanced astronomical activities, and as a model for using processes and databases developed by amateur astronomy as a way of doing useful science at low cost.

Introduction

In any part of the world, *practical* or *laboratory* activities can be an important part of science education at the school and university level, especially when they include a strong element of inquiry and discussion. Carefully-chosen *research projects* can also be used; they are almost obligatory at the graduate level, highly desirable at the undergraduate level, and quite possible at the senior high school level —both because they contribute to effective learning of science processes and skills, and because they motivate students through the excitement of doing real science with real data. Research projects enable students to contribute to research; this is especially desirable in the developing countries, where the supply of research personnel is low.

How can students in developing countries do research? If specialized equipment is available and usable — in their supervisor's lab, for

W. Wamsteker et al. (eds.), Developing Basic Space Science World-Wide, 331-337.

instance — there is less problem. Here, we suggest some solutions when little or no equipment is available. One solution is to make simple but useful measurements of the sky, using processes which have been developed and perfected by *amateur astronomers* and/or to use the data which amateur astronomers have obtained. We shall use the measurement of variable stars as an example. This may lead to the use of processes and data obtained by *professional astronomer*s, as explained below.

1. Amateur Astronomers

An *amateur astronomer* is someone who loves astronomy, and cultivates it as a pastime or hobby. A more stringent definition might be that an amateur astronomer is someone who does astronomy with a high degree of skill, but not for pay (Williams 1988). Amateur astronomers make important contributions to astronomical research (Dunlop & Gerbaldi 1988) and education (Percy 1998a). Astronomical research and education done by amateur astronomers is "democratic" in the sense that it enables anyone to do astronomy, whether they have formal credentials or institutional affiliations or not. It is "science for the masses". It is certainly true of variable star astronomy because, in the words of the song, "the stars belong to everyone". The Internet can also be a powerful tool for those who are "connected", since it provides access to large and sophisticated databases (such as the *Hipparcos* catalogue and epoch photometry) and information and instructions for analyzing and interpreting the data.

2. Variable Stars and the AAVSO

This topic was reviewed by us at the 1996 and 1999 UN/ESA Workshops on Basic Space Science, in Bonn, Germany (Percy 1998b) and Mafraq, Jordan, so only a brief account will be given here.

Variable stars are those which change in brightness. If measured sufficiently carefully, almost every star turns out to be variable. The variation may be due to geometrical processes, such as the eclipse of one star by a companion star, or the rotation of a spotted star, or it may be due to physical processes such as pulsation, eruption, or explosion. Variable stars provide astronomers with essential information about the properties, processes, nature, and evolution of the stars.

In many variable stars, the changes in brightness are large enough to be detected with the eye, using a small telescope, binoculars or (in the case of a few dozen bright variable stars) no optical aid at all. Amateur astronomers and students can make a useful contribution to astronomy by measuring these variable stars in a systematic way. The American Association of Variable Star Observers (AAVSO) was established in 1911. Its purpose is to coordinate variable star observations made largely by dedicated amateur astronomers, evaluate the accuracy of these observations, compile, process, and publish them, and make them available to researchers and educators. As of 2003, the AAVSO receives over 400,000 measurements a year, from 550 observers worldwide. The measurements are entered in the AAVSO electronic database, which contains over 10.5 million measurements of several thousand stars. The demand for these measurements, from researchers and educators, has increased by a factor of 25 in the last two decades —partly as a result of major collaborations in space astronomy (Mattei 2000). To contact the AAVSO, write to: AAVSO, 25 Birch Street, Cambridge MA 02138, USA; email: aavso@aavso.org; web site: www.aavso.org.

3. Hands-On Astrophysics

It occurred to us, several years ago, that the measurement and analysis of variable stars could help students to develop fundamental science and math skills. Variable star measurement and analysis is inherently simple; it can be done by any high school student. The analysis and interpretation of the data involves a wide range of scientific and mathematical skills, some of which would be understood and appreciated by a junior high school student, and some of which would challenge an expert in the field. We therefore developed *Hands-On Astrophysics*, with partial funding from the US National Science Foundation. Its purpose is to bring variable star observing and analysis into the classroom or lab. It includes finder charts for 45 stars, 31 35mm slides of five constellations, 14 prints of the Cygnus star field at seven different times, 600,000 measurements from the AAVSO International Database on 45 stars, user-friendly computer programs to analyze them and to enter new observations into the database, an instructional video in three segments, and a comprehensive manual for teachers and students.

Hands-on Astrophysics is very flexible. It can be used in science, math, or computer classes, or for independent projects. It can be adapted to various levels, from junior high school to university. It can be used as a complete course of study, or in parts. It can be used by *any* individual, of *any* age, to learn the art and science of variable star astronomy. It is self-contained; no previous knowledge of astronomy, or variable stars, is assumed. It is open-ended, and can lead to sophisticated projects which are ideal for science fairs. It actively involves the students in the scientific process, and motivates them by enabling them to do real science, with real data. Students can observe variable stars in the real sky,.4 and send their measurements to the AAVSO, thus contributing to astronomical research. *Hands-on Astrophysics* is available from the AAVSO(address above) from the Astronomical Society of the Pacific (www.astrosociety.org), and from Sky Publishing Corporation (www.skypub.com).

4. Hands-On Astrophysics and the School Science Curriculum

Variable stars are not a central “content" topic in the high school science curriculum in North America (though they relate peripherally to several topics in the US National Science Education Standards (and the Canadian equivalent) dealing with the sun, stars, galaxies, and universe). It should be realized, however, that the science *content* in the curriculum is less important than the science *process* and *skill*s. Variable star observing and analysis develops and integrates a wide range of science and math skills. In the grade 9 (age 14 years) science curriculum in the province of Ontario, Canada, for instance, the following skills expectations are listed:

- Plan ways to model and/or simulate an answer to the questions chosen, for instance ..... how astronomers are able to understand and compare the sizes and distances of objects ..... in the universe.
- Demonstrate the skills required to plan and conduct an inquiry into the.....characteristics of visible celestial objects .....
- Select and integrate information from various sources, including electronic and print resources, community resources, and personally collected data .....
- Gather, organize, and record information using a format that is appropriate to the investigation (e.g. maintain a log of observations of changes in the night sky .....)
- Analyze qualitative and quantitative data .....
- Communicate scientific ideas, procedures, results, and conclusions using appropriate SI units, language, and formats
- Calculate and compare the ..... distances to objects ..... in the universe
- Predict the qualitative and quantitative characteristics of visible celestial objects ….

5. Applications to Astronomically-Developing Countries

5.1 A Prelude to Research

Variable star observing can be a prelude to more advanced astronomical activity. Wentzel (private communication), for instance, has found *Hands-on Astrophysics* very useful in workshops with physics teachers in Vietnam, as part of the IAU's "Teaching for Astronomical Development" program. One of us (JAM) has been using *Hands-on Astrophysics* at the teacher-enhancement workshop TOPS (Towards Other Planetary Systems) held in Hawaii for Hawaiian and Micronesian teachers and students, for the past four years. *Hands-on Astrophysics* helps these science teachers and students, often with no astronomy background, to develop basic science skills of observing, collecting, analyzing, and interpreting scientific data. Several developing countries have acquired (or are planning to acquire) small telescopes which will eventually be equipped with photoelectric photometers or CCD cameras. Is there a way to start doing real science, even before the telescope arrives and is operational? Yes! The solution is to begin doing serious visual measurements of variable stars, using binoculars or a small telescope if available. The AAVSO, either through *Hands-on Astrophysic*s, or by mail, by email, or by its web site, can provide assistance in setting up an observing program. The AAVSO has provided observational material, and *Hands-on Astrophysics* kits, to 16 UN sites with such setups. The measurements, so obtained, can then be contributed to the AAVSO International Database, accessed through the AAVSO website, to be used by researchers and educators.

The next step is photoelectric or CCD photometry with the newly-acquired telescope. These measurements are more precise than visual measurements, but the general principles of analysis are the same. The AAVSO has both a photoelectric program, and a CCD-photometry program. For more information, visit the AAVSO web site (*www.aavso.org*). Other international collaborative photometry programs are listed in Percy (1998b), who pointed out the value of beginning research as part of an international collaboration.

5.2 A Stepping Stone to Other Databases

Databases from space astronomy missions are increasingly available on CD-ROM and/or the Internet, and they provide a practical way for astronomically-developing countries to begin research at very little cost —a PCwith a CD-ROM drive and/or Internet connection. One example is the *Hipparcos* catalogue of astrometry and epoch photometry (Turon 1997, Perryman 1999). AAVSO observers provided crucial support for this mission (Turon 1997). In turn, the mission has provided dozens of new variable stars to be studied by photoelectric or CCD observers (Perryman 1999). It has also provided millions of photometric measurements of "unsolved" stars which require detailed analysis. The *Hipparcos* mission has excellent research and education web pages (astro.estec.esa.nl/Hipparcos/hipparcos.html), with information on variable stars, interactive tutorials on variable star analysis, as well as data. Additional information on variable stars can be found on the AAVSO web site (*www.aavso.org*), and user-friendly software developed by the AAVSO for analyzing variable star observations (the TS11.ZIP program for time series analysis, and the WWZ11.ZIP program for wavelet analysis), can be downloaded from www.aavso.org/software.stm. *Period98*, a powerful period-analysis package, can be downloaded from the web site www.deltascuti.net/period98/ at the Institute of Astronomy, University of Vienna, along with an instruction manual.

5.3 Hands-On Astrophysics as a Model

Hands-on Astrophysics makes use of techniques which have been refined by amateur astronomers over many years, and measurements which they have made for the benefit of research. There are other types of measurements and data which provide a low-cost introduction to real astronomical research.

Sunspot counting is an obvious example which can be done in the daytime with a small telescope. The AAVSO (www.aavso.org) has a well-established solar program, as does the Association of Lunar and Planetary Observers (ALPO:www.lpl.arizona.edu/alpo/); these organizations can provide guidance and coordination. Sunspot numbers, over decades or centuries, can be analyzed with the same software as for variable stars.

Timing of occultations of stars by the moon, planets, and asteroids is a useful scientific activity; among other things, it can provide estimates of the size and shape of asteroids. Since occultations occur along specific geographical paths, observers in the astronomically-developing countries can often provide otherwise-inaccessible measurements. A small telescope, a stopwatch, and access to a time signal, is necessary, along with predictions which can best be provided on the Internet. The International Occultation Timing Association (IOTA, www.anomalies.com/iotaweb/index.htm) coordinates this work. Meteor observation has recently proven to be especially interesting because of the "storms" in the Leonid meteor shower. Meteor storms can occur within a very few hours in time, so it is essential to have observations from all longitudes. No telescope is required for these observations, and only approximate time signals are necessary. The American Meteor Society (www.amsmeteors.org) coordinates this work.

One of the advantages of these "amateur" techniques is that there may be a few local amateur astronomers who are familiar with the sky, with telescopes, and with basic observational techniques. These people can be a useful partners in the astronomy research process.

Acknowledgments

This material was presented at the IAU-COSPAR-UN Workshop on Capacity Building in Basic Space Science by John Percy, who thanks the Natural Sciences and Engineering Research Council of Canada for research and travel support. *Hands-on Astrophysics* was funded in part by a grant to the AAVSO from the US National Science Foundation.

References

Dunlop, S., & Gerbaldi, M., 1988, Stargazers: The Contributions of Amateurs to Astronomy, SpringerVerlag.

Harmanec, P., 1998, A&A, **335**, 173.

Mattei, J.A., 2000, in Amateur Professional Partnerships in Astronomy, ed. J.R. Percy & J.B. Wilson, ASP Conf. Ser. **220**, 35.

Percy, J.R., 1998a, in New Trends in Astronomy Teaching, ed. L. Gouguenheim et al., Cambridge University Press, 205.

Percy, J.R., 1998b, AP&SS, **258**, 357.

Perryman, M.A.C., 1999, Sky & Telescope, **97**, #6, 40.

Turon, C., 1997, Sky & Telescope, **94**, #1, 28.

Williams, T.R., 1988, in Stargazers: The Contributions of Amateurs to Astronomy, ed. S. Dunlop & M. Gerbaldi, SpringerVerlag, 24.

Chapter 6.6

CCD PHOTOMETRY OF KZ HYA FROM THE AAO (PARAGUAY)

F. A. Doncel, A.E. Troche-Boggino†, T. Noguchi
Universidad Nacional de Asuncion
Facultad Politecnica, Observatorio Astronomico
Paraguay
d_fredy@hotmail.com

Abstract A pulsating star of the SX Phe-type, the variable KZ Hya (HD94033) was observed with the CCD camera attached to the 45-cm reflector at Asuncion Astronomical Observatory in Paraguay. We describe here the results of the photometric observations covering 12 maximum phases. A new ephemeris has been obtained, and the result suggests a probable change of the pulsation period of KZ Hya.

Introduction

CCD photometric observations of short period pulsating variable KZ Hya ($\alpha(2000)=10^h51^m54.08^s$, $\delta(2000)= -25^{deg}21^m10.8^s$, 2000) were made during 7 nights from April 18 to July 17, 2002. The observations were made, with the 45-cm Goto reflector (Kitamura, 2003; Brocchi,2003) at the Asuncion Astronomical Observatory in Paraguay. KZ Hya was discovered in 1975 by Przybylski and Bessell (1979) in a photometric survey of early type stars with large proper motion, and was the first known short period Cepheid which clearly belongs to Population II.

Details of the Observatory are given in Trocche (2003) and of the observations are given in Doncel et al. (2004).

† Shortly after the completion of this paper Professor Alexis E. Troche-Boggino passed away.

W. Wamsteker et al. (eds.), Developing Basic Space Science World-Wide, 339-344.

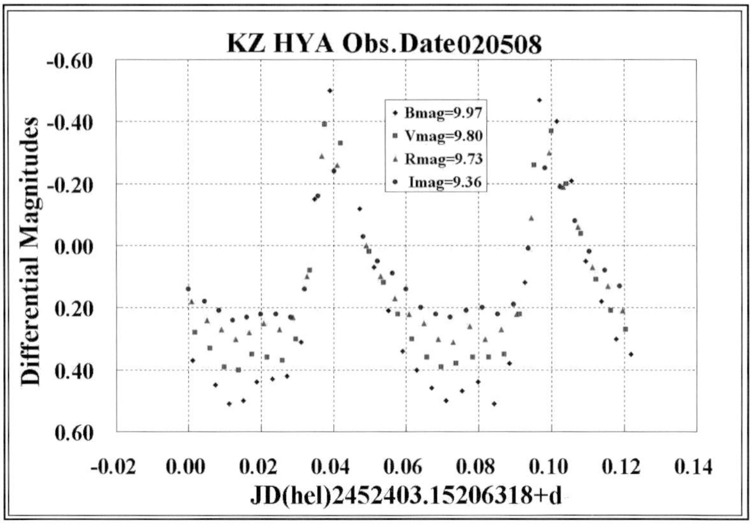

Figure 6.6-1. The observation of CCD photometry of the KZ Hya with BVRI colors which were made during the night of 2002, May 8.

1. Observations and Reduction

Observations of KZ Hya were carried out using the CCD camera with *BVRI* color filters in the Cassegrain focus of the 45-cm telescope. The present ST-8 type CCD camera has 1530×1020 pixels with field of view of about 8.7×5.8 square arc-minutes. The CCD Camera system was made by the Santa Barbara Instrument Group (SBIG).

Figure 6.6-1 shows the result of CCD photometry during the night of 8 May, 2002 as an example. Exposure times for the different bands were 30 seconds in the *B*-band, 10 seconds in the *V*-band, 10 seconds in the *R*-band, and 10 seconds in the *I*-band, respectively. In figure 6.6-1 we show the differential magnitudes with respect to a nearby comparison star. The constancy of this star was checked against a second comparison star with the estimated accuracy of $\sigma \approx \pm 0.02$ magnitude in all colors. The limiting magnitudes in *BVRI* for an exposure of 10 seconds are 13 mag. (*B*), 14 mag.(*V*), 15 mag. (*R*), 15 mag. (*I*) respectively.

Figure 6.6-2 shows the combined differential observations in R-band during the 4 nights covering 7 phases with the maximum phases co-aligned. Figure 6.6-2 shows that the light curve is asymmetric and there is a secondary maximum at phase 0.7. The amplitudes of KZ Hya in the four colors are $\Delta B=0.993, \Delta V=0.792, \Delta R=0.624, \Delta I=0.478$.

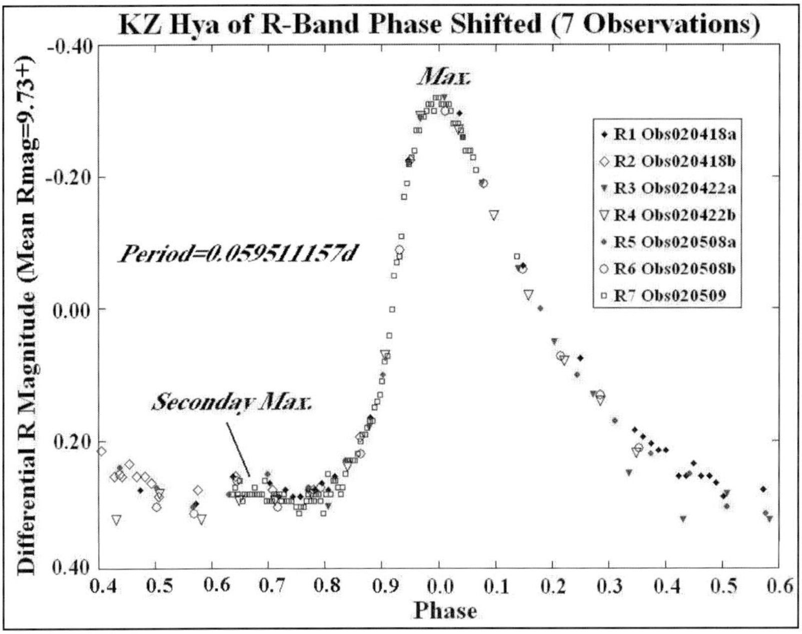

Figure 6.6-2. Differential observations in R color during 4 nights covering 7 maximum phases.

2. Period of Pulsation

From their first observational results coverings 25 maximum phases Przybylski and Bessell (1979), derived the ephemeris

T max(hel) = 2442516.15836 + 0.0595104212 E, (1)
Hobart, Peniche and Pena (1985), later derived
T max(hel) = 2442516.15903 + 0.0591012 E, (2)
and Liu, Jiang and Cao (1991), gave
T max(hel) = 2442516.15576 + 0.05911036 E $+0.5\times2.92^{-12}$ E^2 (3)

The combination of these historical observations with our new results permit the re-determination of the ephemeris covering 102 maximum phases over 27 years. Using the O-C residuals of all maximum phase values obtained a new ephemeris has been derived,

T max(hel) = 2442516.15850 + 0.05911157 E $+0.5\times2.24^{-12}$ E^2 (4)

Two series of O-C residuals corresponding to Eq 3 (LJC) and Eq.4 (DTN) versus epochs for KZ Hya is shown in figure 6.6-3. The accuracy of the ephemeris is ≅ 0.0001 days, and therefore, a change of the pulsation period is suspected because it is found to be about 5 times larger than the accuracy of measurements. Also, it is indicated in figure 6.6-3 that our residuals suggest the existence of a regular variation with a period of about 9 years. Therefore the present result would indicate the necessity of any theoretical consideration.

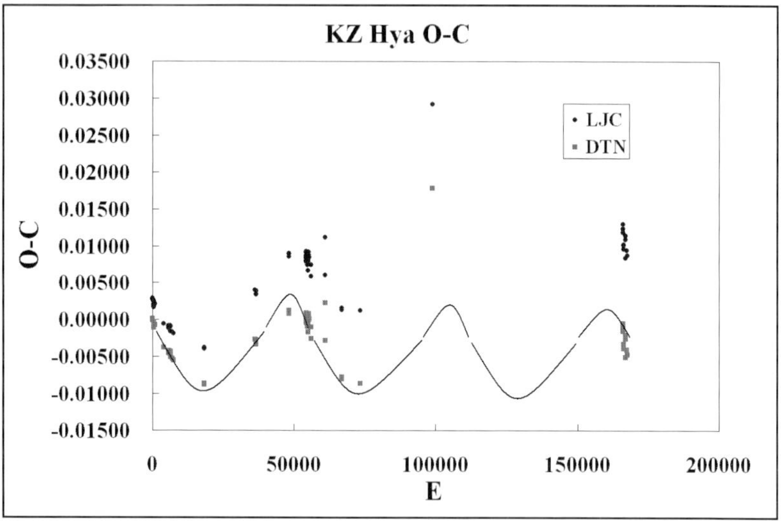

Figure 6.6-3. . O-C residuals versus epochs for KZ Hya

3. Two color Diagram

The *(B-V)* versus *(R-I)* color-color diagram from the present observations is shown in figure 6.6-4 for the smoothed mean color curve. During the rapid rise of bright-up phases and the following darkening phases the star appears to move anticlockwise on this diagram. The separation of the right–hand side and left-hand side around maximum parts are larger than that around the minimum parts. It is found from figure 6.6-4 that the curve at phases of 0.6, 0.7, 0.8 and that at phases of 0.3, 0.4, 0.5 seem to approach each other and eventually twist at the lower part, while the curve at bright up phases 0.9 to 0.0 the *R* magnitude rapidly goes up after 0.0 phase to 0.3 darkening.

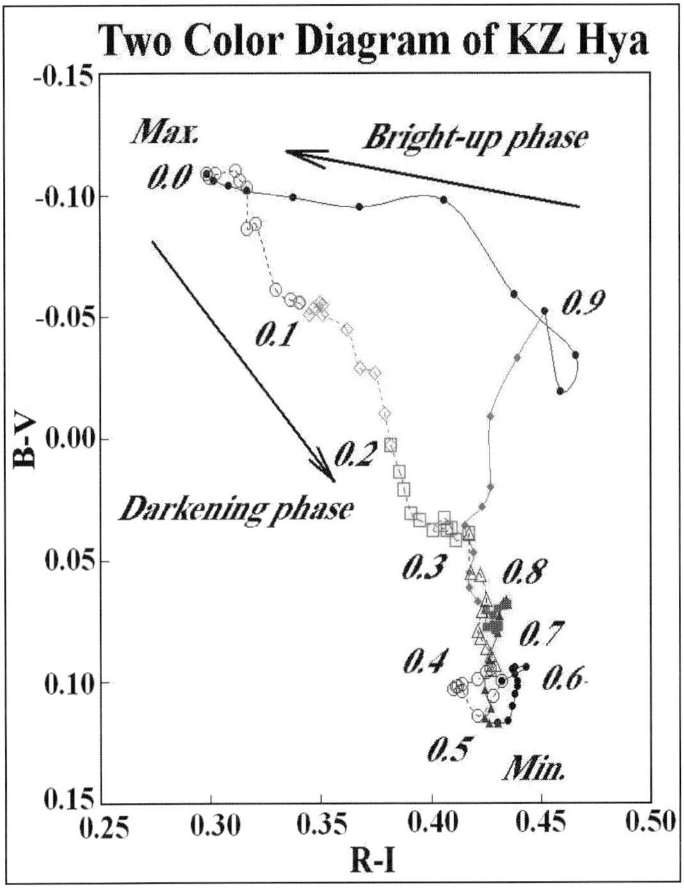

Figure 6.6-4. Two color diagram, B-V versus R-I.

4. Conclusion

This paper presents the first results of CCD photometry using the 45-cm telescope at Asuncion Observatory A new ephemeris has been obtained, and the result suggest a possible change of the pulsation period of KZ Hya.

Acknowledgements

We would like to express our hearty thank to Prof. M. Kitamura and Dr. M. Kondo of National Astronomical Observatory of Japan for their helpful discussions. Thanks are also due to Mr. H. Sato of the same Observatory for his kind help. In the present work, TN acknowledges the partial financial support from JICA (Japan International Cooperation Agency).

References

Doncel, F. et al., 2003, AP&SS, *in press* (DTN)
Hobart,M.A., Peniche, R. and Pena, J.H., 1985, Rev.Mexicana Astron. Astrof., **11**, 19.
Liu, Yanying, Jiang, Shiyang and Cao, Ming, 1991, Inf. Bull. Var.Stars No.**3606**. (LJC)
McNamara,D.H., and Budge, K.G., 1985, Publ.Astron.Soc.Pacific, **97**, 322.
Przybylski, A. and Bessell, M.S., 1979, Mon.Not.R.astr.Soc., **189**, 377-385.
Trocche, A., 2003 this volume.

SECTION:

7. PROJECTS FOR SPACE SCIENCE AND ASTRONOMY IN DEVELOPING COUNTRIES

Chapter 7.1

CHASING THE DREAM

J.R. Percy
University of Toronto
Frindale College
Ontario, Canada
jpercy@erin.utoronto.ca

Abstract A short summary is given, clarifying the reasons why astronomy and basic space science is of importance for all courtiers in the world. Some international efforts in this context are indicated.

Introduction

Why should we in the developed world care about the astronomical needs of developing countries? Certainly there may be tangible benefits if, for in-stance, these countries provide us with good observing sites. There are also many astronomy educators who will be happy only when everyone on Earth understands and appreciates astronomy. The best reason, however, is fairness: we should feel content only when people everywhere are as well off as we are.

Why should developing countries care about astronomy? An obvious benefit is educational. Astronomy is deeply rooted in almost every culture, and its allure can raise public interest in and understanding of science. Astronomy can attract young people into science or engineering, and a cadre of young people trained in science and technology benefits the rest of society. Astronomy captures the imagination and contributes to a sense of shared exploration and intellectual excitement. It shows us our place in time and space. As the mathematician Henri Poincaré said, "Astronomy is useful because it shows how small our bodies, how large our minds." "Astronomy is the stuff of dreams and youthful fascination," said Mazlan Othman of Malaysia " This is true for us in the developing countries as much as in countries like the U.S.A. Our youth are interested in astronomy and space as much as youth elsewhere."

W. Wamsteker et al. (eds.), Developing Basic Space Science World-Wide, 347-350.

Othman is a good example of a common phenomenon in the developing countries: the "lone astronomer"– a lone individual or, at most a small group, who facilitates astronomy at the university, school, amateur, and public levels, 1.2 while teaching mainstream disciplines such as math or physics as well. Their accomplishments are remarkable. What are their needs, and how can they be met? Globally the needs of astronomy are enormous. Nearly 100 countries have professional or amateur astronomical organizations. Only about 60 of these countries are sufficiently involved in astronomy to belong to the International Astronomical Union. Only about 20 countries, representing 15 percent of the world's population, have access to the full range of astronomical facilities and information. This does not include most of the Eastern European, Baltic, and former Soviet countries, whose fragile economies keep them from achieving their full potential, despite the excellence of their astronomical heritage and education.

Even in many developed countries, people feel their education systems desperately need more development. By this reckoning, at least 70 countries could use a hand to enable them to become full participants in the international astronomical community. A primary goal of the IAU is to develop astronomy through international cooperation. Conferences and publications allow astronomy educators to share their experiences, but most IAU programs address the specific needs of specific developing countries. This takes sensitivity: knowing the cultural background of scientists and educators in these countries, the type and level of their previous training, the facilities available to them, even the nature and dependability of their electrical supply.

1. For Want of Books

One important need is to travel to conferences and the developed countries. IAU Commission 38 helps young astronomers to study in the developed countries, and return home afterward. The IAU provides travel grants for participants who need them. IAU Commission 46 designs programs to promote interaction between astronomers from the developing countries and their colleagues elsewhere. There are also programs outside the IAU, such as the Vatican Observatory summer schools for graduate students, that provide full travel and expense support for students who need it.

Astronomers in developing countries also need books, journals, teaching materials, and data. Even when countries have sufficient hard currency to afford the astronomical prices of these materials, their postal services frequently delay or lose them. Several societies and publishers send copies of their journals to selected developing countries. Other organizations, such as the Canadian Astronomical Society, collect and ship surplus books and journals. The need must be identified, a donor found, and a channel of communication established. A contact person in the developing country must be found; otherwise, the material goes astray.

Translation of the foreign material can meet some needs. For example, dedicated local astronomers translate the Astronomical Society of the Pacific's (ASP) teachers' newsletter 'The Universe in the Classroom' into several languages.

As often as not, however, material suitable for teachers and the general public is written by the lone astronomer.

2. Equipment

Teaching and research equipment is needed to break the vicious circle of astronomy in developing countries. Without equipment, young astronomers cannot be trained locally; without trained astronomers, equipment will never be acquired. Japan has donated planetariums and small telescopes to several developing countries in and around Asia. Developed countries can also transfer surplus equipment. Think how many unused telescopes and planetariums there must be in the United States alone. Sending surplus books and journals, let alone a telescope, is not easy. But the rewards are great. A single small telescope at a public observatory can serve the teaching and research needs of professional and amateur astronomers, teachers, students, and the public. The Traveling Telescope is another solution. Traveling research-grade instruments such as photometers and CCD systems may be another.

With small telescopes, developing countries can fill important research niches. Simple, inexpensive instruments enable small telescopes to participate in multi-longitude campaigns on variable stars. The Whole Earth Telescope is a net-work of small telescopes in 10 countries, used for continuous monitoring of rapid variable stars. The French astronomer Francois R. Querci is setting up an Network of Oriental Robotic Telescopes (NORT) for similar purposes (Querci and Querci, 2003). Climatically under-privileged countries could host a crucial antenna in a radio interferometer or concentrate on theoretical studies.

What can one do to support astronomy in the developing countries? First, one should broaden his/her perspective by learning about astronomy worldwide and about the specific needs of the developing countries (Pasachff and Percy, 1990), (Percy, 1996), (Percy, 1996), (McNally, in press), (Batten, 1996). For example, the book, The Teaching of Astronomy, could be a good place to start.

Find out about programs to send surplus books, journals and equipment. If an individual astronomer or group of astronomers produce useful material, those persons should send copies to a few key developing countries. Locate and communicate with the lone astronomers. If a visit to a developing country is made, consider meeting with astronomers, teachers, and students; make arrangements ahead of time by consulting with Commission 46 or the lists of astronomical organizations produced by Strasbourg Observatory. Support the IAU and its programs. Undoubtedly, these efforts will be well rewarded.

References

Batten, A.H., 1996, Ed. "Enhancing Astronomical Research and Education in Developing Countries", Dordrecht: Kluwer Academic Publishers, in Highlights of Astronomy, **11B**, 883.

McNally, D., et al., 2003, Eds. "New Trends in Teaching Astronomy" (Cambridge, UK: Cambridge University Press), in press.

Pasachoff, J.M. and Percy, J.R., 1990, Eds. "The Teaching of Astronomy", Cambridge, UK: Cambridge University Press.

Percy, J.R., 1996, Ed. "Astronomy Education: Current Developments: Future Coordination", Astronomical Society of the Pacific Conference Series, **89**.

Percy, J.R., 1996, Ed. "Triennial National Reports on Astronomy Education" (Paris: International Astronomical Union). Available on the World Wide Web Site at: http://phys.open.ac.uk/IAU46/.

Querci, F.R., and Querci, M., 2003, this volume.

Chapter 7.2

PROVISION OF ASTRONOMICAL EQUIPMENT FOR DEVELOPING COUNTRIES THROUGH ODA OF THE GOVERNMENT OF JAPAN

M. Kitamura
National Astronomical Observatory
Osawa 2-21-1, Mitaka-City
Tokyo 181-8588, Japan

Abstract A summary of the ODA support (1986-2002) in Basic Space Science is given.

Introduction

In order to promote research and education, the Government of Japan started in 1982 to provide high-grade astronomical equipment to developing countries within the scheme of the Japanese Cultural Grant Aid (ODA). Until now, for astronomy, sixteen countries have successfully installed modern planetaria or 45 cm computer-controlled reflecting telescopes through this cultural grant aid (Kitamura, 1999).Currently, similar projects are being pursued for several other countries around the world.

In 1982, the Government of Japan started to provide scientific and educational equipment to developing countries. This programme is called the Japanese Cultural Grant Aid. The provision of astronomical instruments such as research telescopes and planetaria is part of this programme. The applications for aid need to be properly made through the Embassy of Japan in the respective country. The aid provided is not exclusively for astronomy and therefore, astronomical applications must compete with applications for other fields in a country that seeks support through the Government of Japan.

W. Wamsteker et al. (eds.), Developing Basic Space Science World-Wide, 351-353.

Until the end of 2002, sixteen countries have obtained astronomical equipment. This equipment comprises mainly planetaria and 45 cm high-grade reflecting telescopes.

The following lists those countries which received planetaria (1) and telescopes

(2), respectively (CGA: Cultural Grant Aid, GGA: General Grant Aid):

1. **Planetaria**

1986 Burma (Myanmar), Planetarium at Pagoda Cultural Centre (General Grant Aid: GGA)

1989 Jordan, Planetarium at Haya Cultural Centre (Cultural Grant Aid: CGA)

1989 Malaysia, Planetarium at the Space Science Education Centre (GGA)

1990 Philippines, Supplementary Projector for the already existing Planetarium at Manila (CGA)

1993 India, Planetarium at Burdwan University, West Bengal (CGA)

1993 Argentina, Supplementary Projector for the already existing Planetarium at Planetario de la Ciudad, Buenos Aires (CGA)

1994 Uruguay, Supplementary Project for the already existing Planetarium at Planetario de la Ciudad, Montevideo (CGA)

1998 Viet Nam, Planetarium at Cultural Memorial Hall, Vinh City (CGA)

1998 Thailand, Supplementary Project for the already existing Planetarium at Bangkok (CGA)

1998 Sri Lanka, Supplementary Projector for the already existing Planetarium at Colombo (CGA)

1999 India, Goto Planetarium (CGA) at Tamil Nadu Science and Technology Centres, Chennai

2001 Paraguay, Asuncion National Observatory

2002 Ecuador, Planetario Nacional, Cuena City

2003 Dominica, Plaza de la Cultura, Santo Domingo

2. **Reflecting Telescopes and Accessories**

1987 Singapore, 40 cm Reflector at Science Centre

1988 Indonesia, 45 cm Reflector at Bosscha Observatory of Institute of Technology of Bandung

1989 Thailand, 45 cm Reflector at Chulalongkorn University, Bangkok

1995 Sri Lanka, 45 cm Reflector at Arthur C. Clarke Institute for Modern Technologies, near Colombo

1999 Paraguay, 45 cm Reflector at Asuncion National University

2000 The Philippines, 45 cm Reflector at Philippine Atmospheric, Geophysical and Astronomical Services Administration (PAGASA), Quezon City

Provision of the above reflectors and accessories is based upon the Cultural Grant Aid only. The accessories include photoelectric photometers, spectrographs (or CCD set) and computers.

Acknowledgments

In the course of applications to the Cultural Grant Aid, the author would like to thank Mr. N. Jasentuliyana, Office for Outer Space Affairs of the United Nations, for their strong support and assistance. Thanks are due to Prof. T. Kogure, former Director of Bisei Observatory, for his kind collaboration.

References

Kitamura, M., 1999, in Proceedings of the *Conference on Space Sciences and Technology Application for National Developmen*t, Colombo, Sri Lanka, 21-22 January 1999 (Colombo: Ministry of Science and Technology of Sri Lanka), 147.

Chapter 7.3.

THE ASTRONOMY DIGITAL LIBRARY

G. Eichhorn
Harvard Smithsonian Center for Astrophysics
Cambridge, MA02138
USA
gei@head-cfa.cfa.harvard.edu

1. Urania

Astronomy has the most sophisticated on-line information access system of all the sciences (Boyce 1998). At its center is the Astrophysics Data System (ADS), a bibliographic system that provides access to the astronomy literature, as well as a large system of links to other on-line resources like electronic journals and data.

The idea that the Internet could be used to link sources of astronomical information into a unified environment is more than a decade old; it was fully expressed in the planning for the old ADS (Squibb & Cheung 1988) and ESIS (Adorf, et al. 1988) projects. These early attempts were highly data oriented, their initial goals were the interoperability of different distributed data archives, primarily of space mission data. Astronomical data are highly heterogeneous and complex; essentially every instrument has its quirks, and these must be known and dealt with to reduce and analyze the data. This quirky nature of our data essentially prevented the establishment of standardized tools for data access across data archives. The new, hyperlink connected network data system for astronomy is based on the highest level of data abstraction, object names and bibliographic articles, rather than the lowest, the actual observed data in archives. This change in the level of abstraction has permitted the creation of a system of extraordinary power. This new system, still unique amongst the sciences, has been dubbed Urania (Boyce 1996), for the muse of astronomy.

W. Wamsteker et al. (eds.), Developing Basic Space Science World-Wide, 355-361.

Conceptually the core of Urania is a distributed cross-indexed list which maintains a concordance of data available at different sites. The ADS maintains a list of sites which provide data organized on an article basis for every bibliographic entry in the ADS database. The CDS maintains a list of articles and positions on the sky for every object in the SIMBAD database. The CDS also provides a name to object resolver. The possibility for synergy in combining these two data systems is obvious; they have functioned jointly since 1993.

Surrounding this core, and tightly integrated with it, are many of the most important data resources in astronomy, including the ADS Abstract Service, SIMBAD, the fully electronic journals (currently ApJL, ApJ, ApJS, A&A, A&AS, AJ, PASP, MNRAS, New Astronomy, Nature, and Science), NED, CDS-Vizier, Goddard-ADC, and the ADS Article Service. All these groups actively exchange information with the Urania core, they point their users to it via hyperlinks, and they are pointed to by it.

The astronomy journals which are not yet fully electronic, in that they do not support hyperlinked access to the Urania core, also interact with the system. Typically they provide access to page images of the journal, either through PDF files, or bitmaps from the ADS Article Service, or both. Bibliographic information is routinely supplied to the ADS, and the SIMBAD librarians routinely include the articles (along with those of the electronic journals) in the SIMBAD object-article concordance.

While most data archives are not closely connected to the Urania system there are some exceptions. For example the National Center for Supercomputing Application's Astronomy Digital Image Library (Plante, Crutcher, & Sharpe 1996) connects with the ADS bibliographical data via links which are papers written about the data in the archive. SIMBAD connects with the High Energy Astrophysics Science Archive Research Center (HEASARC) (White 1992) archive using the position of an object as a search key, HEASARChas an interface which permits several archives to be simultaneously queried (McGlynn & White 1998), and a new data mining initiative between CDS and the European Southern Observatory (ESO) (Ortiz, et al. 1999) will connect the Vizier tables with the ESO archives. Several archives use the SIMBAD (and in some cases NED) name resolver to permit users to use object name as a proxy for position on the sky, the Space Telescope Science Institute (STScI) Digital Sky Survey (Postman 1996) would be an example. The Space Telescope-European Coordinating Facility archive (Murtagh 1995) allows ADS queries using the observing proposals as natural language queries, and the Principal Investigator names as authors. The Strasbourg Data Center (CDS) has long maintained several of the most important data services for astronomy (e.g. (Jung 1971); (Jung, Bischoff, & Ochsenbein 1973); (Genova et al. 1998)); access to parts of the CDS data via ADS is a key feature of Urania.

The development of the Urania system was strongly supported by NASA through the funding of the ADS and the data centers. Frank Giovane and Guenter Riegler at NASA were the major advocates for this system. Without their vision of a large scale information access system, Urania would not have been as successful as it is. The following describes the center part of Urania, the ADS in more detail.

2. The Astrophysics Data System (ADS)

2.1 History

The first major part of the ADS is the Abstract Service. It was started in 1993 with a custom-built networking software system to provide access to distributed data (Murray et al. 1992). By summer 1993 a connection had been made between the ADS and SIMBAD (Set of Identifications, Measurements and Bibliographies for Astronomical Data, (Egret et al. 1991)) at the Centre des Donn′ ees de Strasbourg (CDS), permitting users to combine natural language subject matter queries with astronomical object name queries (Grant, Kurtz & Eichhorn 1994). The user interface for this first version of the ADS was built with the custom-built software system that the ADS used at that time. The search engine of this first implementation used a commercial database system. A description of the system at that time is in (Eichhorn 1994).

By early 1994 The World Wide Web (WWW) had matured and was widely accessible through the NCSA Mosaic Web Browser. It now was possible to make the ADS Abstract Service available via a web forms interface; this was released in February 1994. Within five weeks of the initial WWW release use of the Abstract Service quadrupled (from 400 to 1600 users per month), and it has continued to rise ever since (Eichhorn 1997). The WWW interface to the ADS is described by (Eichhorn et al. 1995b) and (Eichhorn et al. 1995a).

The second major part of the ADS is the Article Service. It contains scanned full journal articles for most of the astronomical journal literature going back to volume 1 for most journals. The first full article bitmaps, which were of Astrophysical Journal Letters articles, were put on-line in December 1994 (Eichhorn et al. 1994). By the summer of 1995 the bitmaps were current and complete going back ten years. At that time the Electronic ApJ Letters (EApJL) (Boyce 1995) went on-line. From the start the ADS indexed the EApJL, and pointed to the electronic version.

Also from the beginning the reference section of the EApJL pointed (via WWW hyperlinks) to the ADS abstracts for papers referenced in the on-line articles.

With time, other interfaces to the abstracts and scanned articles were developed to provide other information providers the means to integrate ADS data into their system (Eichhorn et al. 1996b). With the adoption of the WWW user interface and the development of the custom-built search engine, the current version of the ADS Abstract Service was basically in place. Combined there are now over 3.3 million abstracts and bibliographic references in the system. The Astronomy Service is by far the most advanced, and accounts for

85% of all ADS use.

A detailed description of the ADS and the CDS has been published in a special issue of Astronomy & Astrophysics Supplements in April, 2000; (Overview: (Kurtz et al. 2000), Search Engine and User Interface: (Eichhorn et al. 2000), System Architecture: (Accomazzi et al. 2000), Data: (Grant et al. 2000)).

2.2 Data Holdings

2.2.1 Abstract Service

The ADS abstract service consists of four semi-autonomous (to the user) abstract services covering Astronomy/Planetary Sciences, Instrumentation, Physics, and ArXiv Preprints. The Astronomy database contains about 880,000 records, the Physics database has 1.4 million records, the Instrumentation database has 660,000 records and the ArXiv Preprint database has 230,000 records. About 60% of the records have abstracts, the rest are table of contents entries (title and authors).

Linking between from the ADS to other on-line information is a central part of the Digital Library in Astronomy. The ADS contains about 5.4 million links, roughly half of them to external resources, the other half to resources within the ADS. Maintaining and expanding this system of links is a major part of the ADS mission.

2.2.2 Article Service

With permission from most astronomical societies world-wide and some commercial publishers, the ADS has scanned a large part of the astronomical journal literature, as well as many conference proceedings. This part of the ADS provides access to full journal articles. We have currently scanned 1.9 million pages. These scanned articles can be retrieved in different formats (gif format for viewing on-screen, and PDF or Postscript for printing).

In collaboration with the libraries at the Harvard-Smithsonian Center for Astrophysics and at Harvard we have included scans of historical observatory publications from the 19th and early 20th century. Currently we have about 300,000 pages from these publications in the ADS. We expect to get another 500,000 pages of these publications by the end of 2003. This part of the astronomical literature is often difficult to access, especially from small colleges and from developing countries.

Having much of the astronomical literature available through the ADS makes it easy to access for anybody world-wide as long as World Wide Web access is available.

2.2.3 Reference and Citations

In addition to the abstracts and scanned articles, the ADS also provides reference lists and citation information. Currently there are 8.8 million references in the database. Some of them were extracted from scanned articles, some were purchased from ISI (Institute for Scientific Information), others were supplied to the ADS by the publishers directly. We currently have over 630,000 records with references, and over 840,000 records with citation information. These numbers are constantly increasing as we process more scanned articles, and receive more references from ISI and the publishers.

***Table** 7.3.-1.* **ADS Mirror Sites.**

Country	Mirror Site	URL
USA	Harvard-Smithsonian CfA, Cambridge, MA	http://ads.harvard.edu
France	Centre des Données Astronomiques de Strasbourg	http://cdsads.u-strasbg.fr
Japan	National Astronomical Observatory, Tokyo	http://ads.nao.ac.jp
Chile	Pontificia Universidad Catolica, Santiago	http://ads.astro.puc.cl
Germany	European Southern Observatory, Garching	http://esoads.eso.org
Great Britain	University of Nottingham, Nottingham	http://ukads.nottingham.ac.uk
China	Beijing Astronomical Observatory, Beijing	http://baoads.bao.ac.cn
India	Inter-University Centre for Astron. and Astroph., Pune	http://ads.iucaa.ernet.in
Russia	Institute of Astr., Russian Acad. of Scie., Moscow	http://ads.inasan.rssi.ru
Brazil	Observatorio Nacional, Rio de Janeiro	http://ads.on.br
Argentina	University of Cordoba, Cordoba	http://ads.unc.edu.ar
Korea	Korea Astronomy Observatory, Taejeon	http://ads.kao.re.kr

2.2.4 Usage Statistics

In a typical month (Jan 2003) it is used by more than 67,000 individuals, who make ~1.2 million queries, retrieve ~30 million bibliographic entries, read ~1.6 million abstracts, access ~300,000 articles, and download ~1.4 million pages. More than 10,000 users use the ADS regularly (10 times or more per month).

According to a survey of 34,000 of our users, 40% of the ADS users are scientists, 30% are students, 8% are teachers, and 7% are amateur astronomers. The rest are librarians, journalists, and other interested people. One interesting part of the access statistics is the type of queries that are done. About 2/3 of the queries are author queries. 20% of the queries use the title field, and another 20% use the abstract text field.

2.2.5 Mirror Sites

The ADS is mirrored world-wide at 12 sites. Table 1.1 shows the current mirror sites and their URLs. Setting up a mirror site is fairly easy. The hosting institution has to provide a server and an Internet connection. Such an abstract mirror site can now run on a Linux PC with 20 Gb of disk space. A partial article mirror site can run on as little as 80 Gb of disk space. If you are interested in having a mirror site, please contact Dr. Guenther Eichhorn at gei@cfa.harvard.edu for detailed requirements.

3. Conclusion

Astronomy has the most sophisticated and interconnected discipline centered information retrieval system of all the sciences. This has changed the way astronomers do their research. The existence of the ADShas provided large time savings for scientists, and it has made possible searches that were not possible before. The extension of the current system into the Virtual Observatory system will further enhance and expand these new capabilities and enable even larger productivity gains in the near future.

References

Accomazzi, A., et al., 2000, A&AS, **143**, 85.

Adorf, H.-M., Albrecht, R., Johnson. M.D., & Rampazzo, R. 1988, in Astronomy from Large Databases, Eds F. Murtagh and A. Heck, ESO Conf. Proc., **28**, 137.

Boyce, P. 1996, Bull.A.A.S, **28**, 1280.

Boyce, P. B. 1995, Bull. A.A.S., **27**, 1333.

Boyce, P. B. 1998, Library and Information Services in Astronomy, **III**, 107.

Egret, D, Wenger, M., Dubois, P., 1991, in "Databases & On-line Data in Astronomy", D. Egret & M. Albrecht, Eds, Kluwer Acad. Publ., 79.

Eichhorn, G, et al.,. 1995, Vistas in Astronomy, **39**, 217.
Eichhorn, G., 1994, Experimental Astronomy, **5**, 205.
Eichhorn, G., 1997, AP&SS., **247**, 189.
Eichhorn, G., et al., 1994, Bull.A.A.S., **26**, 1371.
Eichhorn, G., et al., 1996,:in "Astronomical Data Analysis Software and Systems V", ASP Conf. Ser. **101**, 569.
Eichhorn, G., et al., 2000, Astron.Astrophys. Suppl. Ser., **143**, 61.
Eichhorn, G., et al.,. 1995, in "Astronomical Data Analysis Software and Systems IV", ASP Conf. Ser., **77**, 28.
Genova, F., et al., 2000, Astron.Astrophys. Suppl. Ser., **143**, 111.
Grant, C. S., Kurtz, M. J.,& Eichhorn, G. 1994, , Bull.A.A.S., **26**, 899.
Jasniewicz, G., Lesteven, S., Ochsenbein, F. , Wenger, M. 1998 , in "Astronomical Data Analysis Software and Systems VII", ASP Conf. Ser., **145**:, 7, and 470.
Jung, J., 1971, Bull. Info. Centre de Données Stellaires, **1**, 2.
Jung, J., Bischoff, M., Ochsenbein, F., 1973, Bull. Info. Centre de Données Stellaires, **4**, 27.
Kurtz, M. J., G. , et al., 2000, Astron.Astrophys. Suppl. Ser., **143**, 41.
McGlynn, T. , White, N., 1998, in "Astronomical Data Analysis Software and Systems VII, ASP Conf. Ser.", **145**, 481.
Murray, S. S., et al.,. 1992, in "Astronomy from Large Databases II" , 387.
Murtagh, F. 1995, *http://www.eso.org/hst-prop-abs-search.html*.
Ortiz, P.F., Ochsenbein, F., Wicenec, A., & Albrecht, M., 1999, in "Astronomical Data Analysis Software and Systems VIII", ASP Conf. Ser., **172**:, 379.
Plante, R. L., Crutcher, R. M., & Sharpe, R. K. 1996, in "Astronomical Data Analysis Software and Systems V", ASP Conf. Ser., **101**, 581.
Postman, M. 1996, Distribution to the Astronomy Community of the Compressed Digitized Sky Survey, STScI Report.
Squibb, G.F., & Cheung, C.Y., 1988, in "Astronomy from Large Databases", Eds F. Murtagh and A. Heck, ESO Conf. Proc., **28**, 489.
White, N. E., 1992, in: Astronomical Data Analysis Software and Systems I, ASP Conf. Ser. **25**,52.

Chapter 7.4

THE CHANGING ROLE OF THE TELESCOPE IN ASTRONOMY

R. Albrecht*

Space Telescope European Coordinating Facility
European Southern Observatory
Garching, Germany

ralbrech@eso.org

Abstract This paper argues the point that not only telescopes are important items to conduct research in astronomy. Access to high quality astronomical data is already possible through astronomical science data archives. Virtual Observatories (VOs) will be the most important general purpose research tools in the not too distant future.

Introdution

Starting with Galileo Galilei at the beginning of the 17th century, the terms astronomy and telescope have been synonymous. The astronomer was the stargazer, working through the night, charting the heavens, discovering new objects and finding out new details about known objects. The human eye was the detector, with all the advantages of the built-in neural network based image processing, and with all the disadvantages inherent in a system with a psychological bias.

The quest for a permanent record for the purpose of comparison and classification quickly led to drawings produced right from the eyepiece. Still, the observer had to have the knowledge to evaluate the importance of the actual observation on the spot (Barnard's: "Mein Gott da ist ein Loch im Himmel!) and do his own record keeping as part of the observational process, all during the night, in the dome, at the telescope. When photographic techniques became available they were used for more permanent and more objective recording of the observations.

* Affiliated to the Space Telescope Operations Division, Research and Scientific Support Department, European Space Agency

W. Wamsteker et al. (eds.), Developing Basic Space Science World-Wide, 363-371.

Quantitative measurements became possible. This changed, but only by degree, when photoelectric photometry was introduced in the 1930's.

Only the measurements were performed at the telescope, the data analysis became desk work, occupying the astronomer during the day. This had already been the case with the photographic plates were the development was made during the day, but with electronic registration night work became only a fraction of the astronomer's total work. This trend has since increased. Incidentally, even though the photomultiplier tube was more sensitive that the photographic plate it did not contribute much to the efficiency of the telescope because it only observed one object at a time, discarding the signal from all the other objects in the field of view, but it had the great advantage of being essentially linear!

An enormous increase of efficiency occurred in the 1970es. On- Line A/D conversion allowed direct digital conversion of signals by computers. With these tools the next breakthrough came when CCD detectors were introduced. They were 30 to 50 times more sensitive than photographic plates, and, even though the first devices were not very large, they could observe several objects at the same time. However, this came at the expense of an increase of non-telescope related work. Among it was also computer-related work. It turned out that without computers to aid in the acquisition and in the analysis devices based on CCDs and other panoramic detectors just could not be operated. At the same time the telescopes became larger and more complex, again requiring electronic control devices, i.e. computers, to operate them.

1. The Collapse of the Wavelength Boundaries

As long as astronomical research was essentially being conducted by the astronomer working at, or close to, the telescope, observational astronomy was divided by sharp borders along the wavelengths, which the telescopes and the detectors were able to record. It was evident that the optical observer, working miracles with his telescope and plates, was not going to get involved in, for example, radio astronomy. This in turn was the domain of the wizard who could tweak the optimum gain out of amplifiers and antennas.

As observing facilities and their auxiliary instrumentation became more and more complex, the observers found themselves removed from the telescopes, at best operating them from command consoles through computer based interfaces. This was even more pronounced in the case of space borne instruments. Often the "observer" was not even in the control center when the data were taken, and even if they were, like in the case of IUE, the actual commanding was done by a telescope operator.

While this removed one of the romantic "joys" of doing astronomy, the fascination of being alone with the universe, and being connected to it through an astronomical telescope, it also had its advantages. The wavelength boundaries collapsed. The physics of the objects manifesting their characteristics across the full electromagnetic spectrum became the center of the interest of the researcher. This process started in the mid-1970es and took about 20 years to mature and reach a certain level of respectability in the astrophysical community. Today it is quite common to consider the emissions of, say, an active galactic nucleus across all wavelengths, from gamma rays to radio, in the effort to understand the physics.

2. Big Science

Until well into the 70's, and even the 80's, it was possible to conduct meaningful astronomical research with moderately sized telescopes. A 24-inch telescope, located at a good site and equipped with a good photometer, or, later, with a CCD camera was a formidable research tool. Without claiming that no meaningful work can be done at all with such small facilities, it is in general no longer the case. Whether we like it or not, operating inefficient telescopes at favorable sites may amount to a waste of valuable resources in terms of operational expense for the major observatories, which have consequently discontinued the support of small telescopes. On the other hand it is indeed the location of the major observatories which is driving this trend, since it is relatively expensive to operate an observatory at a mountaintop far from everywhere at high elevations. So this argument is not necessarily valid for observatories with moderate sized telescopes associated with institutes such as Universities and located at well chosen, but not remote sites.

Big science started in nuclear physics because of the obvious military and industrial implications. The impact (literally) of nuclear physics changed the role of the scientist, and at the same time prepared the tax paying public for the fact that science costs money. Big science in astronomy, while possibly having started with the Hale Telescope in 1948, had its roots in radio astronomy, where, owing to the wavelength of the electromagnetic emissions to be recorded, antennas of up to 100 meters in diameter were constructed in the sixties.

However, the real drivers were the space missions, which, regardless of the size of the telescope, require a substantial investment just for the launch. The most expensive space mission is The Hubble Space Telescope, but ground based facilities are not lagging far behind. The ESO VLT total cost over its life time will be in the range of the cost of a small space mission. Nations tend to be protective of their major investments of public money and therefore often it is not easy to obtain access to major facilities for scientists whose government did not contribute financially to the project. An attempt to overcome this issue – although with limited success - has been the organizational structure chosen for the Gemini telescopes.

Although a large amount of front line science in astronomy in today's world relies on telescopes of an aperture of at least 2 meters or more and on space missions, there remains a large volume of the knowledge space which needs to be addressed and can only be addressed with the type of facilities which I classify above as small telescopes. It is exactly this combination of the needs for small facilities and the availability of the Astronomical Virtual Observatories, which make the data from the major facilities available world-wide, which, if carefully chosen and properly applied, can present a wonderful opportunity for scientists in the developing world.

3. Astronomical Data Archives

The idea of data archives is as old as astronomy itself. Observations have always been diligently recorded and served as the basis of later discoveries. The best known case is Kepler, who used the observations which were done by Tycho Brahe to substantiate the laws which he derived. Large plate archives were assembled during the first half of the 20th century. The photographic plate, serving, at the same time, as the detector and as the storage medium could be used by different observers for different types of science. In actual fact, this was hardly ever done, as the plates usually did not find their way into an organized archive from which they could be distributed, and quite often ended up in the desk of the original observer. Also the complexity of the non-linear detection of the photographic plates contributed to this lack of multiple usage. The IAU Task Force for Preservation and Digitization of Photographic Plates is trying to assure that also historical photographic material will be available through the AVO (Griffin, 2002). A special place is taken in this all by the photographic surveys of the Palomar Observatory Sky Survey type, which produced plates of standard format, standard calibration, and which were copied en masse to serve the needs of astronomers.

This situation changed dramatically with the introduction of electronic detectors, together with the advances in digital storage technology and the associated decrease in price. It became easily possible to produce multiple copies of the data without loss of quality. The high cost of the space missions made it obvious that the data had to be exploited to the maximum possible degree. This led to the development of institutional calibration and of procedures and policies, which make it possible for scientists other than the original observer to obtain the data and to re-use them. The Internet revolution of the last 20 years, through which high bandwidth computer connections became easily accessible, finally made it feasible to get access to such data.

Today we have reached the stage that most publicly funded facilities operate science data archives from which data can be obtained with relative ease and at moderate cost. To give an impression of what is available: the archive at ESO holds, at the time of this writing, about 10 Terabytes of data. The HST archive alone contains almost 300,000 images and Spectra. Various other large science projects are being performed with the specific purpose to obtain statistical properties of large populations of specific objects, such as 2dF Galaxy Redshift Survey (Colless,1998 http://www.mso.anu.edu.au/2dFGRS) and the Sloan Digital Sky Survey (Stoughton,2002; http://www.sdss.org/sdss.html) which will have multiple use for the scientists. These will require archival facilities just as large science projects have created their data archives. A good example of such project for a general purpose observatory is the Hubble Deep Field, which was so successful that similar projects quickly followed. The broadest based project so far is the Great Observatories Origins Deep Survey (GOODS), which combines data taken with space facilities and with large ground based telescopes. The data are calibrated by the GOODS team, and they become public almost immediately. While the GOODS team, having designed the observations, has definite advantages, the sheer volume of the data is so large that many collaborators are required to optimally exploit the data (http://www.eso.org/science/goods/).

4. Virtual Observatories

The next logical step is to combine several data archives. Apart from the very practical advantage of not having to familiarize with the different access mechanisms there is also a scientific benefit: different data sets coming from such a combined archive can be easily compared.

The first step was the combination of the European HST archive and the archive of the ESO VLT (http://archive.eso.org/). This was a relatively easy step as the two archives were developed on the same hardware by the same people, so it was only logical to combine the access mechanisms. Trying to do this on a larger scale and combining the archives of different institutes and different projects is a really challenging task for very practical reasons. Not only is it necessary to overcome the differences in technology and conventions, which are not only caused by different developers who are reluctant to give up control over their brainchildren; quite often the reasons are much more mundane, like different generations of technology- hard to avoid as long as Moore's Law, which says that computing power will double every 18 months, still holds. In short, this is formidable task, as anybody who has ever tried it can attest to. Making archives inter operable also requires considerable network bandwidth, plus serious compute power and very advanced computer science concepts and techniques, which have only just become available at affordable prices.

Several initiatives have started recently. On the European side there is the Astrophysical Virtual Observatory (http://www.euro-vo.org/, Quinn, 2003), on the American side there is the National Virtual Observatory (http://www.us-vo.org/, Hanisch, 2003), both projects being collaborations of several leading astronomical institutions and the respective space agencies. There are also several efforts on a smaller, national scale, notably in Australia, China, Canada, Germany, Japan, Russia, and the United Kingdom. We are fortunate that all these projects talk to each other and coordinate their activities (*http://www.ivoa.net/).* This is being paralleled by a development which is happening outside of astronomy; the introduction of the GRID. The GRID will be the successor of the Internet (although the normal users will probably not realize because the transition will be quite smooth). From the point of view of the user, the data will be provided by the Virtual Observatory as a multidimensional hypercube. The planes of this cube will be populated on demand in real time by the different archives. The user can then define subsets of the hypercube, or specify cuts, or threads, through the hypercube to obtain astrophysically relevant information. For instance, a cut along wavelengths will yield the spectral energy distribution of the selected object. Threads along spatial and lambda axes will produce 2-D resolved spectra. Objects can be shifted along wavelength to compare their morphology with objects at different redshifts. Calibration of the data, rebinning to the desired pixel grid and registration of the data in world coordinates is done on the fly. Calibration will have to be performed at the archive sites of the different data sets, as is only there that high quality calibration is possible. Data analysis, on the other hand, can be done at the user site, given the fact that powerful data analysis systems are available in astronomy at essentially no cost. Computations which require considerable computing power will be done through the GRID, using resources which are made available for the purpose at participating institutes.

Tools which make this possible will be provided by the VO organizations. The range of tools spans from very computer science oriented (the so-called middleware which makes it possible for the different components to talk to each other and to the GRID), to astronomy-oriented, such as the user interface and the interactive work area, in which the user assembles the data and analyzes them synoptically. Collaborators can work "shoulder to shoulder", even though they are separated by large distances. That voice and video communication will be part of all this goes without saying, this will just be spin-off from today's web activities. All this will augment and incorporate services which are already available, like the services offered by the CDS and by the ADS (Eichhorn et al.,2002).

5. Consequences for Astronomy

We have, at this point, ten telescopes of an aperture of 8 meters or more. Deducting bad weather and maintenance they are producing about 3000 nights worth of data per year. We also have another 19 telescopes between 3.5 and 8 meters. In addition we have a variety of space missions, past and present, with archives that hold literally millions of high quality data frames.

The inevitable conclusion is that in astronomy we are, at this time, not suffering from a lack of data. We are suffering from a lack of astronomers to analyze the data. And here is an opportunity for astronomers in developing countries. Except for some extremely remote locations reasonably good Internet connections could be made available almost everywhere in the world. At the same time even powerful computers have become very inexpensive. And data analysis systems like IRAF are available for free to bona fide scientists. The following statement is admittedly controversial. I have discussed it with colleagues from many different countries, and while not everybody feels as strongly about it as I do, the statement is still basically correct: The most efficient way to do astronomy for a scientist in a country with limited resources is not to invest all available energy into building a telescope. The most efficient way to do astronomy for an astronomer in a developing country is to get a reasonably powerful computer, a reasonably good network connection, and to install IRAF. Then to pick a subject, which is of interest, get information and data through the net, and start to work. Once first results are available it is possible to contact other people working in the field, and in most cases they will be delighted to find colleagues who are willing to collaborate. As soon as the Virtual Observatories become fully available (which should be around 2004), access to their services should be requested. This can be done at various levels, either through a scientist-to-scientist collaboration, or through institutional arrangements.

While full access to the services of a virtual observatory would be an advantage it must be emphasized that it is not a prerequisite: cutting edge astronomy can be done today with computer equipment which is available and priced for the home market, and with software which comes for free.

6. What about a Telescope?

All the above does not mean that you should not get a telescope at all. For one, it is required for teaching; After all, students must learn how astronomers used to work for centuries, before the introduction of service observing and queue scheduling. They should experience the excitement of being in visual contact with the objects of our science, and they need to realize that at around 4 am observing loses a lot of its glamour. As mentioned above, carefully selected research programs can be meaningfully carried out and can not only provide strong motivation for students, but also lead to important first ranked results (e.g. Golap Krishna et al., 2003).

The telescope could be a commercial product and not a one-of-a- kind type of facility, which is impossible to maintain. The aperture should be in the range of 50 cm, which is large enough to capture a reasonable amount of light, yet small enough so the apparatus can be operated and maintained by one person working alone and without heavy equipment. There should be a CCD camera and a small spectrograph, again commercial products. Such combinations are available essentially off-the-shelf (Williams, 2002). The telescope should be located at a reasonably sheltered site, but not necessarily in a very remote location, which makes support and upkeep difficult and expensive. Beyond teaching and research, the telescope has an important role in conveying the science of astronomy to the public. This also means that the telescope has to look good on photographs and on TV. Whether we like it or not, we live in a world which is dominated by the commercial market, so the public is being inundated with glossy ads, with which we have to compete. We need to portray our science as it is: professional, intellectually exciting, and high-tech. The telescope should convey to the public that meaningful research is being done, and funds are required to do it. The public and the media should have access to the telescope through guided tours and open door days. In this manner the telescope will contribute to astronomical research: by helping to generate the support, which is required to do astronomical research and to attract young people to the excitement of the science of astronomy.

Acknowledgments

The author wishes to thank Dr. W. Wamsteker for his many helpful comments.

References

Colless M.M., 1999, Phil.Trans.Roy.Soc.Lond.A, **357**, 105

Eichhorn, G., et al., 2002, AP&SS, **282**, 340.

Gopal-K., et al., 2003, ApJ, **586**, L25.

Griffin, R.M.E.G. (2002). PDPP Newsletter No. 1

Quinn, P. (2003). Astrophysical Virtual Observatory. In: Proceedings of the ESO/ESA/NASA/NSF Conference "Towards an International Virtual Observatory", ESO Conference Series (Springer Verlag) In press.

Hanisch, B. (2003). The National Virtual Observatory. In: Proceedings of the ESO/ESA/NASA/NSFConference "Towards an International Virtual Observatory", ESO Conference Series (Springer Verlag) In press.

Stoughton, C., et al., 2002, AJ, **123**, 485

Williams, R., (2002). Planning Your OMI Observatory System, *http://www.opticalmechanics.com.*

Chapter 7.5

THE WORLD SPACE OBSERVATORY/ULTRAVIOLET (WSO/UV) PROJECT

W. Wamsteker and B.M. Shustov
ESA/VILSPA
P.O. Box 50727
28080 Madrid, Spain
willem.wamsteker@esa.int

Institute of Astronomy of the Russian Academy of Sciences (INASAN)
48 Pyatnitshaya Str.
109017 Moscow, Russia
bshustov@inasan.rssi.ru

Abstract The World Space Observatory project model was first introduced in the 5th UN/ESA workshop. Over time, the original concept has been further developed to evaluate how such project could become a reality. An assessment study has been made of the WSO-Ultraviolet (WSO/UV) as this was considered the best implementation model since the world-wide need for such project in the astrophysics community clearly exists. Also in the developing world considerable expertise exists through the wide distribution of the ultraviolet spectroscopic archive INES from ESA. Therefore WSO/UV presents the best chance of success for the next stage in the accelerated BSS evolution needed for sustainable development in the pre-industrial countries. The WSO/UV Implementation Committee (WIC), composed of scientists from some 14 countries, is the leading organization in this project. We will here describe the implementation model for WSO/UV and highlight the scientific importance as well as the plans for the future.

1. Introduction and Background

The original concept of the World Space Observatory (WSO) Model is described in the UN General Assembly Document A/AC.105/723. One of the assumptions associated with the idea of a WSO is to avoid the excessive complexity required for multipurpose missions, this

W. Wamsteker et al. (eds.), Developing Basic Space Science World-Wide, 373-387.

would directly make any meaningful participation of the developing countries very difficult. As the history of space science has shown, even the successful completion of a relatively simple space science mission is a complex activity, associated with various obvious risks (e.g. the Launch).

In the past many arguments have been used, beyond the prime scientific driver, in the selection processes for space science missions. Only a few space agencies implemented the limited number of successful Space Science missions. In the period before 1990 all missions were essentially new, so the selection process was to some extent driven by what was technically feasible (especially detector technology). In more recent times the advanced possibilities have made this process evolve into a fierce competition in *an essentially national-only context* for the available resources. Very focused mission goals are then required to justify the expenditures and control the costs. This type of reasoning does however essentially exclude the participation of scientists in developing countries from the experimental aspects of Basic Space Sciences. The often expressed intentions to solve this issue through a ¨small simple mission for the developing countries¨, although well intended, does not really address the issue. The motivation explored in A/C.105/732 has shown that there does exist a valid case to overcome this problem through the WSO concept. The first assessment for a World Space Observatory Project has been done for the ultraviolet domain: WSO/UV (Wamsteker and Bandecchi, 2000; Rodriguez-Pascual, 2000). The Ultraviolet (UV) wavelength domain ($\lambda\lambda$120 nm to $\lambda\lambda$320 nm) for WSO/UV as implementation model for WSO has been chosen for four reasons:

1. The UV domain is **only** accessible through observations with telescopes in orbit.
2. A UV observatory is clearly lacking in the tools available to the interested scientific community (see figure 7.5-1). As this community covers a range extending from all Astrophysics to Planetary Atmospheric and Surface science, a wide application range exists.
3. Considerable experience has been build up in the developing world, through the extensive use of archival data in the widely distributed UV spectroscopic Archive INES (Wamsteker et al., 2000) with local installations in some 22 countries.
4. Astronomy addresses fundamental questions which are of interest to all people of the world.

Before we proceed with the details of the implementation model for WSO/UV and the current status of the Project it might be useful to give short summary of the history and current status of observational UV Astrophysics.

In the early 1970's the first orbiting UV missions were launched with ESA's TD-1 UV photometric mission, NASA's Copernicus mission for high resolution spectroscopy in the UV and extending into the Lyman range, and the Astronomical Netherlands Satellite (ANS). From 1978 through 1996 most of the Astrophysics in the Ultraviolet wavelength domain has been done with the highly successful International Ultraviolet Explorer satellite IUE (e.g. Wamsteker and Gonzalez-Riestra, 1998), a joint project between NASA, ESA and PPARC. From 1984 through 1987 the Russian ASTRON mission was in orbit (Boyarchuk, 1994). In 1990 the ESA/NASA Hubble Space Telescope (HST) introduced a new capability for the UV community. At the time of writing the early instruments with UV capabilities (GHRS; WF/PC2; FOS and FOC) have been exchanged by astronauts for the Advanced Camera for Surveys (ACS), and the Space Telescope Imaging Spectrograph STIS (Woodgate and Kimble, 1999). One further future replacement by the Cosmic Origins Spectrograph COS (Green and Morse,1998), and WF/PC3, is still planned.

The interest of the scientific community in pointed observations in the UV has always been so high that a deep Sky Survey mission has had to wait for a long time. Finally, the first dedicated deep Ultraviolet Sky Survey (NASA´s GALEX mission) has been launched in April 2003. The results from this mission will generate an additional strong demand for an efficient follow-up UV spectroscopic instrumentation. The GALEX survey Catalogue (planned for release in 2006) is expected to contain some 1 million QSO's; 1000 Clusters of Galaxies; 300,000 White Dwarfs; 10,000 Cataclysmic Variables etc. (Bianchi et al., 2000). The majority of these will be newly recognized objects.

The increasing importance multi-$\lambda\lambda$ observations for the interpretation of the high energy astrophysics and the formation of stars and planets, requires the combination of the major ground-based 10-m class telescopes (see e.g. Green, 1999) with the capabilities of the current generation of powerful X-ray telescopes (CHANDRA and NEWTON_XMM) and the UV capabilities of a mission like WSO/UV.

In section 2, a summary of the Space Astrophysics plans of the major Space Agencies is given. In section 3 we give the background of the World Space Observatory concept. In section 4 we highlight the detailed capabilities of WSO/UV as presently agreed in the WIC. In section 5 the current status and implementation planning is indicated. We will here not address the scientific impact of the WSO/UV in detail. For reference to this we refer to the home page of WSO/UV at URL http://wso.vilspa.esa.es and other papers (e.g. Morse, Shull and Kinney, 1999 and Shustov and Wiebe, 2000). Other WSO/UV associated websites in a national context (language, specific national interest, etc.) are available for the following countries:

Argentina: http://www.fcaglp.unlp.edu.ar/wso/indexcas.html
Germany: http://astro.uni-tuebingen.de/groups/wso_uv/
Russia: http://www.inasan.rssi.ru/rus/WSO/
South Africa: http://www.saao.ac.za/space_science/WSO/
Spain: http://www.mat.ucm.es/~wso/
United Nations: http://www.seas.columbia.edu/~ah297/un-esa/wso.html

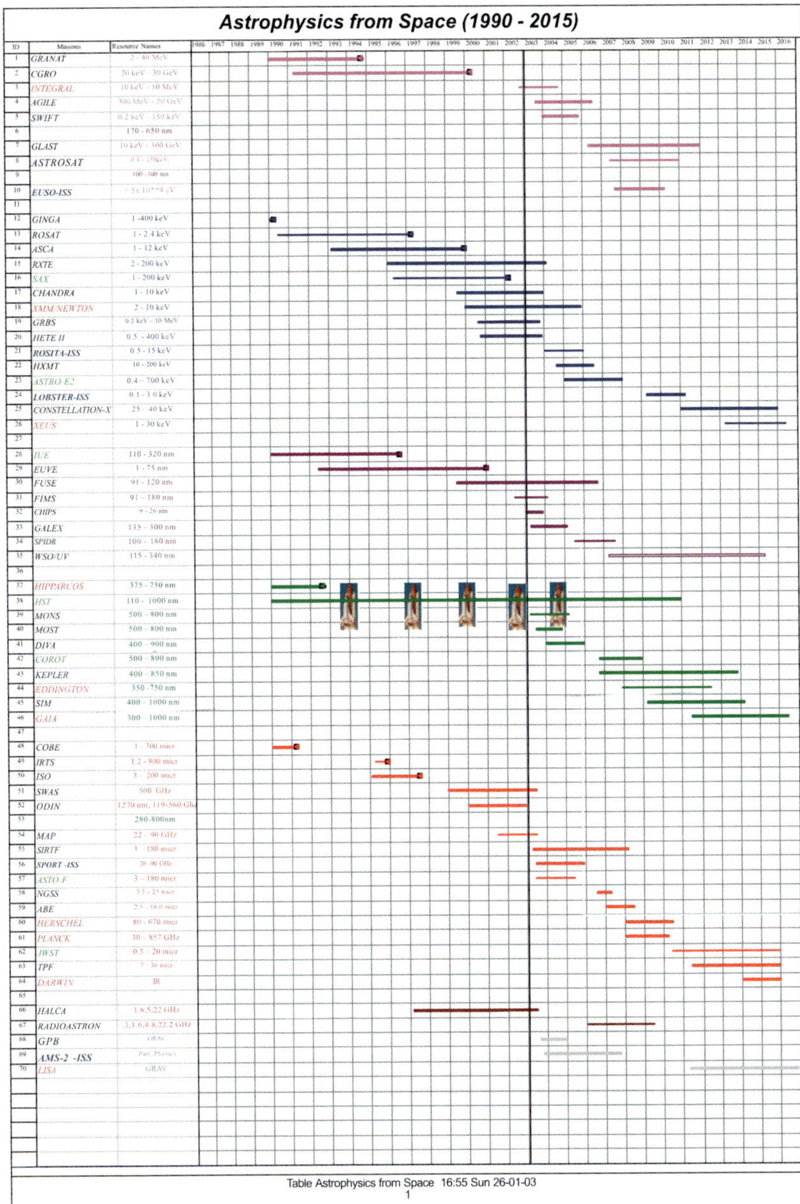

Figure 7.5-1 Space Astrophysics: A 25-year overview of Photon Astrophysics. The time of writing is indicated thick vertical line and the missions color-coded according to energy, wavelength or frequency domain and ordered by date of launch. The color coding is **pink** *for the MeV to GeV range*, **blue** *for the .1 keV to 500 keV range*, **violet** *for the UV range from 1 to 320 nm*, **green** *for the optical NIR range 320 to 1000 nm*), **red** *for the IR-mm range from 1 μm to 1000 μm*, and **brown** *for the radio domain in GHz*, and gray for fundamental physics. The shuttle symbols indicate manned missions with STS to service the HST.

2. Photon Astrophysics from Space

Astrophysical studies from space have been historically driven by the major Space Agencies from the United States, Europe, Russia, and Japan. The table (figure 7.5-1: *Astrophysics from Space 1990-2015*) shows the past and future Astrophysics missions in the quarter century spanning the millennium change.

Over the last part of the 20th Century, each Space Agency developed its own scientific program with priorities mostly driven by local academic and industrial capabilities and other interests. Collaborations in projects usually happened late in the project development, and were often driven mainly by the fact that the defined science missions were found to exceed the available funding. At that time an acceptable contribution was sought to be delivered by a potential partner. As the frequency of launch opportunities for science missions was relatively low, this supplied also a strong impetus for collaborations. The introduction of such collaborations in the later phases of the implementation of a mission - which can last as long as 10-15 years- often gave rise to new complexities through the engineering adjustments required, and the associated cost effects.

As the results of the missions over the past 30 years have been digested by the scientific communities, the needs for more sensitive instruments has become obvious and with this, a considerable growth in size and cost of the missions. An increase in sensitivity can normally only be obtained by three more or less unrelated properties:

- the telescope size,
- optical design and coatings,
- new detector development.

Each of these has significant impact on the mission costs, either through launch requirements or through the necessary costs associated with an industrial development program. The available launch capabilities represent a determining factor in mission size and associated weight. These developments and future plans of the space fairing nations in the Astrophysics are shown in figure 7.5-1. As indicated, this figure is arranged according photon energy. The WSO/UV is indicated in this overview as an open bar. This clearly illustrates the importance of WSO/UV in the context of the currently approved plans for future missions in Astrophysics by the major space agencies. It is interesting to note here that the large collecting area requirements of Astrophysics is already starting to show a change in the approach to fulfill the needs for observing facilities in Astrophysics. The future mission types fall into two categories:

a) very large missions exceeding even the capabilities of the strongest economies, clearly requiring international cooperation, or,
b) small missions with a very specific, narrowly defined, science goal, which could still be accomplished by a single agency. However, even these are, in recent times, found to require already bi- or trilateral agreements. Again here the lateness of this realization brings about its own additional problems.

The WSO concept claims to overcome these limitations and the WSO/UV mission has been designed

- To be in accordance with a realistic evaluation of the current world economy,
- To demonstrate the efficiency of the mission model, and
- To illustrate that, even purely scientific goals, can contribute in an efficient and effective way to the acceleration of global and sustainable development.
- To fulfill an important need for UV data in the scientific community.

3. WSO Concept and Purpose

History has shown that development in a socially peaceful environment is extremely difficult to achieve and that revolutionary changes, driven by intellectually advanced (and *at times extreme*) ideas, can easily become dominant. Therefore sustainable development in the modern world needs a culturally appropriate and sociologically complete development environment. This can only be accomplished when, apart from the essential educational processes and basic sustenance requirements, also *professional outlets* for those motivated to learning and development in a broad sense, are generated. For the post-industrial times this presents an important challenge to the world at large, and requires the creation of complex and fast information distribution capabilities, local mobility support and complexity support structures.

The strong democratization drive in the industrialized world is accompanied by an economic globalization, where regional and cultural identity plays an important, but not well recognized, role. The influence of cultural factors defy quantitative analysis, and the absence of proper consideration of those has been one of the main problems associated with the implementation of sustainable development programmes. It appears now clear that the activation of sustainable development schemes *will have to incorporate original and innovative approaches to the development process, where sharing must be an integral part of the collaborative efforts of all countries involved in the process*

The current development strategies in many developing countries, include a significant investment in education, which does not appear to bear the desired fruits, because of strong emigration pressures on the best educated individuals. One reason for this may be that participation in advanced science and technology can only function efficiently if also access to advanced investigation and innovation tools are accessible. Consequently, investment in education often results only in the creation of a consumer market, without the creation of the professionally well-formed, culturally and intellectually identifiable, and academically oriented cadre of scientists that is necessary for sustainable development. In hindsight, it is very clear that the success of the western industrial revolution was based on a fruitful interplay between the, relatively small, academic community and the commercial sector of the population.

Over the centuries, astronomy has played a major cultural role as the predecessor of all philosophical, scientific and technological development. This is because it uses scientific methods to approach a most fundamental question, basic to many religious as well as non-religious philosophical concepts:

What is the place of (the people of) planet Earth in the Universe?

During the United Nations/ESA workshops on Basic Space Science (Report, 1998), the concept of a World Space Observatory has been recognized as an important tool to bring about the necessary quantum leaps in development. The World Space Observatory embodies a twofold goal:

a) To create opportunities for all countries of the world, to participate in the frontiers of space science, on a sustainable basis and at the national level, *without the need for excessive investment.* In doing so, a WSO will make an important contribution to the development, in a multidisciplinary environment, of an academically mature and competitive cadre in many developing countries within 5 to 10 years after inception of the project by offering equal opportunities to highly trained scientists all over the world, independent of the development level of the country, sociologically and/or economically;
b) To support and stimulate worldwide collaboration and to assure that the study of the mysteries of the universe from space can be shared, in a sustainable way, by scientists from all countries in their own cultural environment.

This will not only maintain the curiosity-driven spirit of discovery that is an integral part of sustainable development, but also make a reality in the scientific world of the visionary principle that "***space is the province of all mankind***".

For scientific reasons upon which we will expand below, the choice for the assessment study of a World Space Observatory (Wamsteker and Bandecchi, 2000) has been made for the UV domain. One obvious reasons being, that the UV domain can only be reached from beyond the atmosphere and secondly, that already considerable expertise exists in the developing world for the associated astrophysical problems. This will incorporate a direct application of the "headstart" principle. Thus it is possible to benefit greatly from a new space observatory to be launched in the 2nd half of the first decade of the 21st Century. Considering the combination of the WSO concept and the needs of the astrophysics community for a dynamic Ultraviolet science program, the WSO/UV mission was conceived.

4. WSO/UV: Astrophysics of the Future

Mission concept WSO/UV

The driving principles behind the design of the mission are:

a) Operation of a 1-2 meter-class telescope in Earth orbit with a spectroscopic and imaging capacity specific to the ultraviolet domain;
b) High throughput and optimized operational and orbital efficiency;
c) Optimum benefit to be derived from the fact that ultraviolet cosmic background radiation is at a minimum around 200 nm;
d) Minimal operational costs without affecting the scientific excellence of mission products;
e) Direct access to a front-line facility for basic space science for the world-wide international astrophysics and planetary science community;
f) Limitation of the technological developments needed for the prime science mission;
g) Data distribution and data rights established as in UN A/AC 105.723 (1998).

4.1 The Instruments

Due to the heavy demand on technological innovation for a SUVO type mission (an 8m UV Space Telescope), there are no plans in the programs for any UV mission until well beyond 2010 (Weiler, 2000; Bonnet, 2000), as can be seen in figure 7.5-1. The earliest opportunity for such mission seems to be beyond the projected end-of-life of WSO/UV. Thus WSO/UV is important to maintain an active science and technology community for the UV, which otherwise is likely disappear due to the lack of intellectual and practical stimulus.

The ESA study (Wamsteker and Bandecchi, 2000) showed that a distributed implementation plan as foreseen for WSO/UV with a technology cut-off in 2002/3, represents a practical feasible mission

An independent review (JPL,2001) showed that the engineering requirements of the WSO/UV mission and associated risk evaluation, are well within the current capabilities of space technologies, with the decision not to incorporate technology which has not reached the TR Level 6 at the end of 2002. The specific mission model will allow a very fast implementation possibly with a launch in the 2007/8 time frame.

WSO/UV mission model

- Telescope: 1.7 m diameter (**T-170** *Russian model; IoA Acad. of Sciences* in Boyarchuk et al., 1998), PSF(550 nm) ≈ 0.2 arcsec
- Spectrograph for the UV only: primary 110 - 340 nm. (**HIRDES** *German model; IAAT* in Kappelmann, 1998) with spectral Resolution R≈5-6 x 10**4; as well as a low resolution capability R ≈ 1000.
- Imaging: 115 - 340 nm with quality ~ 0.1- 0.3 arcsec (**MCP based** *Israelian model; TAU* in Brosch, 1998); 2 UV Imagers: one for Max. spatial resolution; one for Max. sensitivity; and one Imager for visual domain
- UV Spectro-Polarimeter (under feasibility evaluation)
- High Earth Orbit (non-Halo orbit around L2).
- Distributed Mission Operations
- Science Operations fully distributed at level of Nations
- Overall properties of the mission as a whole as described in Report, 1998

4.2 The Science

We will below summarize some of the astrophysics areas for which the WSO/UV defined above will be capable to answer fundamental questions and contribute, in an essential way, to solve currently pending questions with respect to evolution of the Planetary System, Stars, Galaxies and the Large Scale Structure of the Universe.

In *Planetary System Science*, we consider the study of global atmospheric circulation and magnetospheric interaction. The gaseous planets present an excellent laboratory for the understanding of weather patterns less stochastic than the terrestrial patterns which are, in only partially understood ways, affected by human activities. The study of these undisturbed natural phenomena contributes not only to planetary studies per se, but will supply a better understanding of our own planetary environment. Of course a major contribution to the history of our Solar System will be associated with the study of Comets.

For *Stellar Science* the complete life cycle of stars can be studied with many new discoveries to be expected. The project could make major breakthroughs in stellar evolution as a consequence of the multiplicity in star systems, through the detailed studies of the effects of close binary mass exchange and accretion on condensed objects. Also the rapidly changing shock phenomena in Young Stellar Objects and the physical mechanisms driving jets in such objects, form an extremely exciting area of application of the WSO/UV. Also the spectroscopic studies in the ultraviolet of photometrically discovered exoplanet candidates presents an exciting field of studies for WSO/UV.

The studies of *Interstellar matter* and *Galaxy population* will allow the full evaluation of the cycling of Interstellar Matter and the subsequent chemical evolution of Galaxies. The capability to study such phenomena systematically over the full range from zero to high redshifts at the resolution supplied by WSO/UV, presents a very important contribution to these fields, and will connect the early Universe with the current epoch. The important questions and the ways in which these can be addressed are well illustrated in the recent investigations of stars associated with our Galaxy of Christlieb et al. (2002) and the results from the observation of the absorption line system in a Quasar at redshift z=2.7 by Prochaska et al. (2003).

Also *Cosmological questions* associated with the re-ionization phase are well within the capabilities of the WSO/UV. The general problem to establish the nature, location and time of the Galaxy formation will give important results for Galaxy evolution. The study of the Inter-Galactic Matter and its relation to Clusters of Galaxies and other sources of ionization will present a superb challenge for the capabilities of the WSO/UV.

The nature of Astrophysics is that many breakthroughs in the field are the consequence of *unforeseen or unpredicted occurrences.* We mention here discoveries of Comets and their behavior during their passage near the Sun, Novae, Super-and Hypernovae, γ-ray Bursts, OVV's and others, are strong components in the science addressed by the WSO/UV. It is specifically the rapid response capability foreseen in the mission which, together with the extended visibility periods, will present new opportunities and major challenges to the scientists. Of course, in all these fields the capability to make simultaneous multi-$\lambda\lambda$ observations of variable phenomena will add and an important extra dimension to all these studies.

5. WSO-UV Current Status

After the first introduction of a WSO concept at the 5th UN/ESA Workshop the ideas were further expanded by the participants in the 8th UN/ESA Workshop were an overall motivation and concept was developed (Report, 1998). The scientific aspects of these ideas have been broadly addressed in a joint NASA/ESA/PPARC Conference in Sevilla, Spain in late 1997 specifically dedicated to UV astrophysics (Wamsteker and Gonzalez-Riestra, 1998). Afterwards, the combination of the WSO concept and the UV requirements for Astrophysics to bring them together in the WSO/UV model was introduced (Wamsteker, 1999). To define a scientifically sensible model, a meeting was held between interested parties to supply the science requirements (Minutes, 1999). These were used as input for the assessment study made by ESA in the context of its long term planning (Wamsteker and Bandecchi, 2000).

Currently the structure of the project is centered around the WOS/UV Implementation Committee (WIC), in which the interested national scientific communities have centered their efforts through a National representative. The WIC has been constituted under Russian Chairmanship through the Russian Academy of Sciences (INASAN). At the national level Working groups (NWWG) have been formed with interested scientists as members. Each of these NWWG's works independently to obtain support at the national level for participation in the WSO/UV. The WIC acts as the project scientist for WSO/UV and follows and coordinates the presently ongoing phase A activities in various countries at a local level. The WIC members represent some 200 scientists, members of the NWWG´s.

The indications of interest from the scientific community have been presented to funding organizations at the national level in the following countries (WIC membership): Russia (Chair), Argentina, 5 Baltic/Nordic countries, China, Cuba, France, Germany, Israel/India, Italy, Mexico, the Netherlands, South Africa, Spain, United Kingdom, Ukraine, with ESA and UN as observers. The WOS/UV is also incorporated in the tasks of the IAU Working Group for Future Large Scale Projects. Phase A activities are ongoing currently in various countries (e.g. Jena-Optronik, 2001; see also minutes of the WIC Meetings http://wso.vilspa.esa.es Section: Minutes). An independent study (JPL, 2001) has confirmed the results in Wamsteker and Bandecchi (2000).

The WSO/UV mission should be really considered a mission with a "*window of opportunity*". This concept of opportunity is the occasion that at the current space and time a rather unique coherence is available in:

- the obvious need of the astrophysics community world wide for a UV facility, to address important questions, mainly associated with Planetary atmospheres, Star formation and the baryonic content of the universe at redshifts out to $z \leq 2$;
- absence of a prime UV observing facility in the major Agency planning;
- the progress made in many developing countries in observational astrophysics through the availability of both UV Archives and modern smaller telescopes;
- the advances made in the electronically driven communications;
- the benefits which can be generated by a major stimulus in the development of communications infrastructure in the developing world;
- the strong drive for democracy world wide which requires more complex structures on a national level, at the same time requiring better intercommunications between culturally different countries;
- the realization in many developed countries that future space science projects can only be implemented sensibly if early coordination can be organized, resulting in more opportunities and better utilization of the missions in orbit;
- the availability of a new view of the universe in the UV domain through the results of GALEX;
- The combination of all these factors makes that the conceptual aspects of a World Space Observatory can all be brought together in an effective implementation plan resulting in the Launch of WSO/UV in the 2007/2008 time frame.

The challenge is now to reach a mutual agreement in an equal partners environment, where the possible real contributions to the WSO/UV can be demonstrated to show the validity and time scales of the WSO/UV.

To make such **really global** space science activity with strong participation at the national level in a local environment, the preparations for a smooth implementation of WSO/UV are currently being developed under the supervision of the WIC. The direct challenge here is that a Phase A study by a *single agency* would not learn us anything new. A multinational Phase A study is required, and has to clarify the following important issues,

1. Develop an overall WSO/UV management model and legal structure for the WSO/UV implementation and operations.
2. Assure the concurrence of the capabilities of WSO/UV with the scientific requirements for Ultraviolet observations.

3. Definition at a national level where the participant's interest is for Hardware and Software development, fabrication and operations, and astrophysical importance.
4. Reconfirm the technical feasibility of WSO/UV.
5. Definition of total funding need for the WSO/UV project.
6. Definition of participant costing and obtain national funding.
7. Detailed definition of a Science programme for WSO/UV, technically consistent with the mission model.

When these issues have been solved and the necessary engineering documentation has been agreed, industrial activities can start, with a firm timeline. Once this process has been started, also those countries which have currently no space hardware capabilities can join the Project to prepare their direct participation in the operations and the exciting scientific exploitation of the World Space Observatory Ultraviolet. This will allow the national scientific communities to participate directly in front-line science without having to build up first the economic strength to carry the full load of major Space Science missions. Through this a major step forward will be made in the realization of some of the goals envisioned in the UN Outer Space Treaty.

References

A/AC.105/723, 1998, Report and recommendations 8th UN/ESA workshop Jordan, 14-34

Bianchi,L., and the GALEX Team, 2000, Mem. Soc. Astron. It., Vol. **70**, in press

Bonnet, R.M., 2000, Horizons 2000, ESA Space Science Plans,

Boyarchuk,A.A., 1994, Ed. "Astrophysical studies on-board Space Observatory "ASTRON", Nauka, Mocow, 416pp..

Boyarchuk,A.A., et al., 1998, in "UV Astrophysics beyond the IUE Final Archive", ESA **SP-413** ,809.

Brosch,N., 1998, in "UV Astrophysics beyond the IUE Final Archive", ESA **SP-413** ,789.

Christlieb,N.et al., 2002, Nature, **419**, 904.

Green, J.C., Morse,J.A., and COS Instr. Team, 1998, in "UV Astrophysics beyond the IUE Final Archive", ESA **SP-413**, 805.

Green R.F., 1999, in "UV-Optical Astronomy beyond HST", A.S.P. Conf. Ser. **164**, 346.

Jena-Optronik, 2001, HIRDES phase A Study Final report, HIRDES-FR-DJO-001

JPL, 2001, Advanced Projects Design Team "Assessment of World Space Observatory Proposal", ADF, CL# 01-1168.

Kappelmann,N., 1998, in "UV Astrophysics beyond the IUE Final Archive", ESA **SP-413**, 831. Minutes, 1999, Minutes WSO/UV Definition Technical meeting, 25-27 Nov. 1999, *http://wso.vilspa.esa.es/ (Meetings #1).*

Morse, Shull and Kinney, 1999

Prochaska, J.X., Howk, J.C., Wolfe, A.M., 2003, Nature, **423**, 57.

Rodriguez-Pascual, P.M., 2000, UE-CEES PX0116000 [**CDF-05 (C)**].

Shustov, B, Wiebe, D., 2000, Eds. Proceedings "Scientific prospects of the Spectrum-UV Ultraviolet Observatory" GEOS, Moscow, (in Russian).

Wamsteker et al., 2000, Ap&SS, **277**, 155.

Wamsteker, W., Bandecchi, M., 2000, WSO/UV Assessment Study, ESA **CDF-05 (A)**.
Wamsteker,W.,Gonzalez-Riestra, R., 1998, in "UV Astrophysics beyond the IUE Final Archive", ESA **SP-413** , 809.
Wamsteker, W., 1999, in "UV-Optical Astronomy beyond HST", A.S.P. Conf. Ser. **164**, 261.
Wamsteker, W., Gonzalez-Riestra, R., 1998, Panel Discussion WSO, 1998, in "UV Astrophysics beyond the IUE Final Archive", ESA **SP-413** , 849.
Weiler, E.J., 2000, Space Science Strategic Plan, NASA
Woodgate, B.E., Kimble,R.A., 1999, in "UV-Optical Astronomy beyond HST", A.S.P. Conf. Ser. **164**, 166.

SECTION:

8. ASTROPHYSICS FOR UNIVERSITY PHYSICS COURSES

Chapter 8.1

INTRODUCTION

D.G. Wentzel
University of Maryland
US College Park MD 20742 2421
USA
wentzel@astro.umd.edu

The problem

Chapters 8.1 to 8.9 respond to the question posed by many universities: "How can we introduce some astrophysics in our physics courses?" The question is often qualified: "We cannot teach a whole course in astrophysics."

This solution

In these Chapters I present an array of astrophysical problems, any one or a few of which can be selected and used within existing physics courses on elementary mechanics, or on heat and radiation, kinetic theory, electrical currents, and in some more advanced courses. Answers are provided to all problems.

These astrophysics problems are designed to be an interesting and challenging extension of existing physics courses, to test the student's understanding of physics by testing it in new realms, and to stretch the student's imagination. A brief tutorial on the astrophysics is provided with each problem, enough so that the physics professor can present the problem in class. The higher-level problems start with a brief introduction to the physics.

All the problems seek compact algebraic and numerical solutions that can easily be translated into physics. For many problems, the solution is shorter than the statement of the problem. A few mathematical proofs appear in Chapter 8.9.

W. Wamsteker et al. (eds.), Developing Basic Space Science World-Wide, 391-397.

The Physics

The problems on mechanics (Chapter 8.2 to 8.4) are nearly independent of each other, so that any one of them can be used in an appropriate introductory physics course. However, the seven problems in Chapter 8.2, "Mechanics - Orbits and Kepler's third law" can be used together as a mini-course on many of the interesting topics in modern astronomy and astrophysics, ranging from the solar system to black holes in galaxies.

The problems in Chapter 8.5 on thermal radiation can be selected independently of each other. The problems on the lives of stars (Chapter 8.6) are best done in sequence, starting with the introduction, but it is easy to stop at any point without reaching the (academically more challenging) end of the Chapter. It is quite possible to create a mini-course on the Sun by selecting the solar problems in Chapter 8.5 to 8.7.

The problems on the cosmic magnetic fields and high-energy astrophysics (Chapter 8.7 and 8.8) are academically more advanced. However, although their backgrounds involve Maxwell's equations and special relativity, the actual problems deal with quantities that are physically intuitive. A mini-course on neutron stars and pulsars can be constructed using problems in Chapters 8.4, 8.5, 8.7, and 8.8.

The Astrophysics

The format for all the problems is similar. Once the students have been introduced to the physics, the text outlines the astrophysical setting, which is normally to be given as part of lecture. Then time needs to be allowed for the physics problem to be done by the students. Finally, the interpretation of the results of the problem is to be given as part of lecture.

Astrophysics is an attractive science not only because it stretches the imagination but also because it is highly interdisciplinary. Astrophysics involves atomic physics, nuclear physics, fluid and plasma physics, solid state physics, chaos theory, organic chemistry, special and general relativity, and more. But students are trained in solving specific problems, and they acquire a broad view of science largely through solving many kinds of specific problems. Thus, the problems in this booklet provide a focus for the students to which the broader astrophysical challenges can be tied. Most of the text provided with each problem is designed to highlight the broader questions and challenges, which are then crystallized in the given specific problems that are to be solved by the students.

Astrophysics is a frontier science. Even students can formulate good questions suitable for research. Some observations made by the Hubble Space Telescope have been requested and will be investigated by high-school students. But the frontier nature of astrophysics makes teaching it difficult. Even the professional astrophysicist soon learns to admit to some students' questions: "I do not know", or preferably "I do not know, but perhaps we can try to figure it out together." Indeed, the problems in this booklet will be difficult to teach because students will inevitably ask questions that go well beyond the specific problem and the provided tutorial astrophysics. But the value of students formulating questions far exceeds the discomfort of the professor answering "I do not know". Many physics students merely memorize their physics. The astrophysics breaks them out of memorization and gets them to think independently. Students' questions are a sign of their progress.

Interpretation

Physics relies on a mixture of theory and experiments. In physics experiments, one controls the parameters such as temperature or imposed magnetic field. Astrophysics relies on observations that one cannot manipulate. Often the interesting astrophysics is based on observations that are just barely possible (though they might be easy to verify three years later with newer equipment). Observations must be qualified as to their accuracy and theories must be interpreted as to their plausibility.

A by now classical example of astrophysical discovery and interpretation is provided by quasars: Observations of quasars indicate an astronomically enormous energy output from an astronomically tiny source at an astronomically enormous distance. Is the energy perhaps derived from one star per year falling into a black hole that has already consumed a hundred million stars? At first this answer seemed so many orders of magnitude different from anything we knew that it seemed most unlikely. Yet it was the only available answer, and the problem seemed so compelling that scientists pursued it. It took twenty years' observations and their interpretations, but now this kind of answer is generally accepted. It is not proven, but generally accepted.

Astrophysics is often the first frontier science encountered by physics students. Therefore, physics students find it unsettling that one always needs to qualify astrophysical observations and theories. They must gradually learn the meaning of the various qualifying words. There is "compelling evidence for" Newton's laws of motion as observed in our daily lives (even though these laws are not correct in a highly relativistic setting). The source of quasar energy is "probably" gravitational; that means that major aspects of the observations fit, there is an approximate theory, but there are many observational and theoretical details unexplained. The direction of the rotational axis of Neptune "may" be due to a collision; that means there is no direct evidence for that collision (although major collisions clearly are and have been important in the solar system), and alternative explanations are likely (chaos theory?). The description of this manner of thinking, the interpretation and the judgments, the qualifiers such as probable, may be, or perhaps, these all are an essential part of teaching astrophysics, and of teaching any frontier science. The parts of the text labeled introduction and didactics aim to provide some of this description.

Didactics

How does one deal theoretically with a newly observed phenomenon? No, one does not start with a computer! One starts with the question: "What kinds of physics are relevant?" It is essential to select merely a very few physical parameters and construct a minimum of analytical equations that "contain the essential physics". These are often called "back of the envelope" calculations. In astrophysics, one first considers appropriate forms of energy without worrying, yet, about the detailed forces that lead to these energies. Are we dealing with gravitational, nuclear, kinetic, electromagnetic energies or some exchange between two of them? What are the main parameters such as size or mass of an object that influence these energies? Sometimes answers can be found by dimensional analysis. Never mind if the numerical coefficients in these estimates may be off by a factor of two or three. Several of the problems in these Chapters emphasize this kind of analysis. In particular, some problems ask students to solve differential equations by a one-step integration, which explicitly brings out the main physical parameters.

In all problems, the student is encouraged to think about the desired degree of accuracy. A few problems have as their main goal a discussion of likely errors of measurement. Any student who gives an answer to any problem accurate to four decimal places has totally missed the scientific process. Only after "back of the envelope" estimates make sense, and only after the relevant observational quantities are actually known to better than a factor of two or three, then one can think of solving a more complete problem either analytically or, more often these days, using a computer. Successful computing requires a thorough understanding of physics and astrophysics.

In all the physical sciences, students trained to ask "what are the important physics" and "what are the appropriate magnitudes" gain a much wider view of science than the students who can work out prescribed equations to three-digit accuracy. In fact, such a wide view of science is needed for scientists to help the development of their country effectively. Hopefully, students' study of the astrophysics in this booklet helps them toward this lofty goal.

Group collaborative learning

Frontier science is a collaborative venture. Discussion is an integral part of learning and research in astrophysics. If necessary, the problems in these Chapters can be presented and solved as part of a lecture, but they are selected and written so that they can be discussed and solved by small groups of students, preferably during a class period. Groups of 2, 3, or 4 students work well, depending in part on the physical limits of the seating arrangement.

The professor should introduce the physics and the astrophysical problem. Then the student groups should work on the problem, while the professor wanders about the classroom to check on the progress of the groups and asks helpful questions when students are stuck. When most groups are finished, one group should be asked to report the group's answer. Other groups may report that they have a different answer, which leads to discussion. Quite possibly the groups have selected alternative satisfactory paths to the solution. If necessary, finally, the professor should identify either the correct answer, or a reasonable answer for problems where students must estimate input quantities. This answer should then be interpreted in terms of astrophysics and the uncertainties in observational and theoretical quantities.

If students are not used to working in groups, perhaps the problem must be divided into smaller parts. The professor seeks a group's report and discussion of each part after most groups have finished that part. This also gives more groups a chance to contribute their reports. However, not every group needs to report for any given problem. A way to make sure that all students within a group participate is to identify one student who writes down the group answer, and another student who is ready to report to the class. Every class member should have these jobs at one time or another.

Students working in groups take much time. The professor can lecture about three problems in the time that student groups need to do one. Compared to pure lecture courses, some topics of the course must be omitted because the time is no longer available. But assuredly the students will understand the one problem they solved, and the professor will have evidence of it. This is much more useful to the student, in the long run, than some additional material stored incompletely in the students' memory.

Astronomical Numbers

It is amazing that our tiny brains comprehend objects such as stars, galaxies, and the cosmos. But "astronomical numbers" can be rather intimidating. Our understanding requires time! We all, including physics students, need time to assimilate astronomical distances and energies. Mere memorization does not help much. It is necessary to think about the astronomical objects for a while, to become familiar with them.

Most people assimilate astronomical distances best by progressing step by step, from objects we know to new (further or larger) objects; once we know them, we can progress to still newer (further or larger) objects. Therefore, the problems in Chapter 8.2, involving Kepler's third law, follow a progression of ever larger distances, and the problem in Chapters 8.3 and 8.4 follow such a progression at least approximately. The computational units in this book are mks units. But various other units appear because they are convenient, because they help to understand the physics without drowning in huge numerical exponents. Thus students must become used to units of km/s, Astronomical Units, light-years, and the solar mass, radius and luminosity. A table of frequently used values follows. For most problems, only the first significant digits will be needed.

Acknowledgment

I acknowledge extensive and detailed constructive comments on a draft of this booklet by Prof. Francesco Zaratti, Planetario Max Schreier, La Paz, Bolivia, and by Prof. John Wang, University of Maryland, USA.

Numerical values

$G = 6.7\text{x}10^{-11}$ N m^2 kg^{-2}	gravitational constant
$k = 1.4\text{x}10^{-23}$ j/K	Boltzmann constant
$c = 3.0\text{x}10^{8}$ m/s	speed of light
$h = 6.6\text{x}10^{-34}$ j s	Planck constant
$\sigma = 5.7\text{x}10^{-8}$ w m^{-2} K^{-4}	Stefan-Boltzmann constant
$\sigma_T = 6.7\text{x}10^{-29}$ m	Thomson cross section (sometimes s = collision cross section)
$m_p = 1.7\text{x}10^{-27}$ kg	mass of proton
$m_e = 0.9\text{x}10^{-30}$ kg	mass of electron
A.U. = $1.5\text{x}10^{11}$ m	Astronomical Unit = mean Sun-Earth distance
yr = $3.2\text{x}10^{7}$ s	one year = period of Earth's orbit around Sun seen from afar

$ly = 9.5x10^{15}$ m	Light-year = distance traveled in one year at speed c
$M_o = 2.0x10^{30}$ kg	mass of Sun
$L_o = 3.9x10^{26}$ w	luminosity of Sun
$R_o = 7.0x10^{8}$ m	radius of Sun
$M_E = 6.0x10^{24}$ kg	mass of Earth
$R_E = 6.4x10^{6}$ m	radius of Earth
1 radian = $2.1x10^{5}$ arcsec	units of angle
$\mu = 4\pi \text{ x } 10^{-7}$	vacuum magnetic permeability
$eV = 1.6x10^{-19}$ j	electron Volt (KeV = 10^3 eV, MeV = 10^6 eV, GeV = 10^9 eV)

Relevant books (in English)

Two textbooks that have provided much inspiration for this booklet are:

1. M. Zeilik and S. A. Gregory, Introductory Astronomy and Astrophysics, 4th edition, 1998, Saunders College Publishing, ISBN 0-03-006228-4; 515 pages plus a prelude introducing the physics, at the end appendices and glossary, about $60 in the USA. This book is used in the USA for students in engineering, physics, etc. in the 2nd of 4 years toward a B.Sc.
2. B. W. Carroll and D. A. Ostlie, An Introduction to Modern Astrophysics, 1996, Addison-Wesley Publ. Co., ISBN 0-201-54730-9; 1325 pages, plus appendices including a stellar-structure computer code, about $90 in the USA. An exhaustive treatment, but emphasizing those physics that can be done with relatively simple computations. Portions of the book can be selected for individual courses. The academic level is appropriate to physics students in the USA in the fourth year leading to a B.Sc. and to students studying for an M.Sc. in astrophysics.

A more advanced book organized around the physics of astrophysics is:

3. Martin Harwit, Astrophysical Concepts, 2nd edition 1988, Springer Verlag, ISBN 0-387-96683-8; about $60 in the USA.

Chapter 8.2

MECHANICS: ORBITS AND KEPLER'S THIRD LAW

D.G. Wentzel

1. The first human voyage to Mars - Orbital periods

Physics needed:

Kepler's first and third laws. Chapter 8.9 gives a brief derivation of Kepler's first and third laws, starting from constant energy and constant angular momentum.

The Problem:

Calculate, using Kepler's third law, the time needed for a spacecraft to travel from Earth to Mars, stay on Mars a while, and then return to Earth. Assume circular orbits for both Earth and Mars and a spacecraft orbit that just touches the two planet orbits. Given: the radii of the planet's orbits are 1 and 1.6 A.U. respectively. Round off all periods to simple fractions of a year. Sketch the planetary positions at the time of launch and at the time of arrival at Mars, together with the spacecraft orbit from Earth to Mars, and again for the return trip.

The Setting:

In 1969 and for a few years thereafter, humans visited the Moon. The next major goal for human travel is Mars, perhaps by the year 2020. Even now, Mars is being explored intensively, by computer-directed spacecraft, because of indirect (and debatable) evidence that some form of primitive life may have occurred on Mars long ago.

To send humans to Mars is technically much more complicated. A massive spacecraft will be needed to assure human survival. Its launch into an orbit to Mars will need the most powerful rockets available. Given likely rocket power, even twenty years from now, an orbit from Earth to Mars will have to be chosen to minimize the required energy. With the approximation that the orbits of Earth and Mars are circles,

W. Wamsteker et al. (eds.), Developing Basic Space Science World-Wide, 399-421.

which is adequate for our purposes, the minimum-energy orbit is an ellipse (practically a circle) that just touches the orbits of Earth and Mars. To visit and stop at Mars, the orbit will consist of two half ellipses. The spacecraft will be launched from Earth, perform half an orbit in traveling from Earth to Mars, land on Mars, later be launched from Mars, perform half an orbit, and land on Earth. It must stay on Mars long enough so that the Earth will be in the right position for the spacecraft to meet Earth. The question is: how long will the astronauts be absent from Earth?.

The Solution:

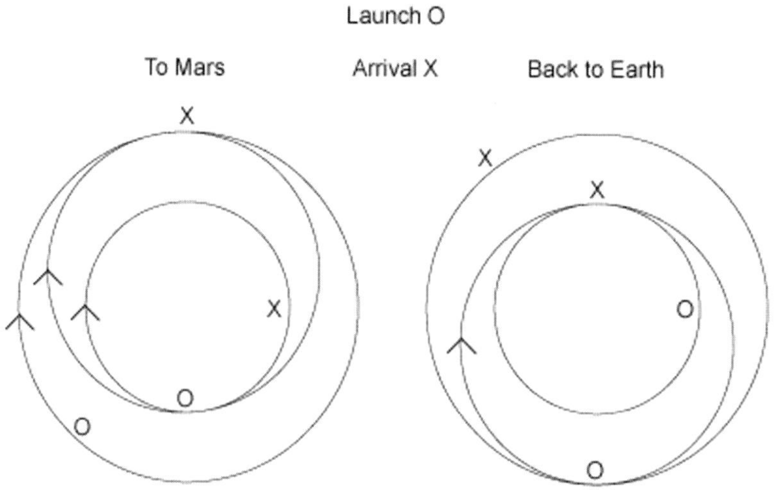

Figure 8.2-1. The semi-major axis of the spacecraft's orbit = (1 + 1.6)/2 = 1.3 A.U.

From Kepler's third law, the period of the spacecraft orbit is 3/2 years. The trip from Earth to Mars takes 3/4 year = 9 months. Now the period of Mars' orbit, again from Kepler's third law, is 2 years. In 3/4 year, Mars moves 3/8 of a full circle, while the spacecraft moves 1/2 of a full circle. Therefore, we must launch the spacecraft from Earth when Mars is 1/8 of a full circle "ahead" of Earth. By the time the spacecraft reaches Mars, Earth will be 1/4 of a full circle "ahead" of Mars and the spacecraft.

The explorers must stay on Mars until, at the time of launch from Mars, Earth is 1/4 of a full circle "behind" Mars, such that the spacecraft can meet Earth 3/4 year later. While the explorers are on Mars, Earth must progress by half a full circle relative to Mars. Since Earth travels a full circle in one year, Mars only 1/2 a circle in one year, the Earth progresses by 1/2 of a full circle relative to Mars in 1 year. Therefore, the astronauts must stay on Mars for 1 year (1/2 of a Martian "year"). The return trip takes 3/4 year. The whole trip takes 2 and 1/2 years.

Interpretation:

The difficulty of this trip resides not only in the engineering. 3/2 years is a long time for humans in space. They will be exposed to cosmic rays and energetic particles from solar flares. So far, we do not know how to assure the health of astronauts in space for so long a time.

The experiences in the Russian space station Mir have begun to solve this question. The explorers will also be exposed to cosmic and solar particles while on the surface of Mars, but they may spend most of their time on Mars underground.

For space probes without humans, it is no longer necessary to use the minimum-energy orbit. Moreover, if there is enough time, spacecraft may obtain extra needed energy by suitably passing near any convenient planet (see problem 8.3.2).

Didactics:

1. This is an excellent problem for students to solve in small groups. The professor needs to walk among the groups, check their progress (especially by checking their diagrams), watch for preconceptions such as item 3) below, and ask questions that will get the groups to progress when stuck. Afterwards, or possibly part way through, (the appointed reporter from) different groups should be asked to report their results. Insist on students drawing the diagrams. Diagrams focus the human mind remarkably well.
2. Depending on class preparation, students should review, again in small groups, the physics leading to Kepler's third law for circular orbits around the Sun. The force balance is given by $mv^2/r = GmM_0/r^2$ where M_0 is the Sun's mass, $M_0 \gg m$. Together with $v = 2\pi r/P$ where P is the period of the orbit, the relation yields $r^3/P^2 = GM_0/4\pi^2$. Conventionally, the right side is set equal to unity upon expressing r in the unit of the semi-major axis of Earth's orbit, 1 Astronomical Unit (A.U.) = 1.5×10^8km, and expressing P in units of 1 year = 3×10^7s. The derivation of Kepler's third law for elliptical orbits (see Chapter 8.9) does not provide any useful additional physical insight.
3. Watch for a preconception: When they are asked to draw an elliptical orbit connecting Earth and Mars orbits, many students will draw an ellipse like this: If they do, ask them to figure out why the sketch is wrong. (see figure 8.2-2: Sun not at focus).
4. Do not allow your students to calculate a semi-major axis or the number of seconds in a year accurate to 3 or 4 decimal places, because the approximations already made (such as taking Mars' orbit to be circular) cause errors even in the second decimal. A sense of

awareness of possible errors must be reflected in the number of decimals one uses.

5. If we insert into Kepler's third law $r^3/P^2 = GM_o/4\pi^2$ the Earth's values r = 1 A.U. = $1.5x10^8$km and P = 1 year = $3x10^7$s, and given G = $6.67x10^{-11}m^3s^{-2}kg^{-1}$ [measured in the laboratory by the Cavendish experiment], then we deduce the mass of the Sun, $M_o = 2.0x10^{30}$ kg. How do we know that 1 A.U. = $1.5x10^8$ km? We can measure the distance to a planet like Mars or Mercury by radar : we measure the time for the radar signal to return to us. (Historically, the distance of Mars was first derived by triangulation.) We know that same distance expressed in A.U. from Kepler's laws of planetary motion.
6. Possible students' problem: Consider a human visit to the largest asteroid Ceres, which travels on a (roughly) circular orbit with r = 2.7 A.U. The asteroids, with sizes of our Moon or less, have low surface gravity and are easier to land on than is Mars. Some asteroids may provide opportunities for mining. To simplify this problem, ignore the motion of Ceres during the time while the explorers are on Ceres. [Answer : The orbital periods of the spacecraft and Ceres are about 2.5 and 4.4 years, respectively. Ignoring Ceres' motion means that Earth must be in the same place at launch and arrival, so the whole trip takes the minimum number of whole years, that is 3 years. Therefore, the humans must spend 0.5 years on Ceres. It is now possible to compute the problem in the next approximation : while the humans are on Ceres, Ceres moves about 1/9 of a circle, and so the humans must wait 1/9 of a year longer while the Earth moves into position, etc.]
7. Possible students' problem: Find the mass of the Earth, given the lunar distance $4x10^5$km and lunar orbital period (as seen by a very distant observer) of 27 days. Answer: $6x10^{24}$ kg.

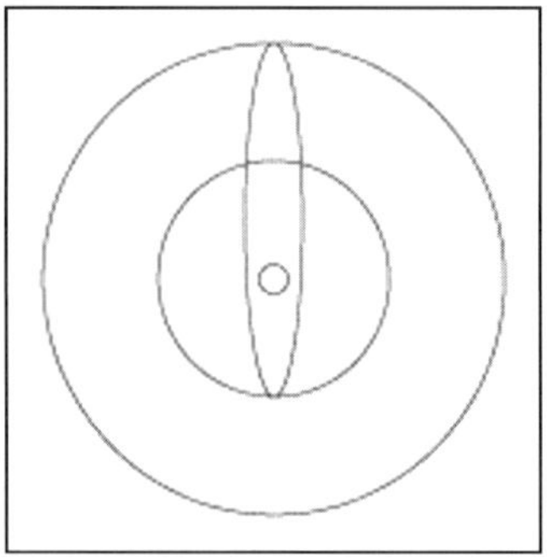

Figure 8.2-2 See Didactics 3

2. The first human voyage to Mars - Energy equation

Physics needed: Energy equation for orbits around the Sun. Chapter 8.9 gives a short derivation starting from constant angular momentum and constant total energy.

The Problem: Using the energy equation for a Keplerian orbit around the Sun, evaluate v^2 in $(m/s)^2$ for the Earth (assuming a circular orbit at r = 1 A.U., and for the minimum-energy orbit from Earth to Mars when the spacecraft is at Earth's orbit (but free of Earth's gravity) r = 1 A.U.. Evaluate the difference in v^2. Compare to the value of v^2 needed to escape the Earth's gravity. Which takes less energy, leaving the Earth's gravity or changing from circular to elliptical orbit around the Sun?

Also evaluate the difference in v^2 between the space probe in its minimum-energy orbit arriving at Mars (r = 1.6 A.U.) and in its circular orbit with Mars (but free of Mars' gravity) at r = 1.6 A.U. Compare to the v^2 needed to escape Mars' gravity (mass of Mars = 0.11 Earth's mass, radius of Mars = 0.53 Earth's radius).

The Setting: The spacecraft taking humans to Mars and back to Earth on a minimum-energy orbit (tangent to the orbits of Earth and Mars, both assumed circular) must use rocket propulsion to escape Earth's gravity, then to accelerate to acquire the minimum-energy orbit to Mars, later to accelerate to match the speed of Mars, and finally to reduce the speed of falling onto Mars; still later, rocket propulsion is again needed to escape Mars' gravity, to slow down to acquire the minimum-energy orbit to Earth, to slow down to match the speed of Earth, and finally to reduce the speed of falling to Earth. This problem shows that all of these maneuvers require significant energies. They are compared to the energy needed to escape Earth's gravity because most students have seen pictures of major rocket launchings and have at least an intuitive feeling that this energy is quite large.

The Solution: The relevant formula for the velocity of a planet or smaller object in an orbit around the Sun, with semi-major axis a and distance r from the Sun, is $v^2 = GM_o[(2/r) - (1/a)]$. For Earth orbit, $v^2 = 8.9\text{x}10^8\ (m/s)^2$, for the spacecraft at Earth orbit, r = 1 and a = 1.3, v^2 is 1.23 times larger, so the extra v^2 that must be imparted is $v^2 = 2.0\ \text{x}10^8\ (m/s)^2$. For comparison, the value of v^2 needed to escape Earth's gravity, is $v^2 = 2GM_E/r_E = 1.25\text{x}10^8\ (m/s)^2$. Conclusion: The energy needed for the spacecraft to escape the Earth's gravity is less than the energy needed, after escaping from Earth's gravity, to place the craft into the correct orbit around the Sun.

At Mars, the energy required to change from the elliptical orbit to the circular orbit moving with Mars is given by $v^2 = (5.5\text{-}4.2)\text{x}10^8\ (m/s)^2 =$

1.3×10^8 (m/s) ^{2}, once again a substantial energy. The energy to escape from Mars is $v^2 = 0.25\times10^8$ (m/s) ^{2}.

Interpretation: The energies needed to change orbit are quite comparable to the energies needed to escape Earth's gravity. The energies are near the limits of technical capability if the spacecraft is heavy enough to carry humans safely to Mars and back. The use of v^2 has in effect focused on the energies needed by the unfueled spacecraft occupied by humans. Additional energy is needed for fuel, which is reduced every time the rockets are fired, and possibly discardable rocket stages.

For space probes without humans, it is no longer necessary to use the minimum-energy orbit. For instance, the spacecraft Ulysses reached Jupiter on a much shorter orbit. It then gained energy while flying past Jupiter (see problem 8.3.2), enough so as to fly over the south pole and then over the north pole of the Sun.

Didactics:

1. The energy equation is an appropriate topic for lecture. However, then the students can solve the problem in groups, with different students serving as recorders and reporters than for problem 8.2.1. If one group reports some part of the problem incorrectly, let other student groups recognize the mistake. The professor need only make sure that the incorrect group figures out what they did not understand correctly.
2. The energy equation simply means: kinetic energy plus gravitational energy equals a negative constant, the total energy. The derivation in Chapter 8.9 merely identifies the constant. Check the constant: It yields the correct result for a circular orbit, $r = a$.
3. Students may be tempted to evaluate v(final)-v(initial) during the time the rocket operates and then square that difference. They will get different answers because they are implicitly placing themselves into different frames of reference. Energy is frame-dependent. All the energies of the spacecraft and planet orbits are evaluated in one frame, stationary with respect to some distant star.
4. Check that your class realizes: no rocket power is needed while in orbit to Mars. Rocket power is used briefly to provide the necessary energy, but then the spacecraft moves without effort along its new Keplerian orbit, until the rockets must again be used briefly upon reaching Mars.
5. Check that your class understands the minimum-energy orbit qualitatively: Once free of Earth's gravity, the spacecraft is in a circular orbit around the Sun. The spacecraft must be accelerated so that it "overshoots" Earth's orbit to reach further out. Once at Mars, it must again be accelerated to achieve a circular orbit at the distance of

Mars. If it were not accelerated, it would continue on its elliptical orbit and begin to fall back toward r = 1 A.U.. You can check that the class understands this : Ask for equivalent arguments about the decelerations on the way back from Mars to Earth. (At Mars, the spacecraft must be decelerated in order to "undershoot" Mars' orbit and begin to fall toward r = 1 A.U.; at Earth, the spacecraft must again be decelerated because otherwise the craft would start traveling outwards again.)

6. Once again, students should work out this problem using no more than two significant digits. The differences in v^2 computed here are very sensitive to the approximations made, especially the approximation that Mars moves in a circular orbit.

3. Planets around stars other than the Sun

Orbits in center-of-mass frame, Doppler shift

Physics needed: Two-body problem, for circular orbits. Doppler shift only for interpretation.

The problem: A star is observed to change its radial velocity, $v^*(rad)$, sinusoidally with time, with an amplitude v^* and period P. Assume that the star moves because it is pulled by a planet orbiting in a circle around that star. Assume that our line of sight to the star is parallel to the plane of the planet orbit. The observed star velocity is relative to the center of mass, and the mass of the star, M^*, is known. Derive an equation for the mass of the planet, M_p, and its distance from the star, r, in terms of v^*, M^*, and P. Then express M_p in units of Jupiter's mass ($M_J / M_o = 10^{-3}$), r in units of A.U., M^* in units of M_o, and P in years. (The velocity v^* remains in m/s; M_o, G, A.U., and the year are among the units given in the Introduction.).

Check that your equations work for Jupiter: its orbital speed is

$v_J = 12$ km/s.

Evaluate M_p and r, in these units, for the star

47 UMa : $M^* = 1.05\ M_o$, $v^* = 44$ m/s, and P = 3 years; and for 51 Peg: $M^* = 1.0$, $v^* = 58$ m/s, and P = 4.2 days.

The Setting: The Sun is a fairly normal star. Therefore, most astronomers expect that there are other stars with planets orbiting around them. But it has been difficult to detect planets. They cannot be photographed because they are too faint compared to the immediately adjacent star. The contrast is smaller in the infrared. With improving infrared technology, perhaps planets can be photographed in the infrared within a few years.

Since 1995, orbiting planets can be detected by the corresponding motion of the star (action and reaction). In the case of the Sun, Jupiter's gravitational pull on the Sun causes the Sun to perform a small orbit around the Jupiter-Sun center of mass, with a velocity of 12 m/s = 43 km/hour. (Given momentum conservation in the center-of-mass frame, this velocity is Jupiter's orbital velocity of v_p= 12 km/s multiplied by the mass ratio $M_J/M_o = 10^{-3}$.) After long technical development, periodically changing Doppler shifts as small as 10 m/s can now be measured for a few dozen reasonably bright stars.

The Solution: We assume that the planet's orbital plane is in our line of sight. Then the amplitude of the observed v^*(rad) equals the actual orbital velocity of the star. Since the observed v^*(rad) varies sinusoidally in time, the orbit is circular. The planet, moving around its star, accelerates the star. The three fundamental equations are $v^*/v_p = M_p/M^*$ (which represents momentum-conservation in the center-of-mass frame), $2\pi r/P = v_p$, and $v_p^2/r = GM^*/r^2$ (force balance for the planet). Elimination of the unknown v_p gives two equations for r and M_p involving the observed quantities v^* and P, and the known M^*. One gets $(M_p / M^*)^3 = v^{*3}P/(2\pi GM^*)$ and $r^3=P^2GM^*/4\pi^2$. The latter should look familiar! For changing units, set $M_p' = M_p/M_J$ and then $M_p = M_J (M_p / M_J) = 10^{-3}M_oM_p'$, use the value of M_o given in the Introduction. Similarly for other quantities. On introducing the asked-for units, one obtains $M_p^3 = 4.4\text{x}10^{-5}v^{*3}PM^{*2}$ and $r^3 = P^2M^*$, where the primes have been omitted.

For Jupiter, $v^* = v_J (M_p / M_o)$ = 12 m/s, P = 12, M^* = 1, and the result is indeed M_p=1, r = 5.2, correct. Given the data in the problem, for 47UMa, M_p= 2.3, r = 2.1; for 51 Peg, M_p = 0.46, r = 0.05.

Interpretation: Given that $v^* < 3$m/s can not be measured, we can not yet detect planets with masses similar to Earth, only planets with masses similar to Jupiter. Nevertheless, the elation is great: We have finally detected planets around other stars! However, two possible objections have to be kept in mind:

First, if the orbit is inclined to the line of sight, the actual orbital velocity is larger than the measured radial velocity and the properly deduced mass of the planet is larger. Thus the above procedure yields only a minimum planet mass, $M_p > 2.3$ for 47UMa, $M_p > 0.46$ for 51 Peg. One argues that, if the inclination angles are random, then most deduced minimum masses will be "close" (by better than a factor two) to the actual masses.

Second, star spots and motions in the stellar surface might cause an apparent Doppler shift. For the stars observed so far no such confusion is expected.

Surprise! No one expected the variety of orbits that have been detected so far.

1. Some planets such as the planet of 47 UMa are indeed similar to Jupiter in minimum mass and orbital radius.
2. Several planets with nearly circular orbits are at a surprisingly small distance from the star, with orbital periods of merely a few days. Examples are the planets of 51 Peg and of τBoo, $M_p > 3.87M_J$, and r = 0.046 A.U. With their Jupiter-like masses, one expects these planets to be gaseous (mostly hydrogen, like Jupiter). How can gaseous atmospheres so close to the star survive the intense heating by the star? Have these planets perhaps formed only recently or perhaps recently arrived from a larger orbit, or both? All the central stars are (selected to be) fairly similar to the Sun. Clearly we do not yet understand the formation of planetary systems such as our solar system.
3. Some derived orbits are very eccentric orbits. Examples: 70VirB, with $M_p > 8.1\ M_J$, P = 117 days, has an orbit reaching from 0.3 to 0.7 A.U. from the star; and 16CygB, with P = 2.2 years, has an orbit reaching from 0.6 to 2.8 A.U. Many astronomers are convinced that planets must form out of a rotating disk of gas and dust. Such planets must form with a nearly circular orbit. Therefore, these astronomers argue, any object with a very elliptical orbit must actually be a star which formed together with the central star in a common cloud of gas. But not everyone agrees.

There are no planets detected with P around 10 years or longer, in part because such planets would cause v^* to be unobservable small, in part because accurate data have been taken and analyzed for only a few years, so that the oscillation in v^*(rad) cannot yet have become obvious.

Didactics:

The conversion of units is essential to recover a sense of the physical quantities such as planetary masses and years that we are familiar with.

The mass of the planet is important in relating v^* to v_p, but it has been neglected wherever it appears together with M^*, in particular, in Kepler's third law.

4. Binary Stars

Activity: measurement of Doppler shift in stellar spectrum

Physics needed: Two body problem; formula for Doppler shift.

Didactic Purpose: An error analysis typical for frontier science.

The Problem: Two spectra are given for the spectroscopic binary Mizar A ((UMa). At the time of spectrum A, the spectrum lines are single. At the time of spectrum B, two days later, the lines are double. Their separation informs us of the radial velocity of the stars relative to each other.

Given the wavelength calibration on the spectrum, measure the relative velocity in km/s using the Doppler formula. Estimate the uncertainty in your value : If your best measurement is x km/s, what is the largest y so that x+y or x-y might be correct values, given this spectrum? Your estimate of the uncertainty will be compared to the range of velocities measured by others. According to Kepler's third law (extended for two objects of similar masses M_1 and M_2), M_1+M_2 depends on the cube of the velocity difference. What is the uncertainty in M_1+M_2?

The two lines in spectrum B are shifted nearly equally to the right and left of the single line in spectrum A. That suggests the two stars have nearly equal masses . In the center-of-mass frame, $M_1v_1= M_2v_2$. Estimate the uncertainty in the ratio of masses, given your estimated uncertainty in radial velocity.

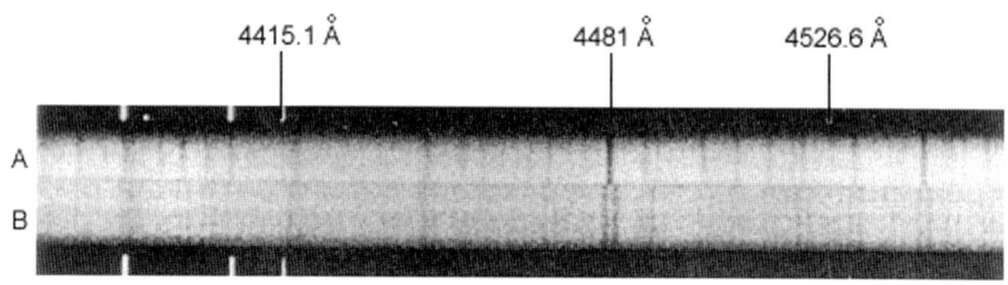

Figure 8.2-3. Spectra of Mizar A. Credit: California Institute of Technology.

The Setting: The Sun is somewhat unusual in that it is a single star. About two thirds of all stars are binaries, that is, members of a binary pair of stars which orbit about each other under their mutual gravitational attraction. The star orbits satisfy Kepler's third law (extended to include comparable masses, see Chapter 8.9). The orbits of the stars around each other are extremely small compared to the distances of both stars from the Sun. Only a few binaries are sufficiently close to us that the two stars can be photographed separately and the orbit can be measured completely. For such binaries, one obtains the mass of each star. For example, bright Sirius has $M = 2.1M_o$ and its very much fainter (white-dwarf) companion has $M = 1.0M_o$.

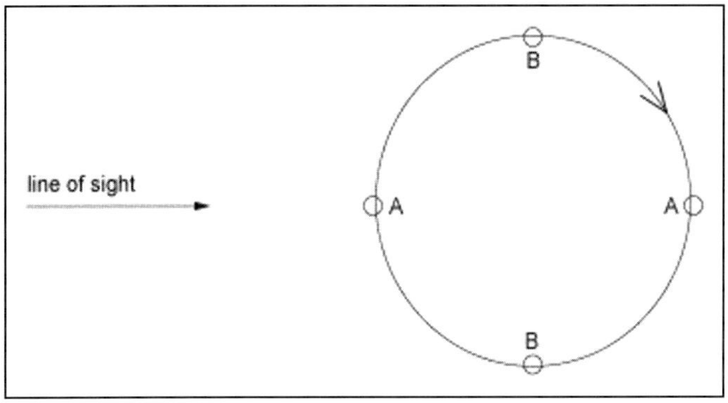

Figure 8.2-4 Observational configuration.

In many more cases, the two stars are so close together on the sky that they appear as a single dot on a photograph. We can recognize that the dot is a binary if we take its spectrum. The spectrum contains the light of both stars. Differences in the radial velocity of the two stars show up as different Doppler shifts of the spectrum absorption lines.

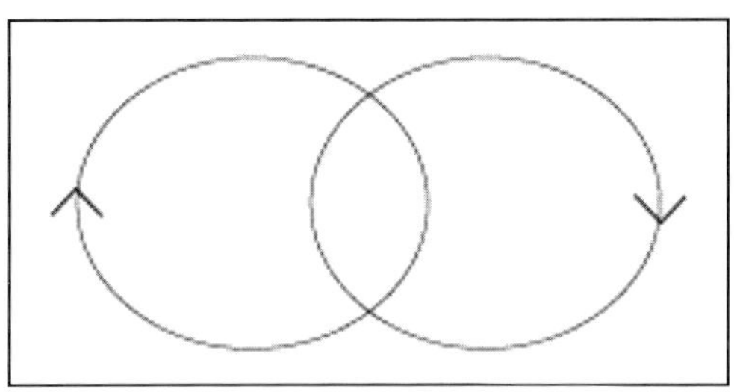

Figure 8.2-5

Because the stars orbit about each other, the Doppler shifts are seen to change from day to day. The graph shows a simple geometry with circular orbits. We see the orbit from the left. The Doppler shifts are identical when the stars are at positions A, but quite different at positions B. Given sufficiently frequent spectra and information about the orientation of the orbital plane, one can deduce the star masses.

This problem concerns Mizar A, a binary with small separation between the stars (P = 20 days, rather short) and elliptical orbits (e = 0.5). The orbits involve unusually high radial velocities, high enough to allow a student's measurement of the radial velocity. Mizar A was the first spectroscopic binary to be discovered, in 1889, because of the high velocities. In figure 8.2-3 are shown two spectra of Mizar A. The two gray bands labeled A and B are the two spectra of interest. Longer wavelengths are to the right (1Å = 1Angstrom = 10^{-10}m). Each spectrum contains some darker vertical lines, referred to as spectrum lines. [The star basically radiates a Planck "black body" spectrum, but atoms in the stellar surface absorb light at specific wavelengths, dependent on their temperature. The wavelengths at which light is "missing" appear as vertical dark lines in the spectrum.] The lines labeled 4481Å are of interest here. In spectrum A, only one line appears at 4481Å. Thus the two stars have the same radial velocity. In spectrum B, taken two days later, the line from one star is Doppler shifted to longer wavelengths, the other to shorter wavelengths. The goal of this observational activity is to measure the difference in radial velocity between the two stars in spectrum B, using the Doppler formula, $\Delta\lambda/\lambda = (v_1 - v_2)/c$, where the v are the components of the two stars' velocities along our line of sight.

The Solution: By comparing the distance between the split lines to the separation between the lines at 4415.1Å and 4526.6Å, the split in spectrum B is about 2.0Å, amounting to about 130 km/s, with the last digit rounded off. But 120 km/s and 140 km/s also are frequent measurements. On this scale, the human eye cannot distinguish 10% on either side of the "true" line center, and about half the measurements are likely to range between 115 km/s and 145 km/s. Thus these data involve a probable error in the radial velocity of about 10%, and M_1+M_2 has an uncertainty of about 30% [since $(1+0.1)^3 \approx 1+3x0.1$]. If one compares the lines in spectrum B to the line of spectrum A, the same measurement error translates to a 20% uncertainty in the mass of each star. This adds to the 30% uncertainty!

Didactics:

1. While these measurement uncertainties appear in astronomical data, students should learn that that concerns about measurement uncertainties pervade physics, engineering, and all the physical sciences. If physics students have little chance to measure things in the laboratory or in a city street, then data like these spectra may be used to help indoctrinate the students so that they will always ask about uncertainties before tackling a real physical problem.
2. These spectra are quite old. More modern data can be much more accurate since this is a bright star. But the goal here is the students' experience with the limitations and uncertainties in measuring data. The experience with uncertain data is an essential part of any scientist's and especially any astronomer's training. Even though error analysis can often be made much more precise than here, a judgment about possible errors based on experience is often the only possible way to inform other scientists about the reliability of new observations.

3. The students themselves should measure the spectrum. Only a personal measurement of the line splitting, and an estimate of its accuracy, gives students a sense of the importance of worrying about accuracy. If possible, a few copies of the spectrum should circulate through the class. Even if only a few students can measure, it is enough to establish the range of measurements. In any case, measurements should be recorded quietly, by each student, so that they will not influence each others' measurements. When all are finished, the accuracy of the measured line splitting can be established by comparing all the measurements and determining the spread in their values. This practical result is much more important than the values given in the "solution".
4. A mm-scale is the most likely measurement tool, but if a different scale (inches?) is available, it is useful to have some measurements done with each scale. If a mm scale is used, there is a strong temptation to round off to whole mm, and that may give a systematic preference to one value.
5. How do we know the wavelengths of spectrum lines? The short white lines surrounding the spectra are made by a spark (usually from an iron arc) created within the telescope. These lines have wavelengths that are known accurately from the laboratory and are used to calibrate the lines appearing in the star spectrum. The entire spectrum shown covers less than 200Å = 20 nm. For the human eye, all the light contributing to the spectrum would have practically the same color. The Doppler shifts could not be perceived by the human eye as a change in color.

The masses of stars are the most fundamental property of stars, because the mass determines the star's surface temperature and size throughout its "lifetime" and the fate of the star at "death". (See Chapter 8.6) Only few of all the stars have masses measured directly through binary orbits. A few very bright stars have their spectrum analyzed so carefully that the surface gravity and radius can be measured, which yields the mass. For example, Vega has $M = 2.0(\pm 0.1)\ M_o$. For most stars, one matches the star's spectrum to the most similar spectrum of a star with a directly measured mass. Stars with relatively rare spectra have large uncertainties in M. For example, the companion to the black hole Cyg X-1 (see problem 8.2.5), has M uncertain by 30%.

5. The Stellar Black hole Cyg X-1

Black hole, Schwarzschild radius

Didactic purposes: A stretch of the imagination. Interpretation of uncertain data.

Physics Purpose: Demonstration of the existence of black holes.

The Problem: Deduce the mass of the stellar black hole called Cyg X-1, given that its visible orbiting companion has a mass of $M_1 = 33\ M_o$ with an uncertainty plus or minus $9M_o$, an orbit with a period of 5.6 days and a velocity amplitude of 76 km/s. Assuming a circular orbit.

1. Assume $M_1 = 24M_o$, the orbital plane is parallel to our line of sight ($\sin i=1$). This yields the minimum mass for the invisible object.
2. Assume a more probable $M_1 = 33M_o$ and orbital inclination given by $\sin i = 0.5$. To do this, first correct the modified Kepler's third law so that it is valid for an orbit of an unknown inclination, and then eliminate one of the velocities in terms of the other (remembering action and reaction).

Physics Background: Stars create their own gravity. If a star becomes smaller, the speed needed to escape from its surface increases. Eventually, when the star becomes sufficiently small, the escape speed exceeds the speed of light, c, and nothing can escape from the surface, not even radiation. Such an object is referred to as a black hole. A black hole occurs once the object shrinks so that its surface is within the "Schwarzschild radius", $r = 2GM/c^2$, where M is the mass of the object . If $M = M_o$, then r = 3 km. [As a mnemonic, this radius follows from the "classical" relation for escape speed c: $1/2\ mc^2 = GMm/r$. Actually the Schwarzschild radius was derived from general relativity by Karl Schwarzschild in 1916.]

Astrophysical background: The first good candidate as a black hole was the brightest x-ray source in the star constellation Cygnus, called Cyg X-1. In 1971, radio astronomers, detecting radio bursts from the same direction of the sky, located the x-ray emitting object much more precisely. They showed that it is in the same position as a hot (surface temperature about $3\text{x}10^4\ {}^\circ\text{K}$) and very large (radius about 20 solar radii) star which orbits about an invisible companion. The theory for the lives of stars (Chapter 8.6) says that the companion might be invisible because it is an extremely small star called a neutron star, with a radius of only 10 km, or because it is truly a black hole. Theory also says: Neutron stars can have a mass no greater than $5M_o$, for otherwise they collapse to become a black hole. Therefore, if we can show that the mass of the invisible object exceeds $5M_o$, then the invisible companion must be a black hole. (Several later problems deal with neutron stars, starting problem 8.4.1)

The Solution: We start with the modified form of Kepler's third law, $M_1+M_2 = (v_1+v_2)^3\ (P/2\pi G)$. Here the v are the actual orbital velocities, but v_2 is not observed.

a) From momentum conservation in the center-of-mass frame, we know $v_2 = v_1 M_1/M_2$. Kepler's third law becomes $v_1^3 (P/2\pi G) = M_2^3/(M_1+M_2)^2$. With P = 5.6 days and v_1= 76 km/s, the left side becomes 0.25 Mo. With $M_1 = 24\ M_o$, one gets $M_2 = 6\ M_o$.
b) Now we need to realize that the actual orbital velocities are larger than the observed radial components, $v = v_r/\sin i$. Kepler's third law now becomes $v_{r1}^3 (P/2\pi G) = (M_2 \sin i)^3/(M_1+M_2)^2$. The left side is still 0.25 M_o. For M_1= 33 Mo and sin*i*=0.5, one gets M_2= 15 M_o.

Interpretation: The mass derived for the invisible object is high enough for the object to be a black hole, even in the improbable case a). In fact, sin*i* =1 is not possible because the x-ray source attached to the invisible object is not eclipsed by the star and, also, the computed separation of the two objects would be no larger than the sugergiant's radius, which would imply a huge flow of gas from the supergiant to the black hole. There is some mass flow, but much more modest. To explain some observed light variations of the visible star with the same period as the orbit, sin*i*=0.5 has been deduced. The mass of the visible supergiant is uncertain in part because the spectrum of such a rare star is normally difficult to translate to a mass, in part because the visible star cannot be quite normal, situated so close to a sink for its gas and a powerful x-ray source.

To summarize, the evidence for this black hole is very good, though by itself it does not convince all astronomers. There are now other stellar objects with very good evidence for a black hole. The existence of stellar-mass black holes is no longer in doubt.

How can we say that x-rays and radio waves are associated with stellar black holes when black holes do not radiate? The answer is that gas flows from the surface of the visible star, falls toward and begins to swirl around the black hole. Indeed there is optical evidence for this gas. As it swirls ever closer to the black hole, it reaches temperatures on the order of 10^7 °K, emitting x-rays. Only then does it fall into the black hole and disappear. Some of the x-rays, flickering on time scales less than one second, tell us that the gas falls into the black holes in chunks.

1. Didactics:
2. Most students readily accept the notion of black holes. It is worth pointing out that astronomers are a conservative and skeptical crowd (as is necessary to avoid mistakes in this frontier science). Black holes with stellar masses were known theoretically for decades, but many astronomers doubted seriously that they exist in nature. The general acceptance of the reality of stellar-mass black holes had to wait until the necessary x-ray technology came along. Similarly, the reality of

long-predicted neutron stars had to await the development of radio technology and the discovery of pulsars (see problem 8.4.4).

3. Since the derivation of Kepler's third law in Chapter 8.9 is rather cumbersome, it may be useful to provide to the class (or have them derive) the same law for circular orbits. The two objects have a common period P, so $v_1 = 2\pi r_1/P$ and $v_2 = 2\pi r_2/P$. Action, reaction, and the force balance yields $M_1 v_1^2/r_1 = M_2 v_2^2/r_2 = GM_1M_2/(r_1+r_2)^2$. If the velocity amplitudes are both measured, the ratio of the masses derives from $M_1v_1 = M_2v_2$. Using $a = r_1+r_2 = r_1\ (1+M_1/M_2)$, and $v = v_1+v_2 = v_1\ (1+M_1/M_2)$, one obtains most directly $M_1 + M_2 = av^2/G$, from which the usual forms follow by using $2\pi a = vP$.

6. The Mass of our Milky Way Galaxy

Gravity within a spherical mass distribution

Relevant Physics: Gravity within a spherical distribution of mass, at a distance r from the center, is given by $g(r)=GM(r)/r^2$, as if all the mass M(r) out to a sphere of radius r were placed at the center. An outline of the proof appears in Chapter 8.9. The relation is analogous to that found for some electrostatic problems.

Physics Implication: Used for our Galaxy, g(r) leads to the inference of much invisible, "dark" matter in our Galaxy.

The astronomical setting: The stars that we see in the Milky Way constitute part of a large rotating disk of stars and gas that we call the Milky Way Galaxy or usually simply the Galaxy (written with a capital G). In the part of the sky containing the star constellation Andromeda, one can see (with binoculars) a spiral galaxy, called the Andromeda galaxy or M31, that probably looks similar to our Galaxy seen from a far. (M31 is $2x10^6$ light-years from us.) If the Sun were placed into the picture of the Andromeda galaxy, it would be near the outer part of the disk of stars. Most of the stars are a few light-years apart. The Sun and the stars near us rotate about the center of the Galaxy with a speed of about 220 km/s, at a distance from the center of about $2.7x10^4$ light-years. (Newer data might change that distance by 10%.) Judging from the brightness of the center of the Andromeda galaxy, most of the stars of that galaxy are relatively near the center and distributed roughly spherically about the exact center. Although the center of our Galaxy is largely obscured from our vision by interstellar dust, we can tell that in our Galaxy also there is a roughly spherical distribution of stars around its center and distributed roughly spherically about the center. If we take this observation literally, then the Sun is outside most of the mass distribution and we can treat all the mass of the Galaxy as being at the center.

Figure 8.2-6. The Galaxy M31. Credit: National Optical Astronomy Observatories

The old-fashioned Problem: Estimate the mass of our Milky Way Galaxy (concentrated near the center) given that the Sun orbits around the center of our Galaxy approximately in a circle with a velocity about 220 km/s. The center of our Galaxy lies at a distance, D_o, of about $2.7x10^{17}$km.

The Solution: The assumptions allow use of the original Kepler's third law (solar mass << galaxy mass). Use D_o = $1.8x10^9$ A.U., P = $2.5x10^8$years, hence M = $1.0x10^{11}M_o$.

Interpretation: The result, $10^{11}M_o$, is the reason that our Galaxy is often said to contain some 10^{11} stars. But this is a very rough number, because most stars we know of have a mass less than $1M_o$, so there must be more stars. In fact, the result M = $10^{11}M_o$ seems to be quite good : If we add up the mass of all the stars we see (or infer behind the obscuring dust) in our Galaxy, we obtain about $0.9x10^{11}M_o$. However, a first problem arises if we add up the stars in the central, roughly spherical region: they are only a fraction (~0.2) of the stars in the disk. So the assumption of a central mass comes into question.

If Kepler's law really applied, then the circular velocity of stars, and of the gas between the stars, would decrease with distance from the center in proportion to $r^{-1/2}$. Instead, the circular velocity of the gas is observed to remain roughly constant from the Sun's distance outward as far as we can measure. (The observations mainly use the Doppler shift of the radio emission from neutral hydrogen at a wavelength of about 21cm.) The constant velocity means that stars and gas further out than the Sun feel the gravity from more matter than the Sun does. At some distance, the velocity must decrease, but we cannot detect stars or gas there to measure the velocity. In any case, there must be more mass than the mass in stars we see (or infer behind dust). We speak of much "missing mass" or "dark matter".

The constant rotation curve extends to at least $r = 3D_o$ and probably (judging from the optical or radio rotation curves v(r) of other galaxies) several times that distance. That means the dark matter must extend at least that far, well beyond the disk with its visible stars. One assumes that the dark matter is physically independent of the visible stars in the disk, and that it is distributed roughly spherically about the center. The visible stars and gas, in this model, contribute rather little of the mass, but their rotation about the center, v(r), traces the gravity g(r) caused by the dark matter.

Up-to-date Problem: Assume a spherical mass distribution and that M(r) at the Sun is $M(D_o)=10^{11}M_o$. Let the rotation velocity v(r) be constant from D_o to $5D_o$. Deduce the form of M(r) as function of r for $D_o < r < 5D_o$, and find $M(5D_o)$. What fraction of this mass is dark matter?

The solution: Force balance $v^2/r = g(r)=GM(r)/r^2$ makes M(r) proportional to r. Therefore, $M(5D_o) = 5M(D_o)$. Compared to luminous matter of 0.9M(Do), dark matter constitutes about 80%.

Interpretation: A mass estimate based on v(r) = constant yields only a minimum mass, since more mass must exist out where v(r) decreases toward zero. Estimates of the mass of the Galaxy are now about 7 (at least 4 and perhaps 10) $\times 10^{11}M_o$.

What is the dark matter? Perhaps there are unrecognized white dwarfs or black holes left over from former stars, or big planets or very cold gas. Indeed some star-like objects that we cannot detect by their radiation are now being detected because they gravitationally focus onto us the light from more distant stars. After a search of some three years, these newly detected star-like objects, probably white dwarfs with $M \sim 0.5M_o$, apparently make up only part (half?) of the "dark matter", but a firm conclusion must wait.

In cosmology, inflation theory and the search for the density which closes the universe has led to the suggestion that most of the missing mass in galaxies (and clusters of galaxies, see problem 8.4.5) is non-baryonic. This has led to a search for other forms of matter, such as neutrinos or axions.

Didactics:

1. Most students have little trouble getting used to distances of stars we see in our sky, and to the scale of light-years, because the Sun and the solar system still have some role to play as we learn about the stars. But students need time to adjust to the scale of our Galaxy. Photographs of the Milky Way do not help, because we see in the Milky Way only stars within some $3x10^3$ from us, and these appear to be centered on the Sun and us. The visible light from most of our Galaxy is obscured by dust. Maps based on radio observations appear very abstract to most students. But students can relate to photographs of other galaxies, such as the Andromeda galaxy, which probably resembles our Galaxy. And they can imagine the Sun places into the outer parts of the stellar disk of the Andromeda galaxy, and gradually revolving around its center.
2. It should be obvious here that there is no value in evaluating answers to this problem beyond one significant digit. It is important to point out that the uncertainty is not just due to measurement errors, but due to a real physical uncertainty in interpretation of the observations.
3. How do we know that the Sun revolves at about 220 km/s around the center of the Galaxy? Doppler shifts of galaxies in the "forward" direction have average apparent radial velocities about 440 km/s different than galaxies in the "backward" direction.
4. The result $g(r)=GM(r)/r^2$ states, somewhat counter-intuitively, that there is no net influence of the external gas. This is valid only for spherical symmetry. The formula reduces to $g(r) = GM/r^2$ if r is outside the object of mass M. If r equals the radius of the object, g(r) is the surface gravity.
5. This gravitational problem can be used as a test that students have understood the analogous electrostatic problem.

7. The 10^9-solar-mass black hole at the center of the galaxy M84

Didactic purpose: A further stretch of the imagination. Some actual space data.

Physics needed: Kepler's third law, thin-triangle relation $d = r\theta$.

The Problem: Calculate the mass of the black hole at the center of the galaxy M84 if gas observed 0.1 seconds of arc from the black hole circles the black hole with a speed of 400 km/s. Assume a distance of $5x10^7$light-years.

The Solution: Since 1 radian = $2x10^5$ seconds of arc, the gas is observed at a distance ($5x10^7$light-years)/$2x10^6$= 25 light-years from the center. Using Kepler's third law and the units used for the solar system, $P = 2\pi r/v = 4x10^{12}s = 1.3x10^5$years, $r = 1.6x10^6$A.U., $M = r^3/P^2 = 3x10^8M_o$.

The setting: In 1963, quasars (short for quasi-stellar objects) were discovered to be among the most distant objects in the Universe, giving off hundreds of times as much radiation as do large galaxies, and yet, judging from their light variations, as small as light weeks. These strange properties led to the suggestion that the power must come from gravitational energy, specifically from material falling into a very massive and very tiny object, a black hole. A black-hole mass of 10^8M_0 was suggested, but this value seemed quite incomprehensible at that time. Now we know that many galaxies contain very powerfully radiating centers, involving stellar masses up to 10^9M_0. But are these centers really black holes? They might be merely dense accumulations of stars.

In 1997, the Hubble Space Telescope (2.4 meter diameter, in orbit around Earth) observed the center of a galaxy (not a quasar) named M84. This galaxy is a member of a "nearby" cluster of galaxies, located merely some $5x10^7$light-years from us. The galaxy is so nearby that the telescope could resolve light from merely 25 light-years away from the center of M84 and measure the orbital velocity of the radiating gas. The deduced mass of $3x10^8M_0$ is not particularly unusual, but the resolved small distance from the center is. If this mass consists of $3x10^8$stars distributed through a sphere of radius only 25 light-years, then the stars are so densely packed ($5x10^3$ per cubic light-year) that they would have collided by now, and the result would be a black hole.

Interpretation: The evidence for a very massive black hole at the center of the galaxy M84 is very good. Telescopic evidence for most other black holes is not as good. But the consensus now says that many galaxies indeed contain a black hole at the center.

The range of black-hole masses is large. The galaxy M87 has a black hole of $2x10^9M_0$! One of the two small companion galaxies to the Andromeda galaxy (picture in problem 8.2.6) probably has a black hole of about $3x10^6M_0$. Our Milky Way Galaxy, although one of the larger galaxies, probably contains only a modest black hole with about $3x10^6M_0$ (see didactic 3, below). Perhaps the smallness of our black hole is fortunate for our existence on Earth. Quasars are still believed to contain massive black holes, but they are too far away for good telescopic evidence.

Why do bright galaxy centers and quasars with massive black holes shine so powerfully? The energy can be derived from roughly one star per year (or as much mass in clumps of gas) spiraling toward the black hole. A star is broken up by gravitational tidal forces, and its gases heat up and shine powerfully just before crossing the Schwarzschild radius. At this rate, a black-hole mass of 10^8M_0 is accumulated in merely 10^8years. Many quasars are in galaxies that are colliding with other galaxies. The complicated gravity within the colliding galaxies sends stars falling toward the black hole. After some time, no more stars (or gases) are available from the surroundings, and the powerful radiation stops. On the scale of the age of the Universe (roughly $13\text{x}10^9$years), these are transitory phenomena.

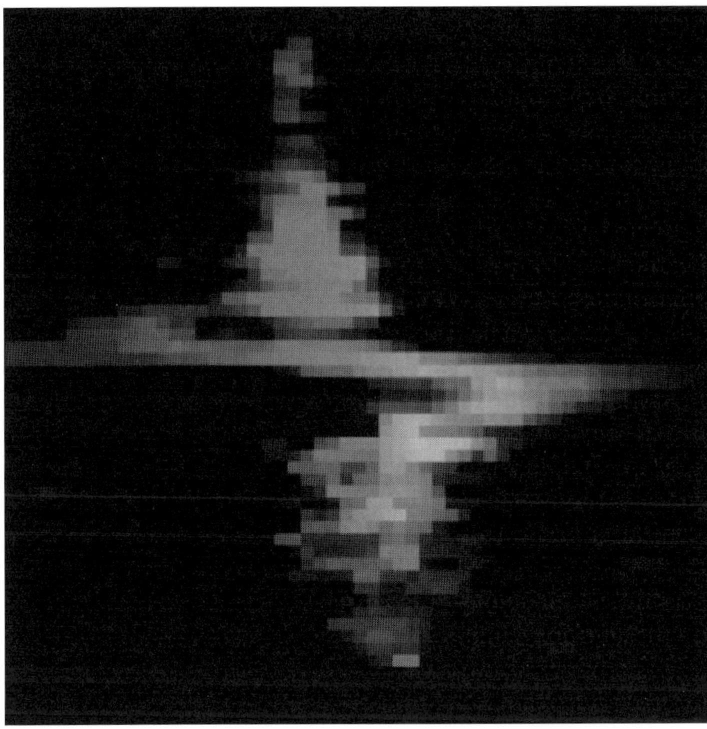

Figure 8.2-7. HST Spectrum of Central parts of M87. Credit: NASA.

The data: The photograph in figure 8.2-7 represents the spectrum obtained by the Hubble Space Telescope. The diagram is intended to help interpret the spectrum. When the light from the galaxy falls onto the image plane of the telescope, where a photo of the galaxy might be taken, a thin rectangular slit is placed on that plane which transmits only the light from a slit-shaped region that includes the exact center of the galaxy. The length of the slit is about 600 light-years at the galaxy. In the photo of M31 in problem 8.2.6, this length would extend over roughly 1% of the size of the galaxy in figure 8.2-6.

The light from each part of the slit was dispersed into a spectrum. Stars practically form a continuous spectrum. But gases between the stars emit their specific wavelengths. The wavelength range observed was chosen to include radiation from hydrogen atoms and nitrogen and sulfur ions. It is assumed that the stars move like the gas moves. In case the gas all has the same velocity, no matter where it appears in the slit, then the spectrum is a straight image of the slit, at the wavelength emitted by the gas.

The spectrum, sketched in the middle, does not look like that at all. Part of the gas shows velocities up to 400 km/s toward us, part up to 400 km/s away from us (relative to the average of all the velocities). The sketch on the right shows a disk of rotating gas, rotating very rapidly near the center (400 km/s), less rapidly further from the center (100 km/s). If we look from the right along the top of the disk, we see gas orbiting toward us at velocities up to 100 km/s. This corresponds to the top part of the spectrum. If we look from the right just above the center of the disk, we look through some gas orbiting at merely about100 km/s, but we also look through gas orbiting toward us very rapidly, up to 400 km/s. This corresponds to the spectrum just above its center. Similarly, just below the center of the spectrum, the gas recedes from us, with some gas receding at 400 km/s, physically near the center, plus other gas receding more slowly, relatively further away.

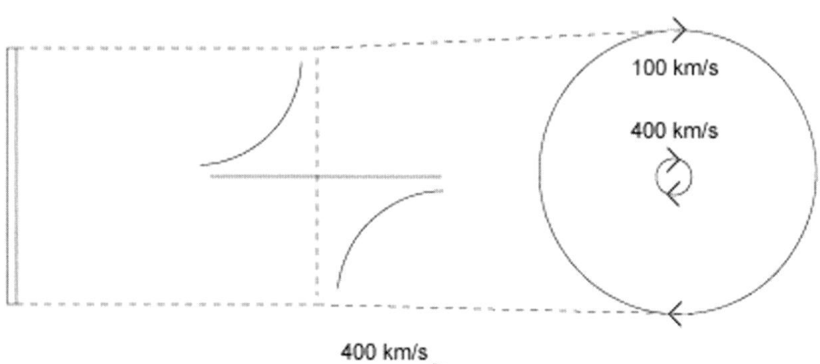

Figure 8.2-8. Viewing Configuration for figure 8.2-7

Didactic:

1. The angular resolution of 0.1 second of arc by Hubble Space Telescope is unprecedented for optical astronomy. With this resolution, one can read the text of a newspaper held at a distance of 1 km (resolution 0.5mm)! Most ground based telescopes are limited to a resolution of about 0.5 second of arc by our atmosphere. For bright objects one can do better by using optical interferometers and automatic corrections for atmospheric blurring.

2. The fastest gas is observed at 0.1 second of arc from the center because that is the limit of angular resolution available from HST. With higher resolution, we could (probably) recognize gas nearer the center and moving even faster. The Schwarzschild radius for M $=3x10^8$ M_o is 10^9km = 10^{-4}light-year. Clearly we cannot see even near to the Schwarzschild radius.
3. Students' problem: The Sun (and Earth) is $3x10^4$light-years from the center of our Galaxy. Various observations have suggested a black hole at the center with a mass $3x10^6M_o$. Predict the orbital period of stars circling the center at a distance (seen from Earth) of 0.1 second of arc. . (The answer: r=10^3A.U., still much larger than the Schwarzschild radius of 0.07 A.U., v = $2x10^3$km/s, P = 17 years.)

Interpretation: P = 17 years means that we can observe the displacement of stars near the galactic center over just a few years! Indeed, photographs taken two years apart with an infrared detector at the European Southern Observatory in Chile show such motion, and the ten-meter Keck telescope in Hawaii has confirmed these observations. These observations thoroughly confirm that there is indeed a black hole at the center of our Galaxy.

Once again, the spectrum used for measuring a Doppler shift shows only a tiny range of wavelengths. Not recognizable in this spectrum is that all this radiation is Doppler shifted somewhat to the longer wavelengths, because M84, as a member of a "nearby" cluster of galaxies, participates in the expansion of the Universe. Since the speed is only $1.2x10^3$km/s$\ll$ c, the Doppler shift does not cause a change in color recognizable to the human eye. For quasars, however, with enormously greater distances, the cosmic expansion causes a very high rate of recession from us, and that causes a significant change in color. Visible radiation detected from the most distant objects actually left those objects (and the emitting atoms) as ultraviolet.

Chapter 8.3

MORE MECHANICS OF THE SOLAR SYSTEM

D.G. Wentzel

1. A walk on Asteroid Icarus

Surface gravity, Escape velocity

The Problem: The asteroid Icarus has a diameter of 1.4 km, a density typical of rock, 3 x the density of liquid water on Earth, and rotates once in 2.3 hours. Determine the weight on Icarus of an astronaut of 200 kg, including space equipment. Determine the escape speed from Icarus and the speed of an object in circular orbit around Icarus, just above the surface of Icarus. How fast must an astronaut progress around Icarus such as to remain on the night-side of Icarus? Discuss whether an astronaut can walk around Icarus at the appropriate speed.

The Astronomical setting: The famous asteroids have diameters of several hundred km. The largest, Ceres, has a diameter of about 1000 km. Much smaller asteroids are known only if they happen to approach Earth closely, i.e. if they cross r = 1 A.U. during their orbits. These are known as "Earth-crossing" asteroids and are of interest because one or another may collide with the Earth in the future.

Icarus is one member of this class. It came within 10^7km = 0.07 A.U. of the Earth on June 14, 1968. Icarus is one of only two known asteroids that approach the Sun closer than the planet Mercury. It has a very eccentric orbit, between 0.2 A.U. and 2.2 A.U. Clearly, the surface on Icarus becomes very hot when it is only 0.2 A.U. from the Sun. Arthur C. Clarke wrote a short science-fiction story "Summertime on Icarus" in which the asteroid is used as a platform for machinery used to survey the inner solar system. At a time when Icarus is near the Sun, an astronaut is dropped by a space ship on the nightside of Icarus, in order to repair some equipment. The rotation of Icarus carries him toward the hot dayside. His heat shield fails. He experiences nearly lethal heat when the

W. Wamsteker et al. (eds.), Developing Basic Space Science World-Wide, 423-436.

Sun rises, but fortunately the space ship rescues him just in time. We shall compute the temperature on Icarus in problem 8.5.1. Here the question is: since Icarus is so small, can the astronaut walk fast enough to remain on the night side permanently?.

The Solution: The surface gravity on an object of mass M and radius R is GM/R^2. With $M_I = (4\pi/3)\rho R_I^3 = 4x10^{12}$kg, the surface gravity is $5x10^{-4}m/s^2$, compared to $10m/s^2$on Earth. An astronaut with space suit totaling 200 kg would weigh on Icarus what merely 10g weigh on Earth! [The ratio of gravities on Icarus and Earth can also be obtained from $(M_I /M_E)(R_E /R_I)^2$ with Earth mass and radius given in the introduction or, more elegantly and with less chance of making an error, from $(\rho_I R_I)/(\rho_E R_E)$ where Earth's density is ρ_E = 5.5 times liquid water.].

The escape speed is $(2G M_I / R_I)^{1/2}$ = 0.9 m/s = 3.2 km/hour, the speed of an object in circular orbit just above the surface is $(G M_I /R_I)^{1/2}$ = 0.63 m/s = 2.3 km/hour. The astronaut needs a speed $2\pi R_I /P$ = 1.9 km/hour to keep up with the asteroid's rotation. This is almost the orbital speed!.

Discussion: The danger of walking on Icarus is that the astronaut accidentally will push off from the ground enough so as to escape the asteroid, thus getting into the full sunshine that is so dangerous. Since the speed with which the astronaut must walk is almost the speed of the circular orbit, each step is really a sub-orbital flight. Each step must be calculated carefully. It seems like a risky maneuver, but perhaps OK when the alternative, staying in place, is lethal.

Regarding the Earth's seasons: If you are at moderate latitudes on Earth and experience "summer" and "winter", ask your students to explain these seasonal temperature differences. If they argue that summer occurs because Earth in summer is closer to the Sun, ask them to explain why summer occurs in July in the northern hemisphere but January in the southern hemisphere. This problem is a good opportunity to make sure that students know the correct explanation of Earth's summers and winters, in terms of the inclination of the Earth's axis of rotation. Icarus, of course, is just the opposite, in that the hot season is caused by proximity to the Sun (see problem 8.5.1).

2. The Planetary Slingshot Effect

Frames of reference-The question of accuracy

Physics Goals: change of reference frames; discussion of appropriate measurement accuracies.

The Problem: Consider a spacecraft approaching Jupiter. Jupiter's orbit is circular, at r = 5.2 A.U. The spacecraft is in a minimum-energy orbit from Earth to Jupiter. See the left diagram below. Using the energy equation, evaluate Jupiter's velocity V and the spacecraft velocity v_{arr}

when it reaches r = 5.2 A.U. and arrives near Jupiter. Which moves faster? Transform to the frame of reference of Jupiter. In this frame, what is the initial velocity of the spacecraft, that is, the velocity of the spacecraft relative to Jupiter before gravitational attraction by Jupiter? Now assume that, after acceleration and deceleration by Jupiter, the final velocity is reversed relative to the initial velocity. See the right diagram below. Transform back into the solar system. What is the new velocity, v_{dep} of the spacecraft, in the frame of the solar system, when it departs from Jupiter? Using the energy equation for the new orbit, estimate a and describe the new orbit.

Write down a formula for the change in v^2 as evaluated in the frame of the solar system, in terms of Jupiter's velocity V and the velocity of the spacecraft just before arrival near Jupiter, v_{arr}

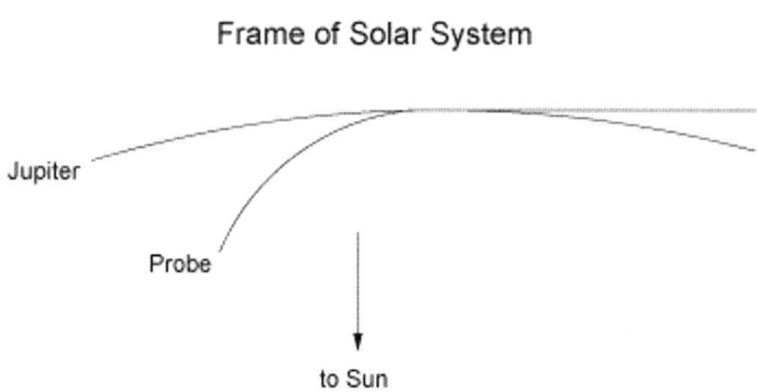

Figure 8.3-1.

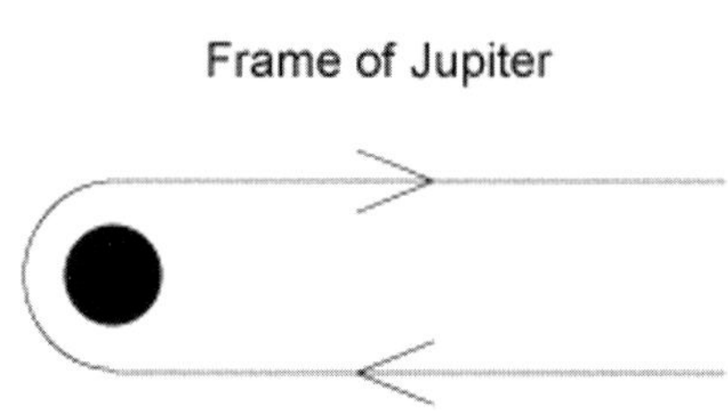

Figure 8.3-2.

The Setting: Once a spacecraft is free of Earth's gravity, it is in an orbit around the Sun similar to the orbit of Earth. Energy is needed so that the spacecraft moves on an orbit toward its destination in the solar system (see problem 8.2.2). Given limited rocket power, the further we want to send a spacecraft, the less equipment it can carry. However, it is possible to boost the orbital energy by having the spacecraft fly past a planet. For instance, the spacecraft Cassini, launched on October 15, 1997, will arrive at Saturn on July 1, 2004, after flying twice past Venus, in April 1998 and June 1999, once past Earth in August 1999, and once past Jupiter in December 2000, with each encounter changing the orbital energy.

The physics of the problem is quite simple: Kinetic energy depends on the frame of reference. It is often worthwhile making a simple computation in a frame in which the energy is constant, and then transforming to the frame of reference of the observer. The brief orbit of the space probe around a planet can be described ignoring the gravity of the Sun. In the frame of reference of the planet, the orbit is a hyperbola. The total (kinetic plus potential) energy is constant. Also, the kinetic energies before and after the encounter are the same, but the direction of the velocity has changed. Upon converting into the inertial frame of the solar system, the change in velocity means that the orbital energy is changed, and thus the orbit about the Sun is changed. This process is called gravity-assist and, more colloquially, the slingshot effect.

The Solution: The spacecraft orbit in the solar system is given by the energy equation $v^2 = GM_o(2/r-1/a)$. For Jupiter we have r = a = 5.2 A.U. and V = 13 km/s; for the arriving spacecraft we have r = 5.2 A.U., a = 3.1 A.U., and v_{arr} = 7.5 km/s. The relative velocity is 13 - 7.2 = 5.5 km/s. Jupiter is faster. It catches up to the spacecraft. After the fly-by, v_{dep} = 13 + 5.5 = 18.5 km/s. For the new orbit, again r = 5.2 A.U. Then v^2_{dep} is very nearly $2GM_o/r$, and a must be large. To within the accuracy of these computations, the probe achieves the escape speed from the solar system! Mathematically, $v_{dep} = v_{arr} + 2\,[V- v_{arr}] = 2V - v_{arr}$, $v^2_{dep} - v^2_{arr} = 4V[V-v_{arr}]$.

Interpretation: This example shows the effectiveness of the planetary slingshot. The main assumption is the deflection by 180° in the frame of Jupiter. Mathematically, it applies only in the limit of an infinitely long hyperbola approaching the focus arbitrarily closely. Physically, it is a good approximation if the velocity of approach to the planet is much less than the escape speed from that planet. For our example, the velocity of approach is 5.5 km/s, Jupiter's escape speed is 60 km/s, and 5.5 <<60 is not a bad approximation. Indeed, when the Pioneer and Voyager spacecraft encountered Jupiter, they experienced significant energy changes and deflections (though with encounter geometries more complicated than the example here). In the opposite limit, a very high velocity of approach means little time is available for gravity to act and a small deflection results. This problem is quite analogous to the electrostatic deflection of electrons by protons.

The terrestrial planets generally produce rather small deflections. But these deflections can still be very useful, especially if one spacecraft experiences several planetary deflections. The spacecraft NEAR (= Near Earth Asteroid Rendezvous), launched February 16, 1996, came past Earth on January 23, 1998 and will arrive at the asteroid 433 Eros in October 1999. A direct orbit to Eros would have required a larger launch rocket, adding US\$ 5×10^7 to the actual cost of about US\$ 1.3×10^8 (which is considered relatively cheap).

Didactics: The criterion for a large planetary deflection, (velocity of approach)/(escape speed) «1, is "obvious" to most experienced scientists because there are no other dimensionless ratios in the problem. But such judgments are not obvious to students and must be substantiated a few times before they get in the habit of looking for such dimensionless ratios. In this case, the proof starts with the hyperbolic orbit, $r = a(e^2-1)/(1+e\cos\theta)$, $r(\min) = a(e-1)$ with e close to 1, and $L = r^2 d\theta/dt = r(\min)v(\max)$. Then $dr/dt = r^2 e \sin\theta \, d\theta/dt \, [a(e^2-1)]^{-1} = 1/2 \sin\theta \, v(\max)$. Far away, a highly elongated orbit requires $\sin\theta$ «1. Therefore, the velocity of approach must satisfy dr/dt «$v(\max)$ ~escape speed.

Astrophysical versus astronomical accuracy: The previous exercises have stressed "astrophysical accuracy", in which 10% accuracy is considered quite good and a factor of two is often no cause for worry. In contrast, the sending of spacecraft through the solar system is a process of "astronomical accuracy". Enormous accuracy was needed for the successful orbits of the two Voyager spacecraft launched in 1977. Both reached Jupiter in 1979 (not on a minimum-energy orbit), both reached Saturn, in 1980 and 1981 respectively, and one continued to Uranus in 1986 and Neptune in 1989. The tracking of the spacecraft is accomplished by carefully measuring the Doppler shift of the radio signal from the spacecraft. Even for the older Pioneer 10, the Doppler frequency shifts of its 2.1×10^9Hz signal can be measured accurate to 10^{-3}Hz.

The spacecraft NEAR, on the way to its Earth fly-by and later to asteroid 433 Eros, flew past the small asteroid 253 Mathilde, approaching within 1200 km on June 27, 1997. The measured Doppler frequency shift showed that the velocity of the spacecraft changed by only 0.23mm/s! This was enough to measure the mass of the asteroid, 10^{17}kg. Given its size of about 46x48x66 km, the density turns out to be only 1.3 times that of liquid water. This is surprising since rocks (such as on our Moon, on two of Jupiter's moons, and in the outer part of the Earth) generally have a density of about 3; even mixtures of rock and ices (such as the two other Galilean moons of Jupiter and the largest asteroid Ceres) have a density of about 2. The asteroid's lower density suggests that it is composed of rather loose rubble, perhaps created or accumulated from past collisions. Indeed, the asteroid shows one truly enormous (relative to the asteroid) crater 30 km wide, 6 km deep, and four more craters at least 20 km wide. The five craters cover nearly half the surface. The craters suggest collisions in the past that nearly shattered the asteroid. Yet the craters are sharply edged and do not seem to consist of rubble. Thus we have both a surprise and a puzzle. Both are made possible by measurement of a change in velocity of only 0.23mm/s.

Figure 8.3-3. Asteroid 433 Mathilde. Photo Credit: NASA.

3. Can a Dust grain destroy a Space Probe?

Collisions, Kinetic energy

The Problem: Estimate the kinetic energy of a dust grain that might hit a camera on the space probe Giotto flying past the comet Halley in 1986. Compare your result to the energy of a stone thrown by a human. Is the camera likely to survive such a collision? (Assume the comet is in a highly elongated orbit, releases dust grains that nearly continue to move with the comet, and this dust is like the dust that pollutes Earth's cities when it is not raining. For all needed parameters, use your general knowledge of the solar system.)

The Setting: From Earth, we cannot photograph the actual body of a comet because it is shrouded in gases and dust that reflect sunlight to us. In 1986, the European spacecraft Giotto flew past comet Halley as near as was technically possible, in order to photograph the actual cometary body and transmit the picture back to Earth by radio. The precise choice of Giotto's orbit was aided by several other spacecraft, also watching Halley.

The comet is made of a mixture of solid particles and ices; on approaching the Sun, the evaporating ices escape from the comet in the form of gaseous jets, and solid particles are also blown off the comet by these jets. This explains the two kinds of tails, made of gas and "dust", respectively, both of which reflect sunlight to us. The released dust grains have orbits through the solar system similar to that of the parent comet. Comet Halley is on a highly elliptical orbit, and so are its dust grains. The orbit of Giotto and its camera is quite different. Is an impact of one of these dust grains likely to destroy the camera on Giotto before it can take the desired photograph? Before Halley and Giotto, we knew very little about the range of sizes and masses of the dust grains, and it was difficult to answer this question.

A Reasonable solution: Regarding the dust grain that might hit Giotto : The dust in our cities is made of small grains that are barely visible. Imagine a spherical grain of radius 0.1 mm. Use the density of rocks, say 4 times the density of liquid water, to get a grain mass of $1.6x10^{-8}$kg.

What is the likely relative velocity of the grain and the camera? The grain, pushed off Halley by thermal forces, has a small velocity relative to Halley, so the grain's orbit through the solar system is almost that of Halley. Halley is in a highly elliptical orbit. Giotto meets it not very far from 1 A.U. Let us use for Halley the speed it would have if it fell from infinity to Earth's orbit. That speed is $2^{1/2}$ times the Earth's orbital speed, about 42 km/s. (Halley actually "falls" from a maximum orbital distance of 35 A.U.) If the spacecraft is relatively slow, say 30 km/s or less, the major component of the relative velocity is the velocity of Halley. So let us use as relative speed 50 km/s. Then the kinetic energy of the grain hitting the camera is 20j.

Is 20j important? What does it take to damage a camera on Earth? Perhaps a thrown stone. How fast can a man throw a stone? If you visualize the thrown stone, it may have the speed of a car moving at, say, 50 km/hour (note the unit: hour, not seconds.). With the stone's mass about 0.1 kg, you get an energy of 10j. The energy of the stone is similar to the kinetic energy of the grain hitting the camera. The stone would damage the camera. Conclusion: The impact of the comet grain on the Giotto camera is likely to damage the camera.

The actual situation: Giotto met Halley on March 13-14, 1986 at a distance of merely 0.9 A.U. from the Sun, when the comet was moving away from the Sun. Giotto came to within 600 km from the comet. The camera took its last picture at the distance of about 1700 km. Then the spacecraft was violently tilted, presumably from the impact of large dust particles. Many instruments became inoperable after the sand-blasting by the dust during the encounter. Equipment on Giotto measured about

12×10^3 impacts of dust grains with masses between 10^{-20} and 10^{-7} kg. Satisfactorily, this range includes our estimate of 10^{-8} kg. We do not know the actual mass of the grain that hit the camera. Giotto's velocity relative to Halley actually was 68.4 km/s, and that was also very nearly the velocity of Giotto's camera relative to the dust grain.

Figure 8.3-4. Nucleus of Comet Halley. Credit: ESA.

Interpretation: The last picture of Halley taken by the Giotto camera showed the comet to have the shape of a peanut, with a length of about 16 km, width about 8 km, covered with a very dark (presumably carbon-rich) surface. Jets of gases reflecting bright sunlight were escaping from sunlit parts of the surface. The jets imply the vigorous evaporation of ices within the comet. In the future, once Halley has approached the Sun several hundred times more, these gases will be gone and Halley's spectacle every 76 years will cease. The grains that now are still encased in ices then will be loose and acquire their own orbits. The orbit of each grain will be similar to the orbit of the comet. But the grains will spread out along this orbit until they constitute an elongated cloud of particles with the shape of Halley's orbit. Whenever the Earth passes through this cloud, it will intercept some of the particles. At those times, we observe a meteor shower. This process is already beginning: Meteor showers around Oct. 21 and May 4 are associated with Halley.

Recent radar observations of the 1996 comet Hyakutake imply that as much as one sixth of all the material ejected from that comet (comet diameter no more than 3 km) is in dust grains larger than 1 mm. Grains with cm sizes are probably loose assemblies of smaller grains, not solid rocks.

Didactics: This problem is supposed to be an enjoyable exercise, giving some opportunity to the students' imaginations. Definitely, the students should estimate the needed parameters on their own, preferably by discussion within groups of 2 to 4 students. Of course, if the students are not used to making estimates based on their general knowledge, then this exercise takes time. When they seek the authoritative knowledge of the professor, they should receive only suggestive questions. This is the same kind of exercise that the experts went through only a few years ago. The time used by this discussion imparts more science to the students than would some additional part of a lecture imposing a few more facts.

The most uncertain parameter in the problem is the size of the dust grain. Class members might well choose a much smaller grain, for instance the size of the wavelength of visible light. In that case, the grain does no damage whatsoever. But then class members should be reminded that we do not know the numbers of dust grains at larger sizes, and danger should be evaluated taking into account "worst cases". Thus they might consider larger grains. At the other extreme, our physical intuition (albeit based on city winds rather than gaseous comet jets) suggests that grains larger than some value (1 mm? 1 cm?) are unlikely to be blown out of the comet. Thus it is rather surprising to detect cm-sized grains from comet Hyakutake.

4. An Asteroid Impact on Earth

Collisions-A new scientific judgment

General science goal: A realistic example of the uneven progress of frontier science.

The Problem: Estimate the energy of an asteroid of radius 1 km hitting the Earth, and compare with the energy released by the largest bomb exploded on Earth. (Assume the asteroid is just a large rock. Estimate whatever parameters you need. A "100 Megaton" bomb yields about $4x10^{17}$joules.)

A reasonable solution: With a density of rock four times that of liquid water, the mass of the asteroid is $1.6x10^{13}$kg. Since most asteroids orbit the Sun in the same sense as the Earth, the relative velocity is probably less than the Earth's orbital velocity. If we take 20 km/s, the energy of the impact is $3x10^{21}$j, comparable to the energy from several thousand of the largest bombs.

Didactics: As in the previous problem, students benefit from discussing the problem and making their own estimates of the density and impact velocity of the object. What matters is the physics, and that the students figure out what parameters need to be estimated. Numerical factors of two are not important.

The Setting: Scientists often participate in fads, which are more politely called paradigms. In the last decade, collisions in the solar system have caught the attention of many scientists and of the public.

For most of the past century, astronomers knew that objects from space occasionally hit the Earth, but these impacts were not considered important. Then, in 1978, came the claim that the dinosaurs became extinct because of the impact of an asteroid about 65 million years ago. The evidence for the asteroid was a layer of iridium-rich material in rock strata, about 65 million years old, found in several places all over the Earth. Iridium is a very rare element, and especially so on Earth because the Iridium originally distributed throughout the Earth has largely sunk down to the Earth's core. Asteroids contain relatively more iridium than the Earth's surface. If the impacting asteroid had a diameter about 8 km, it probably contained 2×10^8kg of Iridium, enough to explain the observations. This argument created much debate.

Now there is compelling evidence for the asteroid impact 65 million years ago, namely an impact crater of the correct age, found on the edge of Mexico's Yucatan peninsula with a diameter of 200 km. There is general agreement that this impact caused the extinction of roughly 70% of all then living species. (There is not yet complete agreement about the dinosaurs, because they might have died out slightly earlier.). At least one additional major extinction event is probably caused by an impact (probably a comet rather than an asteroid, 2.5×10^8 years ago), but some extinctions may be caused by episodes of intense volcanism. The evidence is still debated.

Major international attention focused on comet Shoemaker-Levy 9. This comet was broken into roughly twenty pieces by Jupiter's gravity in 1992. The pieces fell onto Jupiter in July 1994. During the year before these impacts, theoreticians tried to predict the effects of the collisions on Jupiter. Their predictions depended strongly on the unknown mass and solidity of each of the comet pieces. The actual events, especially plumes of gases rising far above Jupiter's surface, were more energetic and violent than had been predicted. Their effects can be seen on Jupiter even five years later.

A new scientific judgment: All the old evidence of impact craters on Earth and Moon is now seen with a different scientific attitude. Possible collisions of celestial objects with the Earth have now become a major scientific topic. An object about 1 km in diameter passed Earth at the distance of only 725×10^3 km, less than twice the Moon's distance. Yet we detected it only after it passed. We are really much more ignorant of the problem than we realized merely ten years ago.

An estimated two thousand near-Earth asteroids exist larger than 1 km. Efforts are beginning to detect systematically all such asteroids and to find which travel on orbits that might some day intersect with the Earth. Unfortunately, such a prediction is not possible with newly arriving comets.

The emphasis will be on searching for asteroids of size 10 km or larger. When they impact on Earth, they tunnel into the Earth and then explode so powerfully that much material is blown into the high atmosphere, above all rain clouds. This material remains up there long enough to be dispersed over the entire Earth. It returns to Earth surface only after months or even years later. The effects of a global cloud of dust, acid rains, and secondary effects on the oceans can cause an extinction episode such as occurred 65 million years ago.

What happens if we actually determine that a large asteroid or comet will hit Earth? The most common suggestion is that we send rockets to the asteroid or comet and use the rocket or bombs to deviate the comet's path. Since Earth is so tiny, even a small deviation will be sufficient if we do it early enough. But this technology is still many years away.

Discussion of large Earth impacts raises strong emotions. To many people, such impacts seem dreadful, indicating a possible future disappearance of humans from the Earth. We are learning, however, that past extinctions spurred the subsequent rapid evolution of new species. The impact 65 million years ago spurred the development of mammals and thus, indirectly, of humans.

5. The Clean up of the Solar System

Collision cross sections-Mean free path - Kinetic Theory

The problem: Estimate the number of years needed for an asteroid flying freely and randomly throughout the inner solar system ($r < 2$ A.U.) before it is likely to crash into a terrestrial planet. Assume there are three planets in this volume, each with radius $2R_E$ (to include captures by the planet's gravity). You may use one (or both) of two analogies: 1) A person is given a ball, blind-folded, turned around many times, and then is asked to throw the ball at a target $1 m^2$ in area, at a distance of 10m. How often does this experiment have to be repeated before the target is likely to have been hit once? How frequently might asteroids orbiting through the solar system "try" this experiment? 2) Use the concept of mean-free-path from kinetic theory.

The astronomical setting: Our Moon shows enormous numbers of craters from past impacts. The lunar dark areas (the "mare") now consist solidified lavas. The lavas long ago rose out of the lunar interior and flowed into giant craters (several hundred km diameter) caused by impacts even longer ago. The radioactive dating of rocks brought back from the Moon (by the U.S. Project Apollo, 1969-1972) shows that most of the impacts, especially the very large ones, occurred in the first 0.5×10^9 years of the Moon's life, roughly in the period from 4.5 to 4.0 x 10^9 years ago. Why did the collisions stop? Apparently, because most of the rocks roaming the solar system at that time either had collided with some planet or had been deflected into orbits such that they left the inner solar system.

A few impacts still happen in the inner solar system even today, that is, in the last 10^8 years. The impact 65 million years ago (problem 8.3.4) is a very energetic example. Some recently fallen meteorites can be analyzed for the amount of cosmic rays that have hit these meteorites while they were in space. (One measures the isotopes ^{3}He, ^{21}Ne, and ^{38}Ar which result from the collision of cosmic rays with rock.) The result is a "space-exposure" age. Typically this age is of the order of 10^8 years. Before the meteorites were exposed to space and the cosmic rays, these objects must have been part of a larger asteroid body. Presumably a collision with the parent asteroid released the objects into space, onto an orbit that ultimately reached the Earth. The exposure ages of 10^8years inform us that the parent asteroids suffered a collision at that time in the past.

Indeed there is evidence for such asteroid collisions. The meteorite from Mars that may carry evidence for long-ago primitive life on Mars (highly debatable) was knocked off Mars 16 million years ago. The composition of the asteroid Vesta (diameter 530 km) led to the prediction that Vesta is the source of a distinctive class of tiny asteroids and some meteorites. Now the Hubble Space Telescope has observed a suitably large crater on Vesta with a diameter of 459 km.

Here we estimate whether a collision time with the terrestrial planets is reasonably on the order of 10^8 years.

One reasonable solution: The target analogy: The area of a sphere 10 m in radius is 400π m^2. If the ball is thrown truly randomly in all directions, a target of 1 m^{2}at 10 m is likely to be hit once in 400π attempts. In the solar-system, the target is Earth at a distance of, say, 1 A.U. from the randomly moving asteroid. The Earth at that distance occupies a fraction of the sky $(\pi R_E^2)/\{4\pi(1\text{A.U.})^2\} = 4 \times 10^{-10}$. Given three planets and gravitational radius $2R_E$, the probability becomes about 5×10^{-9}. Hence about 2×10^8 attempts are needed before there is a reasonable chance of success. If there is one attempt per year, it takes roughly $2x10^8$years.

Kinetic theory: The mean free path is $1/n\sigma$ (ignoring factors of order unity), where n = number of objects per unit volume = $3 / [4\pi/3 \times 2^3]$ A.U. $^{-3}$ = 0.1 A.U. $^{-3}$ and σ is the cross section of an object in the same units, $\pi(2R_E)^2 = 2x10^{-8}$ A.U.^{2} resulting in a mean free path $1/n\sigma$ = $5x10^8$A.U. The mean collision time is $1/n\sigma v$, where v is the relative velocity, say 3 A.U./year (half the Earth's speed), which yields a collision time of about $1.5x10^8$ years.

Interpretation: Obviously the individual numbers are very approximate. One of the uncertainties for the very earliest period of the solar system is the number of terrestrial planets. At least one additional Mars-sized object probably existed for a while, until it hit the Earth and created our Moon. Another uncertainty is the effective cross section of the planet, here taken four times the geometrical cross section. The effective cross section depends on the speed of the asteroid relative to the planet: the lower the relative speed, the more the planet's gravity can deflect the asteroid such as to hit the planet, and the larger is the cross section. There is an additional cross section, of similar magnitude, for the gravitational deflection of an asteroid by a planet that is sufficient for the asteroid to leave the inner solar system.

The main physical uncertainty is the "random" path of the asteroid. During any few thousand years, the orbit of an asteroid is constant and may cross the plane of the Earth's orbit only well inside or well outside 1 A.U., so there is no chance of collision. However, we are learning from "chaos theory" that the orbits of asteroids are changing more often than we had expected, due to "resonances" with the planets, mainly Jupiter. Resonances with Jupiter occur when the ratio of the object's and Jupiter's orbital periods is a ratio of two small integers, so that Jupiter affects the object's orbit repeatedly in the same way. When an orbit is resonant, it may be nearly constant for millions of years, then change "suddenly", perhaps in a mere few thousand years, and then become nearly constant again. Such sudden, large orbital changes may occur every few millions years. They are called "chaotic" changes in orbit because even slightly different initial conditions will lead to different sudden orbital changes at different times. Since we do not know the initial conditions, we must deal with probabilities of sudden orbital changes. After each such change, there is a chance that the new orbit sweeps through Earth's orbital plane at 1 A.U. for many thousands of years, permitting a collision when Earth is in the way. It is this chance that is simulated by the calculation.

The terrestrial and Jovian planets will maintain their orbits indefinitely. The orbits of the Trojan asteroids, situated at Jupiter's distance from the Sun and making an equilateral triangle with Jupiter and the Sun, appear to be stable. Pluto is in a stable resonant orbit with Neptune, that is, Pluto orbits the Sun twice while Neptune orbits it three time. When Pluto crosses Neptune's orbit, Neptune is always a quarter orbit away from Pluto.

Didactics: This should be an enjoyable exercise for the students, allowing some freedom to the imagination. If students come up with parameters that yield 10^9 years, it is more important for them to have chosen some reasonable parameters than being numerically "correct". Certainly, this kind of calculation cannot be expected to agree with reality to better than a factor of, say, three. The two methods used here involve somewhat different assumptions about the "random" velocities and object separations and could easily differ by a factor of two. The answer achieved here, $2\text{x}10^8$ years, is to be considered attractively similar to the observational data regarding the first $5\text{x}10^8$ years and the most recent 10^8 years of the inner solar system.

Chapter 8.4

MORE MECHANICS: NEUTRON STARS AND CLUSTERS OF GALAXIES

D.G. Wentzel

1. Gravity on a Neutron Star

Density of nuclear matter, centrifugal force

Purpose: A stretch of the imagination.

The Problem: Estimate the density of a neutron star assuming that it consists of idealized neutrons that are hard balls with radius of about 0.7×10^{-15}m. If the mass of this star is $2M_o$, compute its radius (compare to the size of the main city of your country), the gravity on the surface (compare to that on Earth), and the escape velocity (compare to the speed of light, c, and ignore relativistic effects).

How high a "mountain" on the surface of the neutron star would you have to climb to gain the same gravitational energy as climbing one of the high mountains on Earth, say Kilimanjaro (Africa), Aconcagua (South America), or Mount Everest (Asia)? Identify some object on Earth with about as much mass as resides in a hand full of neutron star matter.

Finally, compute the centrifugal force acting on gas at the equator of the neutron star for an arbitrary rotational period P. Find that value of P for which the centrifugal force equals the surface gravity (evaluated assuming a spherical star). Explain why one would not expect to observe periods shorter than this P and compare it to observed pulsar periods P > 0.0016 s.

The Setting: The lives of stars are the subject of 8.6. Here we summarize the life of stars only to introduce neutron stars.

Stars are generally static objects. Each layer is supported against gravity by a gradient in the pressure. In a normal star like the Sun, the pressure is that of an ideal gas (p = nkT). The pressure at the center is maintained sufficiently high because nuclear reactions maintain a high

W. Wamsteker et al. (eds.), Developing Basic Space Science World-Wide, 437-447.

temperature. Once nuclear fuel is exhausted, the star contracts, either until some new kind of pressure supports the star or until it becomes a black hole. For a star like the Sun, the new pressure is provided by (degenerate) electrons, and the Sun becomes a white dwarf. For stars with mass between $1.4M_0$ and about $4M_0$, electrons provide insufficient pressure, the star shrinks until electrons are squeezed into protons, the new pressure is provided by the resulting neutrons, and the star becomes a neutron star. For stars with mass above about $4M_0$, even neutrons provide insufficient pressure, no new pressure is available and the star contracts forever, becoming a black hole.

Why can neutrons support a star against its own gravity? The simplest model says that neutrons act like extremely hard balls. That is the model of the present problem. The values of density, surface gravity and escape speed truly stretch everyone's imagination. Neutron stars were theoretically predicted, but the astronomical community remained extremely skeptical that they would actually exist.

Now neutron stars have actually been observed, though only indirectly. Within about 10^7years of the formation of a neutron star, a powerful beacon of radiation rotates with the neutron star and is observable as a pulsar (see problem 8.4.4). Also, when a neutron star is a member of a binary, gases falling from the normal star toward the neutron star begin to revolve around the neutron star. As they gradually spiral toward the neutron star they become very hot and become observable because of their x-ray (and other) emission (see problem 8.4.2). In 1997, the Hubble Space Telescope discovered directly the first non-pulsating "lone" neutron star (see problem 8.4.3).

The Solution: The volume per neutron is $4\pi/3\ r^3 = 1.5x10^{-45}m^3$. If we ignore the small amount of space between the neutrons, their number density is $0.7x10^{45}m^{-3}$ and the mass density is $10^{18}kg\ m^{-3}$. The mass in a "handful", say ten cubic centimeters, has the mass of rock (density 4 x water) on Earth with a volume of 2.5 km^3, a small mountain.

Given the density and the mass, the volume of the neutron star is $4x10^{12}m^3$, and the radius is 10 km. This is smaller than the sizes of many capitol cities of the Earth's countries. The surface gravity is $GM/R^2 = 2.7x10^{12}m\ s^{-2}$, huge compared to 10 m s^{-2} on Earth. A climb of 8000 meters on Mt. Everest is equivalent to climbing $8x10^3x\ 10\ /\ 3x10^{12} = 3x10^{-8}m$ on the neutron star, a height less than the wavelength of visible light. The escape velocity is $(2GM/R)^{1/2} = 2.3x10^8m/s$, about 3/4 c .

When the centrifugal force equals local gravity, $v^2/R = GM/R^2$. With $2\pi R = Pv$, $P^2 = 4\pi^2R^3/GM = 3\pi/G\rho$, and $P = 4x10^{-4}s$. If a neutron star ever had such a period, the star must have lost matter from the equator so rapidly that we would not observe it. The fastest observed pulsars are "safe" from rotational disruption, but only by a factor of 4.

Didactics:

1. The parameters of neutron stars are truly astounding. No wonder that most astronomers doubted their existence until pulsars required it. The escape velocity is near the speed of light. Indeed, the neutron star's radius is only slightly larger than the Schwarzschild radius of 6 km (see problem 8.2.5).
2. The neglect of the space between the spherical neutrons is a simplification no worse than assuming that neutrons are hard spheres.
3. Mathematically, the critical rotational period P can be expressed either in terms of M and R or in terms of ρ. In this case, ρ is actually the primary quantity known physically and, therefore, writing the critical P in terms of ρ is much to be preferred. It not only minimizes the carrying of errors that arose in computing R and M, but shows physically much more comprehensibly what the critical P depends on.

Interpretation: Neutrons are not really simply hard balls. They are made of more fundamental particles. The appropriate equation of state is still uncertain in detail, but one expects most neutron stars to consist of a liquid interior, a crystallized crust, and a gaseous (non-degenerate) atmosphere a few centimeters thick. If the star contains more than about 4 M_o (uncertainty 3 M_o to 5 M_o), then the pressure due to neutrons is inadequate and the star collapses to become a black hole.

2. Accretion disks around neutron stars

Activity: the orbital period of Her X-1

Educational purpose: An example of measurement uncertainties and simple error analysis.

Astrophysics purpose: Introduction to the notion of an accretion disk, a common phenomenon around young planet-forming stars, white dwarfs, neutron stars, and black holes.

The Observational problem: The graph shows the x-ray flux versus time for an x-ray source called Her X-1. The measured data points are shown by the small crosses. The vertical lines show the uncertainty of the individual data points, based on the Poisson statistical error from photon counting. The horizontal lines represent the widths of the detector energy channels and are not strictly error bars. A computer program has connected the data points.

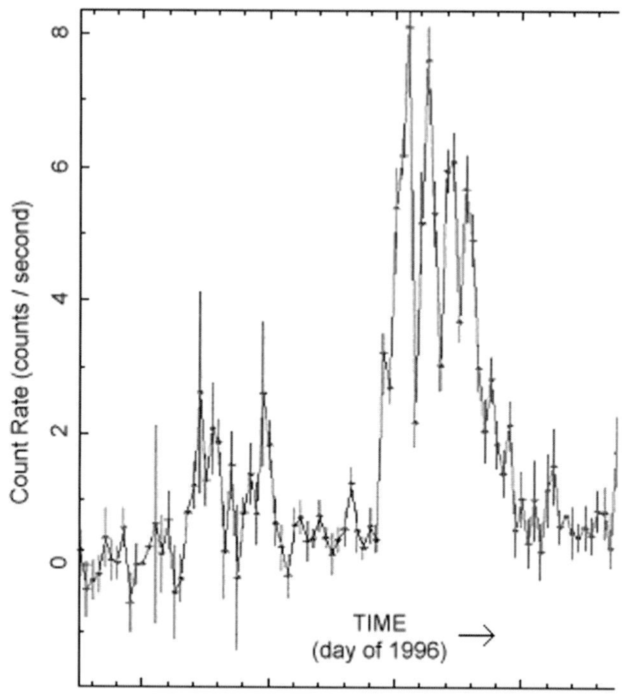

Figure 8.4-1. Data for Her X-1 measured by the Earth satellite Rossi X-ray Timing Explorer (NASA).

These data, looked at from afar, clearly show an oscillation with a period of roughly two days. Measure the period as accurately as you can and estimate the uncertainty in the deduced period: the correct value probably lies between 1 and 2 days. Explain how you arrive at this conclusion.

The Solution: Much judgment is involved in determining the period, and that is the main point of this exercise. The most straightforward attempt is to count the number of periods over the entire interval. But obviously there are some intervals such as at t = 121 days when no maximum is observed. Somehow one has to "correct" for such problems by mentally adding maxima to be counted, based on the rest of the plot. That way students may well determine there are 27 to 30 peaks in 40 days, for a period of 1.3 to 1.5 days. To avoid the "correction", perhaps it is better to select a part of the plot where maxima are very clear, such as between days 129 to 136. But then the time interval for a whole number of periods is difficult to measure accurately. For instance, the high data points with counts of 6 to 8 per second are probably not exactly at the midpoint of the entire peak when one takes into account the data points just before or afterwards. Similarly it is difficult to find the midpoint of a steep slope. Different individuals or groups tend to find values of the period in the range 1.4 to 1.8 days. Which method is better? Should the resulting P simply be averaged? It is not obvious.

Didactics: It is useful to give one copy of the graph (and a tool to measure lengths) to each group of three to five students. Let them work out the period before they receive any description of the radiation or the object that emits the radiation. The groups should talk quietly enough so that neighboring groups cannot hear the answers discussed. When (almost) all groups are finished, ask for the periods from one group, then ask whether any groups have different periods. Let the students suggest reasons for the differences in period derived by different groups. The end result should be the realization that the derived period has an inherent uncertainty, and an estimate of the magnitude of the uncertainty.

Once the discussion is finished, students may be told: If one Fourier-analyzes a much longer data set, one obtains a "signal" in a small range of periods centered on a period of 1.7 days.

The Astrophysical setting: (not really needed for this practical problem): The object Her X-1 is so named because it was the first x-ray source observed in the part of the sky that includes the star constellation of Hercules. It was first carefully observed in the early1970's by the X-ray satellite with the Swahili name Uhuru. The satellite was so named because it was launched off the coast of Kenya on December 12, 1970, on the anniversary of the country's independence.

Her X-1 is a binary star: a normal star and a neutron star orbit around each other. The neutron star is an x-ray pulsar with a rotational period of 1.24 sec. (problem 8.4.4 includes a diagram for a pulsar.) The x-ray data in the graph are taken with low time resolution, so that the pulsar signal is averaged out. Clearly, there is a period of roughly two days. This is the period at which the two stars orbit about each other.

Why the x-ray emission? A binary pair of stars necessarily has angular momentum. If gas from a normal star escapes toward a neutron star, it has angular momentum relative to that neutron star, and cannot fall directly toward it. Instead, the gas swirls about the star, gradually spiraling inward. The gas becomes part of an "accretion disk", centered on the neutron star. In this disk, each part of the gas moves very nearly at the Keplerian velocity around the star, but there is also a very small component of velocity inward.

The small inward velocity has a major physical implication: the gas must lose angular momentum. Viscosity is far too small. Probably turbulence within the gaseous disk acts like a viscosity, gradually transporting angular momentum outward, thus allowing the gas to migrate inward. (The turbulence, in turn, may be caused by magnetic fields.) Inevitably, as the gas sinks deeper into the gravitational well of the neutron star, some of the orbital kinetic energy is turned into heat. The detailed process by which this occurs is not known, though some kind of viscous heating is probably involved. The gas in the disk is heated to tens of millions of degrees and so radiates largely in x-rays ($h\nu$ of the order of kT, see problem 8.5.4).

We observe the orbital period in the x-rays because we view different parts of the accretion disk as it revolves, with the neutron star, around the other star. Sometimes we see the "top" of the disk, sometimes the edge, sometimes the "bottom" of the disk.

Finally, when the gas has swirled very close to the star, the star "grabs" the gas (probably by magnetic forces) so that the gas falls the remaining distance onto the neutron star. The orbital kinetic energy which the gas had just before infall is released in the form of heat on or near the surface of the neutron star, and this heat is radiated away as x-rays. Probably the gas is funneled onto the neutron star near its magnetic poles. Then we see a "pulsating" x-ray source because the rotation of the neutron star repeatedly carries the x-ray emitting region into and out of our view. An x-ray pulsar is thus analogous to a radio pulsar.

3. Radiation from Neutron Star, Accretion Disk

Energy in Keplerian orbits

The Problem: Gas spirals slowly toward a (non-magnetic) neutron star, forming an accretion disk, and then falls a short distance (by much less than a neutron star radius) onto the neutron star. What is the ratio of the energy radiated away by the gas while in the accretion disk to the energy released by the gas upon falling onto the neutron star? (Assume that gas in the accretion disk at any one time is in a practically circular orbit satisfying Kepler's third law.)

The Theoretical answer: One.

Interpretation: Why "one"? Every Keplerian orbit involves a kinetic energy equal to half the gravitational energy released while the gas approached from far away, starting approximately at rest. Therefore, each element of gas in a Keplarian orbit has converted half the gravitational energy to heat and radiation. Specifically, this is valid for those elements of gas just ready to fall onto the star. Once they have fallen onto the star, the entire gravitational energy must have been converted to heat. Since half of that heat was accounted for in the accretion disk, the second half must be heat released on the neutron star.

Interpretation: This situation may be approximately true in the class of x-ray sources called Low Mass X-ray Binary Sources. However, in many other accretion disks, including objects like Her X-1, magnetic fields are important. The physics of these accretion disks is still a very active subject of research.

4. Crab Nebula and Pulsar Slow-down

Rotational energy, moment of inertia

The Setting: The Crab Nebula is the gas ejected by a supernova explosion seen on Earth in the year 1054 A.D. Even now, centuries after the explosion, the nebula is still filled with highly relativistic electrons and with electrical currents and their magnetic fields. The energies inherent in these phenomena cannot be left over from the explosion. Apparently, energy is supplied continuously, at the rate of about $1.2 \times 10^5 L_0$ (L_0 = solar luminosity). What is the source of this enormous energy?.

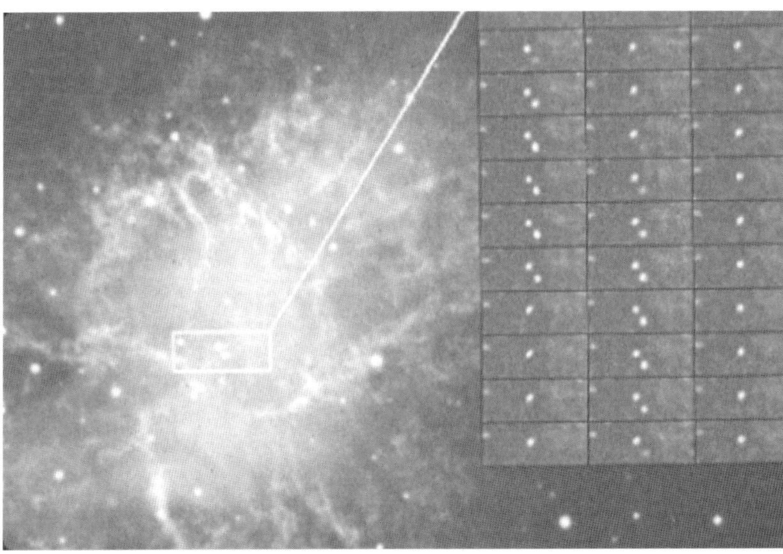

Figure 8.4-2. A sequence of images showing the flashes at visible wavelengths from the Crab pulsar, located at the center of the Crab Nebula. Credit: National Optical Astronomy Observatories.

F. Pacini (of Italy) suggested that the energy needed for energizing the Crab Nebula might come from the slowing-down of a rotating neutron star at the center of the nebula. But how would one detect a neutron star? One year later, in 1967, pulsars were discovered. Pulsars are beacons of radiation that sweep past us. Presumably the beacons are attached to some rotating compact object, such as a neutron star. Initially the pulsars were detected in the radio range, and one pulsar was promptly observed at the center of the Crab Nebula, sweeping past us 30 times per second.

Pulsars have a very precisely defined period, but over many years the period increases, indicating that the rotation of the central object gradually slows down. The energy made available by the slow-down of the pulsar in the Crab Nebula, if the central object is a neutron star, neatly accounts for the energy needed in the Crab nebula. This agreement is one of the main reasons that pulsars were promptly identified with neutron stars, thus providing good evidence that neutron stars exist.

The problem: Write down the kinetic energy of rotation, E, of a neutron star in terms of M, R, and the rotational period P. Assume uniform density. Then write an equation for the rate of change of rotational kinetic energy, dE/dt, in terms of M, R., P, and dP/dt, where dP/dt measures the slow-down of the rotation (with $dE/dt \ll E/P$, R = constant). Evaluate dE/dt for the pulsar in the Crab Nebula using observed values P = 0.0333 s and $dP/dt = 4.21 \times 10^{-13}$ s/s, and theoretical values for a neutron star R = 10 km and $M = 1.4 M_o$. Compare dE/dt to the power needed to energize the Crab Nebula.

The Solution: The moment of inertia of a homogeneous spherical object is $I = 2/5\, MR^2$. The rotational energy is $E = 1/2\, I\, \omega^2 = 2\pi^2 I/P^2$, and $dE/dt = -8/5\, \pi^2 MR^2\, (dP/dt)\, /\, P^3 = 1.2 \times 10^5 L_o$.

Interpretation: The decrease in rotational energy matches the energy needed by the Crab Nebula remarkably precisely! Of course, we assumed a mass of $1.4 M_o$ for the neutron star, and there is no independent measurement of M. But this value is in the correct range expected from theory.

5. Clusters of Galaxies and Cosmology

Integral conditions

Physics Purpose: Theoretical integral conditions provide useful information when detailed observational information is lacking. Integral conditions suggest that clusters of galaxies consist mostly of invisible "dark matter".

Didactic Purpose: When we deal with clusters of galaxies, we are truly dealing with "astronomical numbers": huge distances, sizes, and time scales. And yet the human mind can deal with such phenomena!

Physics Introduction to the integral condition called the Virial Theorem. Consider any number of planets circling around the Sun. For each, the kinetic energy is -1/2 x their gravitational energy. Add them up, and you find that the total kinetic energy in the system $K = -1/2\, \Omega$, where

Ω is the total gravitational energy. More generally, consider a star circling around the center of a cluster of stars, experiencing gravity $g(r) = GM(r)/r^2$. For each star, the kinetic energy is -1/2 times its gravitational energy. For the total system of the stars, $K = -1/2\ \Omega$.

In fact, $K = -1/2\ \Omega$ is true for any system of stars or galaxies, held together by their own gravity, and kept from collapsing by orbital motions. It also holds for a star, held together by its own gravity, and kept from collapsing by thermal motions (gas pressure). (See didactics of problem 8.6.2). The relation $K =-1/2\ \Omega$ is called the Virial Theorem. Its general proof is given in Chapter 8.9, but it is not very useful for most students.

Often in astrophysics, we have insufficient information to make detailed models of observed objects. But the Virial Theorem provides important constraints. Here it is used for the establishing the "missing mass" in clusters of galaxies.

The Problem: A spherically symmetric cluster of galaxies, diameter $3x10^6$ light-years, contains 10^2 galaxies, distributed uniformly through the cluster, traveling at a (root mean square) speed of 10^3km/s, each containing stars with a total mass $10^{11}M_o$.

1. Calculate the gravitational and the kinetic energies residing in the cluster of galaxies. (For the gravitational energy, assume uniform density, so that $\Omega = -3/5\ GM^2/R$.)
2. In view of the Virial theorem, are the galaxies likely to be kept within the cluster by the gravity due to the stars in the galaxies?
3. Compare the time needed for a galaxy to cross the cluster to the probable age of the cluster, roughly the age of the Universe, $1.3x10^{10}$years. Might this cluster of galaxies simply be moving apart if it was created early in the age of the Universe?

If "dark matter" is distributed uniformly in the cluster, with the same random velocities, how much matter would be needed to satisfy the Virial theorem? Compare to the amount of matter observed in galaxies.

The Solution: The kinetic energy per galaxy (on average) is 1/2 $2x10^{41}$ $(10^6)^2 = 10^{53}$j . For the cluster $K = 10^{55}$ j. The gravitational energy of the cluster is 3/5 $6.7x10^{-11}$ $(2x10^{43})^2/1.5x10^{22} = 10^{54}$j. This gravitational energy cannot constrain the random motions. The crossing time is $3x10^{22}/10^6 = 3x10^{16}$s $= 10^9$years. Galaxies could easily have escaped by now. Since they have not, the Virial Theorem must be satisfied. The total mass in the cluster must be about twenty times the mass of stars observed in galaxies.

Interpretation: Typical clusters of galaxies may be 10^7 light-years in diameter and contain roughly 10^3galaxies with a size of 10^5 light-years, separated by 10^6 light-years. Typical motions among the galaxies are

10^3km/s. These clusters have presumably lasted for over 10^{10}years, and so the galaxies cannot merely be flying apart. Yet, if we measure the mass of each galaxy according to the amount of starlight we receive, the total mass of the galaxies does not yield enough gravity. For a while there was the possibility that much gas resides in the vast space between the galaxies. The gas is now observed by its x-ray emission. Its mass is not enough.

There must be much "dark matter" in clusters of galaxies. What is this matter? We know nothing more than we know regarding the "dark matter" in our Galaxy (problem 8.2.6.) However, the problem is more significant for clusters of galaxies: It is estimated that the dark matter in clusters of galaxies amounts to roughly 0.2 of the density that closes the Universe (that is, 0.2 of the density for which the Universe will expand forever with forever decreasing velocity of expansion). For comparison, one can compute the amount of baryonic mass (protons, neutrons) in the Universe that was needed to create, during the Big Bang, the now observed deuterium, ^{3}He and ^{7}Li. One finds an upper limit of baryonic mass of 0.2 times the density that closes the Universe. Therefore, it is possible that the dark matter resides in galaxies in the form of white dwarfs, neutron stars, or black holes left over from earlier stars, or large planets or very cold gas. However, many astronomers take this upper limit on baryonic mass as an additional argument that the universe contains much non-baryonic matter, such as neutrinos or axions. For instance, neutrinos might have a finite mass such that they are gravitationally confined to clusters of galaxies.

Didactics:

1. Students may have heard that we look back in time as we look at ever further objects. They may ask: Perhaps the clusters of galaxies are so far away that they are young and have not had time to expand? No, these clusters of galaxies are within 10^9 light-years from us.
2. The gas between galaxies has been observed to emit primarily x-rays, indicating a temperature of the gas about 10^8 °K. Why is the gas so hot? At this temperature, the thermal speed of protons and the sound speed are about 10^3 km/s. Perhaps, when the gas was less hot, the galaxies moved through the gas supersonically and created powerful shock waves. They churned up and heated the gas until the galaxies were no longer supersonic relative to the gas.
3. The Virial theorem must hold true for equilibrium, but it is not sufficient to guarantee equilibrium. Indeed, galaxies of a uniform density throughout the cluster would not be a distribution in equilibrium. Therefore, the use of the Virial Theorem in problems such as this one only yields estimates, surely no better than a factor of

two. But that is quite satisfactory for exploratory problems such as this one.

4. The gravitational energy of a self-gravitating spherically symmetric object can be derived if one considers the object having been built up to a radius r with mass M(r) and then adding mass m: the gravitational energy released is -GM(r)m/r . [See problem 8.2.6 and Chapter 8.9 for discussion of g(r).] The whole object has the gravitational energy $\Omega = -\int[GM(r)/r]dM(r)$. With $dM(r) = 4\pi\rho r^2 dr$ and uniform density ρ, the result is $\Omega = -3/5\ GM^2/R$.

Chapter 8.5

THERMAL RADIATION

D.G. Wentzel

1. Temperatures on Icarus, Moon and Mars

Stefan-Boltzmann Law

The Problem: First, derive an equation for the surface temperature on the Sun-facing point of an asteroid, as a function of the distance, r (in A.U.), from the Sun. Assume as given that the energy flux from the Sun at 1 A.U. is F, in w/m^2, that the surface reflects a fraction A of the incident cnergy, that a thermal surface emits $\sigma T^4 w/m^2$, and that the Sun-facing point faces the Sun forever.

Second, evaluate your equation for the temperature (in °K) on the asteroid Icarus (A = 0.1) when it is 0.2 A.U. from the Sun, on the Moon (also A = 0.1) at 1 A.U. from the Sun, and on Mars (A = 0.16) at 1.6 A.U. from the Sun. (Assume negligible atmosphere on Mars). Parameters needed: the energy flux at 1 A.U. is F = 1367 w/m^2, and $\sigma = 5.67 \times 10^{-8}$ $w/m^2/K^4$. Speculate on a design of a space suit or heat shield that might allow a human to survive in sunshine on Icarus when Icarus is 0.2 A.U. from the Sun.

Third, suppose the asteroid rotates so rapidly that the temperature is the same all over the asteroid, so that the energy arriving from the Sun is re-radiated uniformly in all directions. By what factor is the temperature lower than your first computation? According to what (qualitative) criterion would you choose which method of estimating the temperature is better?.

The Solution: The energy flux received at distance r is F/r^2 w/m^2 where r is in A.U. and F is given in the problem for r = 1. A fraction (1-A) of this energy is absorbed. Since the point of interest continually faces the Sun, this amount of energy must also be radiated away, at the rate $\sigma T^4 w/m^2$. Therefore, $T^4 = (1\text{-}A)F/\sigma r^2$. Given the numerical values of F and σ, the result is $T = 395\ (1\text{-}A)^{1/4} r^{-1/2}$ °K . The surface of Icarus (see

W. Wamsteker et al. (eds.), Developing Basic Space Science World-Wide, 449-457.

8.3.1), under the given conditions, has T = 863 °K. Perhaps an astronaut could survive if the space suit or heat shield were extremely reflective and if the heat absorbed at the Sun-facing side were to be conducted to a much larger shaded area to be radiated away. On the Moon, with r = 1, T = 385 °K. For Mars, the temperature is 300 °K.

For a rapidly rotating body with uniform temperature, energy is absorbed over an area πR^2 but radiated from an area $4\pi R^2$. Therefore, T^4 is lowered by a factor of 4, and $T = 279\ (1\text{-}A)^{1/4} r^{-1/2}$ °K. This value requires that the period of rotation is much less than the time it takes for the surface to cool off.

Interpretation: For the Moon, the temperature estimated for the Sun-facing point is quite realistic. The space suits of the Apollo astronauts on the Moon had to be cooled when the men were in sunshine. But if the astronauts stepped into the shadow of a big rock, they radiated away enough heat so that their space suits needed heating. Away from the Sun, the lunar surface cools to 110°K.

On Mars, T = 300 °K = 27 °C is really a theoretical maximum if there is no atmosphere, but the surface becomes this warm occasionally because of a slight greenhouse effect. At night the surface temperatures may drop to 130 °K. Where the Mars Pathfinder landed in July 1997, the surface temperatures were much less extreme, 197 °K to 263 °K. The atmosphere there is much cooler than the surface: if you were standing on Mars, your nose would be 20 °C cooler than your feet. Humans on Mars will need space suits in any case because the atmosphere has very low pressure and lacks oxygen.

When a part of the Moon is eclipsed by the Sun, the surface cools off significantly in about two hours. Assuming that the surfaces on the Moon and asteroids are similar, an asteroid surface needs about two hours without sunshine to cool off. Therefore, an asteroid rotating with a period less than about one hour (and tumbling so that all parts are heated) more nearly satisfies the second version of the problem. The surface on Icarus, with a rotation period of 2.3 hours, more nearly satisfies the first version.

Didactics: The solar energy flux at the Earth must be measured from satellites above the Earth's atmosphere. Space experiments are usually very hard to calibrate accurately. It was a major technical achievement to build an instrument that could measure the solar energy flux accurate to about 0.1% . The measured value permits us not only to determine the solar luminosity but to detect that the solar luminosity in fact varies by roughly 0.1% over the years. (See problem 8.7.4)

Since F is known so accurately, the temperatures were evaluated accurate to three decimals, but variations in A (called the albedo), changes in r (the Moon and Earth together change solar distance by about

1% during a year) and variations in the direction of the surface relative to the Sun make even the second decimal inaccurate. The quoted observed temperatures, uncertain by several degrees, were determined from the infrared emissions of the lunar and Martian surfaces, observed by satellites orbiting the Moon and Mars, respectively.

Effect of the Earth's atmosphere: The formulae derived in this problem assume radiation directly from the surface into space, without interference by an atmosphere. For the Earth, A = 0.35, averaged over the Earth and over the year. 35% of the solar energy is reflected back into space, mostly by oceans, clouds, and ice near the poles. The visible light that does reach the Earth's surface is re-radiated in the infrared (according to Wien's law). The Earth's atmosphere (largely its water vapor and carbon dioxide) absorbs the infrared. The radiation is re-emitted and re-absorbed many times, but it gradually wanders upward to cooler layers of the atmosphere, from where it finally escapes into space. The more water vapor or carbon dioxide reside in the atmosphere, the warmer must be the surface so that the radiation migrates upwards and escapes as fast as solar energy is absorbed. This warming is known as the greenhouse effect.

An often quoted measure of the natural greenhouse effect is the following: The Earth's average temperature at the surface is 15 °C = 288 °K. If we did not have the greenhouse effect, but still had enough atmosphere and oceans to makes the temperature uniform over the Earth, then the second version of the problem would apply and our temperature would be –20 °C = 253 °K. The Earth would probably be frozen and without life.

The greenhouse effect is a completely natural phenomenon which has occurred for many millions of years. Since the amount of water vapor and carbon dioxide change in the course of millions of years, due to changes in Earth's volcanism and in the oceans, the Earth's surface temperature has also changed slowly. For instance, at the time of the dinosaurs, 100 million years ago, our atmosphere contained more carbon dioxide and the climate was warmer globally.

Since the beginning of the industrial era, about 150 years ago, human activity has added carbon dioxide to the atmosphere, thus increasing the greenhouse effect, and forcing a gradual warming of the atmosphere . This process has accelerated in the last three decades. The global warming measured during the last two decades is probably due to the human-produced carbon dioxide. But there are alternative explanations for the measured global warming, such as a small increase in the average energy received from the Sun (see problem 8.7.4).

2. The radii of Stars

Stefan-Boltzmann Law-Selection effect in data interpretation

The Problem: The diagram is a plot of the luminosity L and surface temperature T of the twenty nearest stars (symbols ●) and of the twenty apparently brightest stars (symbols x). Identify the largest and the smallest star shown on the diagram. Determine their radii (in units of the Sun's radius R_o), either from $L = 4\pi R^2 \sigma T^4$ or by drawing lines of constant $r/R_o = 10^{-2}$, 1, and 10^2 R_o into the diagram and interpolating.

Which are larger, the stars on the upper or on the lower main sequence?

On the basis of this diagram, is the following text written by a young student correct? "Astronomers observing the apparently brightest stars have learned that most stars are much larger than the Sun." Explain your answer.

The Setting:

On observing a star with a telescope, we can measure three quantities. One is the color of the star, which tells us its surface temperature, T. (Stellar radiation fits a thermal Planck "black body" radiation curve fairly accurately). A second quantity is the distance to the star. For a sufficiently nearby star, we measure its parallax, that is, its apparent shift relative to distant stars caused by the Earth's motion around the Sun. A third quantity is its apparent brightness (units w/m^2). Distance and apparent brightness separately tell us nothing directly about the star, but the two can be combined to yield the stellar luminosity, L = (surface of sphere at Sun's distance from the star)x(apparent brightness). For ease of thinking, all values of L will be expressed in terms of the solar luminosity, L_o.

On a photograph, the size of a star image is determined by our atmosphere (twinkling) and by the size of the telescope (diffraction). The size of the star image tells us nothing about the actual size of the star. The radius of nearly all stars is obtained theoretically from $L = 4\pi R^2 \sigma T^4$ or, relative to the Sun, $\log(R/R_o) = 1/2 \log(L/L_o) - 2 \log(T/T_o)$.

Do stars come in all sizes and temperatures? No, as shown by the adjacent "L-T" diagram. (Diagrams like this are often called an HR diagram after the astronomers who first recognized the usefulness of the diagram, Hertzsprung and Russell. For such historical reasons, T increases to the left in HR and L-T diagrams.) Most stars fall along a diagonal band called the "main sequence". But there are also other stars, notably the largest "supergiants" in the upper right and the smallest "white dwarfs" in the lower left. (The "white" in white dwarf is a misnomer, they tend to be blue.) The physical reason for the distribution of stars in the diagram is outlined in Chapter 8.6.

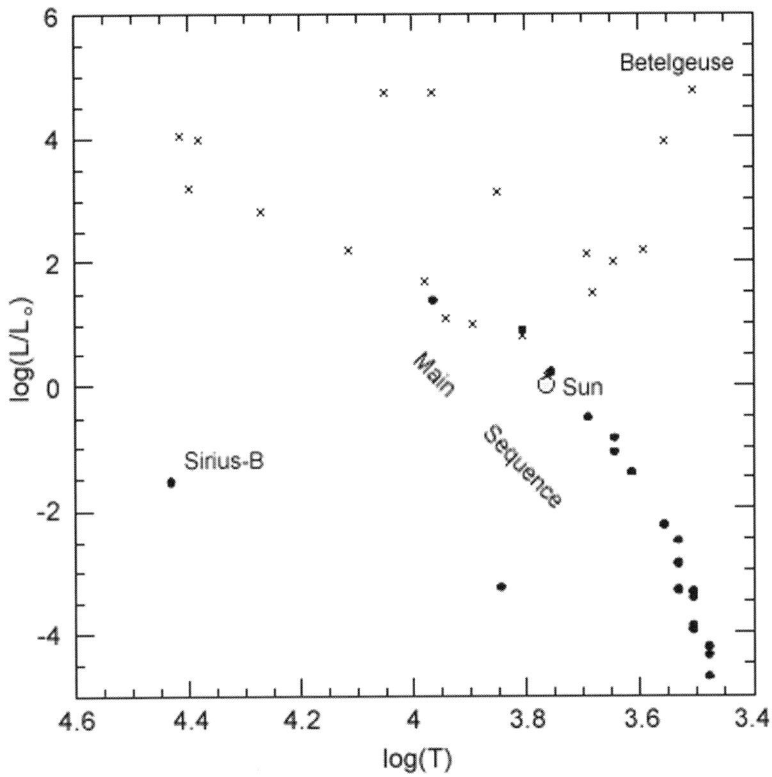

Figure 8.5-1 Hertzsprung-Russell diagram for the 20 nearest and 20 Brightest stars.

The Solution: The largest stars are in the upper right of the diagram. Betelgeuse is some 800 times larger than the Sun. If Betelgeuse were in the place of the Sun, it would reach past Mars and almost to Jupiter. No wonder stars like Betelgeuse are called supergiants. The smallest stars are the white dwarfs, roughly 100 times smaller than the Sun, similar to the size of the Earth.

Since the axes of the diagram are logarithmic, lines of constant radius are straight lines. Their slope of 1/4 is less than the slope of the main sequence. Thus blue main sequence stars are larger than the Sun, which is larger than red main sequence stars.

The student's sentence is wrong. "Most stars" are more like the nearest stars. Each star in a given nearby volume has been listed. The apparently brightest stars are generally not nearby. Being very luminous, according to the diagram, they can appear bright even if seen at large distances. They must be spread over a very large volume. In any given volume, these luminous stars are only a small fraction of all the stars. The apparently brightest stars are not "most stars".

Didactics:

1. This astronomical problem gives a good example of a selection effect, which affects many aspects of the physical sciences and of everyday life. An analogy in towns of developing countries might be: headlights from the infrequent cars are seen from far away; and they attract attention ; the much weaker lights and reflections from the bicycles are seen only if the bicycles are nearby ; yet the most common mode of transportation is the bicycle. An analogy for air travelers might be that distant bright lights from city shopping centers are much more visible from airplanes than are the ordinary lights on ordinary streets almost underneath the plane. Newspapers often report on "surveys" of opinions which show strong selection effects.
2. Each student or group of students should receive a copy of the diagram so that they may read off L and T and/or draw constant-R lines.
3. Sirius A is the only star (other than the Sun) that is in both groups, both nearby and very bright. Its companion, Sirius B, is one of the white dwarfs. Many luminosities are at least as uncertain as the sizes of the symbols in the diagram. Distances have recently been revised, for some stars significantly, by the observations of the European satellite HIPPARCOS. Temperatures of white dwarfs are also uncertain by more than the sizes of the symbols.

3. The Surface temperature of a Neutron Star

Energy conservation-Wien's Law

The Problem: An isolated neutron star moves through the interstellar gas and gravitationally accretes on the order of $dM/dt = 10^7$ kg/s of interstellar hydrogen. Assume that this material falls onto the surface of the neutron star, distributed uniformly over that surface. The gravitational energy released appears as thermal energy that must be radiated away. What is the surface temperature of the star? Assume a uniform density inside the star $\rho = 10^{18}\,\mathrm{kg/m^3}$.

In what wavelength band (radio, infrared, visible, x-ray or gamma-ray) does the star radiate most intensely?

Compare the total luminosity to that of the Sun. Use a neutron star radius R = 10 km

The Setting: Supernova explosions may leave behind a neutron star (introduced in problem 8.4.1). Some neutron stars remain observable for many thousands of years, either because they are pulsars (problem 8.4.4) or because they acquire gas from a companion star (problems 8.4.2,

8.4.3) But if the neutron star is old and isolated, the only source of energy may be the interstellar gas falling onto the star. Can we detect old lone neutron stars by the radiation from infalling interstellar gases?

The solution: Energy is released on the surface of the neutron star because matter falls from infinity to the surface. The gravitational energy released per unit of mass is GM/R. The rate at which mass is accreted is dM/dt. Therefore, the total rate of energy release or luminosity is L = GM (dM/dt)/R . The matter is accreted over the entire surface and thermalized, so we have for the total luminosity $L = 4\pi R^2 \sigma T^4$, where $\sigma = 5.7 \times 10^{-8}$ w/m^2/K^4. Equating the luminosities yields $\sigma T^4 = GM (dM/dt)/(4\pi R^3) = 1/3\, G\rho\, dM/dt$. The result is $T = 2.5 \times 10^5$ °K. By Wien's law, $\lambda(\max) = 2.9 \times 10^{-3}/T = 10^{-8}$ m . This is in the far-ultraviolet and "soft" x-ray range. The total luminosity is about $0.7 \times 10^{-3} L_O$.

Interpretation: The rate of infall, dM/dt, is extremely uncertain because it is proportional to the interstellar gas density and is very sensitive to the velocity of the star relative to the gas. Furthermore, if the star is strongly magnetized and rotating rapidly, the centrifugal forces may give much smaller infall rates. The assumption that the gases fall onto all parts of the star is also debatable. If the star's magnetic field is still strong, the gas may be funneled onto the magnetic poles, a smaller area and thus a higher temperature. Fortunately, these uncertainties enter T only to the power 1/4.

It is not obvious that the infalling gas really creates a Boltzmann velocity distribution within the already present gas at the surface of the star. Thus it is not obvious that the spectrum will have a "black body" Planck character. Rather, the gas may radiate at the moment of impact and first collision, before it has mixed with surface gases. The collisions and radiated photons then are more powerful (problem 8.5.4). In the extreme, if the energy of an infalling proton is converted directly into escaping photons, then the energy emerges as gamma rays.

Several x-ray sources have been suggested to be isolated neutron stars accreting gas from their surroundings. In 1997 the Hubble Space Telescope identified an extremely faint optical object with one of these x-ray sources, thereby establishing with virtual certainty that the x-ray source is an isolated neutron star. However, it is not yet clear that the x-rays are caused by gas accretion onto an old neutron star. They may merely represent the cooling of a hot young neutron star. A few more sources are awaited eagerly in order to better learn how neutron stars evolve and possibly interact with their surroundings.

Didactics: In a rough approximation, gas flowing past a star can be captured with an impact parameter a such that $v^2 < GM/a$. Thus the largest impact parameter of gas that can be captured is proportional to v^{-2}, and the capture cross section is proportional to v^{-4}. The mass capture rate is proportional to the mass flux of gases past the star, ρv, times the capture cross section, so dM/dt is proportional to ρv^{-3}.

4. The Solar Corona and Clusters of Galaxies

Wien's Law, Bremsstrahlung

The Problem: Assume that "black-body" radiation is emitted by thermal electrons colliding with protons. Evaluate $h\nu$ in terms of kT at the peak of the Planck spectrum $I(\nu,T) = (2h\nu^3/c^2)\exp(-h\nu/kT)$. What does this value of $h\nu$ imply about the minimum energy necessary for the colliding electrons that emit this photon? Compare to the mean electron energy 3/2 kT. Given that the electrons have a Boltzmann velocity proportional to $\exp(-mv^2/2kT)$, estimate a likely maximum energy for the colliding electrons.

Now assume that you have a gas, with a Boltzmann velocity distribution at some temperature T, that is transparent to its own radiation. Therefore, the intensity of the Planck spectrum definitely does not apply. But you can still make a prediction about the shape of the spectrum: Given what you have just learned about colliding electrons and the resulting radiation, predict the typical energy of the photons emitted by this transparent gas. For T in the range of 10^6 to 10^{8}°K, what is the wavelength band emitted most intensely (radio, infrared, visible, ultraviolet, x-ray, γ-ray)?

The Setting: Many astrophysical gases have temperatures of millions of degrees, for instance the solar corona (about 2×10^6 °K), accretion disks around neutron stars (10^7 °K), and the gases between galaxies in clusters of galaxies (about 10^8 °K). (See problems 8.6.1, 8.4.2, and 8.4.5, respectively.) These gases are clearly not opaque to their own x-rays, so they do not radiate like "black bodies". At what photon energies might one expect these gases to radiate? The radiation is presumably due to collisions of electrons with protons, a process usually called Bremsstrahlung. This problem asks the students to interpret Wien's law in terms of collisions, with the conclusion that the most intense emission from such transparent gases also follows Wien's law. Therefore, these very hot gases radiate mainly x-rays.

The Solution: The maximum of the Planck spectrum occurs at $h\nu$ = 3kT, twice the mean particle energy. The energies of colliding electrons emitting photons with $h\nu$ = 3kT must have been greater than the photon energies, $1/2\ mv^2 > 3kT$. Yet the energies cannot have been much greater, because much faster electrons are rare. The Boltzmann velocity distribution falls off sharply at about $1/2\ mv^2 > 6kT$. Thus the peak of the

Planck spectrum is due to electrons with kinetic energies roughly between 3kT and 6kT.

The same collisions will occur even if the medium is transparent. Therefore, the x-ray spectrum from a hot transparent gas should have a maximum near $h\nu = 3kT$, just like the Planck spectrum does. For T in the range of millions of degrees, the photon energy is in the x-ray range. The x-rays must be observed from above the Earth's atmosphere.

Interpretation: The estimate made here for the peak of the spectrum is quite good for $T > 10^7$ °K. At these high temperatures, the x-radiation is nearly a continuum, and one determines the gas temperature by the ratio of x-ray intensities in two wavelength bands, much like one determines the temperatures of stars by measuring the relative intensities in red and blue. For the less hot solar corona, the details of the spectrum are quite complicated because atomic transitions cause emissions at many wavelengths with intensities that exceed the collisional continuum.

Chapter 8.6

THE LIVES OF STARS

D.G. Wentzel

Introduction

The Relevant physics: Problem 8.5.2 introduced the kinds of stars we observe. The physical parameters deduced directly from observations are luminosity and surface temperature, and indirectly the size of a star. These refer to the radiative properties of the surfaces of stars. We encounter a variety of other physics when we try to explain the luminosities and sizes of stars. The problems are arranged to isolate various aspects of physics:

- Hydrostatic equilibrium : The isothermal atmosphere
- Energy conservation: Conversion of gravitational energy into heat and radiation
- Energy conservation including $E = mc^2$: The conversion of matter into energy by nuclear fusion
- The equation of hydrostatic equilibrium: a simple solution and an integral limit for a star
- The diffusion equation applied to radiation: a simple solution
- The diffusion of radiation: the Thomson scattering cross section
- Quantum theory: Degenerate electrons and their pressure

The Astronomical setting:

The several physics problems each treat one episode within the lives of the stars. A framework for these problems is provided by the following outline how the Sun was born, how it lives, and how it will die. The L-T diagram, used in problem 8.5.2 to describe existing stars, can also be used to represent the life of the Sun.

W. Wamsteker et al. (eds.), Developing Basic Space Science World-Wide, 459-476.

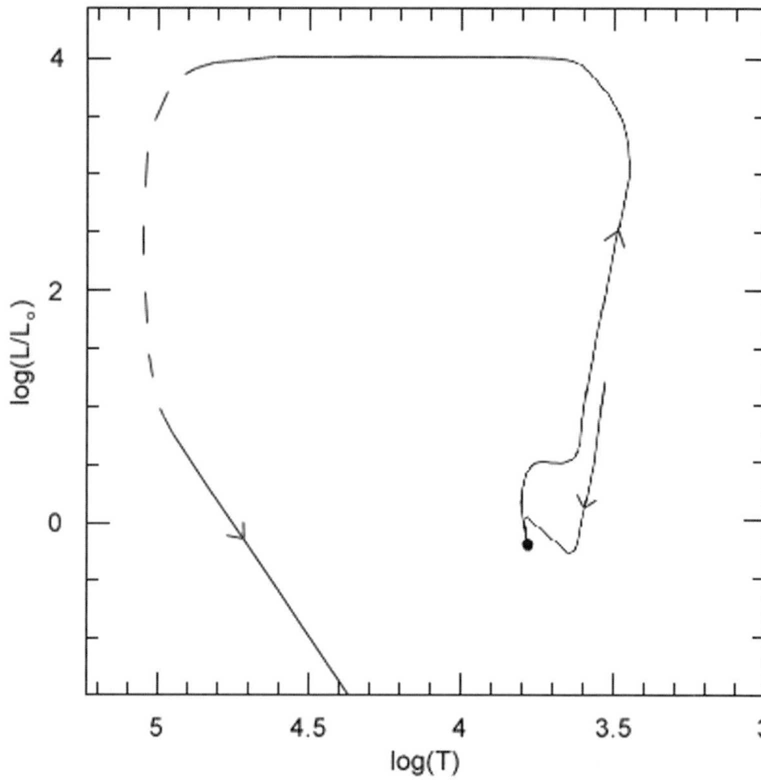

Figure 8.6-1. The evolutionary track of the Sun

Imagine a long-lived extraterrestrial being watching the Sun over the years. Each time it observes the Sun, it places a dot for the observed L and T on an L-T diagram. When the dots are connected, a curve results, which we call the "evolutionary track" or simply the track of the Sun in the L-T diagram. The adjacent diagram shows the track of the Sun. The Sun has four distinct stages in its life:

1. Youth, as a "protostar" (problem 8.6.2): After the Sun's gases have collapsed from an interstellar cloud, it becomes nearly static, every layer supported against gravity by the gradient in the pressure of the gas. Inevitably, heat wanders outward through the star, and radiation escapes from its surface. As radiation escapes, pressure support decreases. The protostar star must shrink and, by compressing the gases, the lost heat and more is re-generated. The central temperature rises.
2. Adult, as a "main sequence star" (Problems 8.6.3, 8.6.4): When the central temperature becomes sufficiently high, the nuclear fusion of

H⇒He starts at the center. The Sun becomes an extremely steady star for some 10^{10} years. Currently, the Sun is about halfway through this longest stage of its life.

3. Senility, as a "red giant": Once hydrogen is exhausted at the center, the outer layers of the Sun expand while the center is further compressed and heated until a new nuclear fuel arises, 3He ⇒C. The Sun becomes a red giant in this relatively brief stage.
4. Death, as a "white dwarf" (problem 8.6.5): After the Sun ejects its outer layers of gas, the rest of the star very rapidly shrinks to become a white dwarf, gradually cooling off forever.

Stars with several times the Sun's mass have similar life stages, but these stages pass much more quickly and energetically (problems 8.6.6, 8.6.7). The stars end as supernovae and leave behind either neutron stars or black holes.

1. Sun and Betelgeuse

Hydrostatic equilibrium-Isothermal atmosphere

Physics introduction - The isothermal atmosphere: For a plane gaseous atmosphere with uniform gravity g, each layer of gas is supported against gravity by the pressure gradient: the pressure p at the bottom of a layer is slightly larger than at the top. Mathematically, $dp/dz = -g\rho$, where ρ is the gas density. For an ideal gas, $p = nkT$, where n is number density of particles. For stellar physics, we relate n to ρ by $\rho = nm_p/\alpha$, where m_p is the mass of a proton. Clearly, $\alpha = 1$ for neutral hydrogen. This applies to the surfaces of cool stars like the Sun and Betelgeuse. For ionized hydrogen, each proton contributes mass, but protons and electrons contribute to the pressure. Therefore, $\alpha = 2$ for ionized hydrogen. This applies generally for $T > 10^4$ °K, throughout the interiors of normal stars, in the surfaces of the hotter stars, and in stellar coronae, here specifically the solar corona. Assuming uniform temperature and gravity, $dp/dz = - p/H$, where the scale height is $H = \alpha kT/gm_p$. Therefore, p is proportional to $\exp(-z/H)$.

Figure 8.6-2. Solar eclipse 11 June 1983. Credit: High Altitude Observatory / National Optical Astronomy Observatories

The problem: Compute the gravity on the surfaces of the Sun, with radius R_0 and mass M_0 given in the Introduction, and of the "supergiant" red star Betelgeuse ($R = 800R_0$, $M = 16M_0$). Compute the atmospheric scale heights at the surface of the Sun ($T_0 = 5.8 \times 10^3$ °K), in the solar corona just above the surface ($T = 2 \times 10^6$ °K), and at the surface of Betelgeuse ($T = 3400$ °K). Compare the results to the stellar radii. [Assume neutral hydrogen for the stellar surfaces, ionized hydrogen for the solar corona.] ,

The setting: Stars are entirely gaseous, mostly hydrogen. The gas density is highest at the center and decreases continuously outward. The star is in hydrostatic equilibrium : Each layer is supported against gravity by a pressure gradient. Once nuclear fusion occurs near the center, radiation travels outwards through the star, being absorbed and re-emitted millions of times on the way outward. Finally, the radiation reaches a layer from which it can escape into space. This layer, called the photosphere, is what we consider the surface of the star. Slightly outside the photosphere of many stars, including the Sun, starts an extensive very hot layer, the corona, seen prominently during a solar eclipse.

This problem is a simple introduction to hydrostatic equilibrium. It assumes an isothermal atmosphere, which is a good approximation for both the photosphere of a star and the solar corona

Betelgeuse, the red star in the "shoulder of Orion" in the Orion star constellation, is an enormous star, which would engulf Mercury, Venus, Earth and Mars if it were in the Sun's place. Not surprisingly, its other parameters are also quite different from the Sun.

The solution: The surface gravity on the Sun is $GM_o/R_o^2 = 2.7\text{x}10^2 \text{m/s}^2$, that on Betelgeuse $6.7\text{x}10^{-3}$ m/s^2. For the surfaces, α=1. At the solar surface, H = $1.8\text{x}10^2$ km = $0.3\text{x}10^{-3}R_o$ «R_o. At the surface of Betelgeuse, H = $4\text{x}10^6$km = $8\text{x}10^{-3}$R= $6R_o$. For the solar corona, α=2, H = $1.2\text{x}10^5$km = $0.2R_o$.

Interpretation: At the solar surface, H < 10^{-3} R_o. That is why we observe an apparently sharp rim of the Sun. In the solar corona, H = $0.2R_o$ explains why we see the corona as a very extended object during solar eclipse. Further out, at lower density, the coronal gas accelerates and becomes part of the solar wind, which extends to roughly 150 A.U. and there merges with the interstellar gases.

For Betelgeuse, the thickness of the "surface" layer alone is about six solar radii, nearly 1% of the stellar radius. If we were to travel near to Betelgeuse, the star would look quite "fuzzy". The reason, of course, is that Betelgeuse is enormously bloated, and the surface gravity is relatively low. The low surface gravity allows gas to escape as a stellar wind. Betelgeuse pulsates : Its radius changes regularly between about 700 and 1000 R_o. Betelgeuse is in its last stages of its life, ready to explode violently within the next 10^5 (or perhaps only 10^3?) years.

2. The Sun in its Youth

Energy conservation: heat and radiation from gravity

The Problem: When stars first start shining, they shrink slowly, converting the released gravitational energy into heat, half of which is radiated away as the star's luminosity.

1. Make an order of magnitude estimate for the time, τ_o, the "youthful" Sun needed to shrink to its present state, using the information in the previous sentence. This τ_o should contain only G and quantities known about the present Sun, namely L_o, M_o, and R_o. Evaluate τ_o in years.
2. Construct a differential equation for dR/dt assuming that the Sun shrank with a constant surface temperature, T. Assume that half the gravitational energy released by contraction appears as radiation, L = 1/2 d $|\Omega|$/dt. For the gravitational energy of the Sun use $|\Omega|$ = 3/2 GM_o^2/R. Eliminate σ appearing in Stefan's law by expressing all quantities in units of R_o, T_o, and τ_o. Integrate the equation. Assume that the star begins with initial luminosity L_i » L_o. Evaluate the time, in terms of τo, needed for the star to reach a final luminosity $L_f = L_o$.

3. Similarly, construct a differential equation for dR/dt assuming that the Sun shrinks at constant luminosity $L = L_o$. Integrate the equation and evaluate the time needed to change from an initial T_i (the same as in the previous differential equation) to the final T_o.
4. Using $T_i/T_o = 0.5$, evaluate the entire time the Sun has needed to reach its present state, in terms of τ_o. Did the order-of-magnitude estimate yield a reasonable value?.

The Setting: This problem seeks to estimate how long the Sun's "youth" lasts, simplifying the track in the L-T diagram (see introduction to Chapter 8.6) into two straight lines. On the vertical track, the temperature $T \sim 3x10^3$ °K is determined by the atomic physics of hydrogen: The degree of ionization of the hydrogen determines how radiative energy can escape.

The Solution:

1. Order of magnitude: The gravitational energy of the Sun is about GM_o^2/R_o. The rate of energy loss is about L_o. Therefore, the appropriate time scale is $\tau_o = GM_o^2/R_oL_o = 10^{15}$s $= 3x10^7$years.
2. Exact integration at constant T: We use $L = 1/2\ |d\Omega\ |/dt = -3/4\ (GM^2/R^2)dR/dt = 4\pi\sigma R^2T^4$. This is a differential equation for R(t). But if we integrate in this form, with σ appearing explicitly, it is easy to make a mistake and difficult to see the physics. We eliminate σ via $L_o = 4\pi\sigma R_o^2T_o^4$. The resulting dimensionless differential equation is $\tau_oR_o^3R^{-4}dR/dt = -4/3\ (T/T_o)^4$, which integrates to $R_o^3\ (R_f^{-3}-R_i^{-3}) = 4(T/T_o)^4\ (t_f-t_i)\ /\tau_o$, where subscript i and f stand for initial and final values. This is to be evaluated for initial and final luminosities $L \gg L_o$ and L_o, respectively. Consequently, $R_i/R_o \gg 1$ so that R_i^{-3} may be neglected, and $R_f/R_o = (T_o/T)^2$. The desired time interval becomes $1/4\ (T/T_o)^2$, that is, $\tau_o = 1/16\ \tau_o$, about 2 million years.
3. Exact integration at constant L: Now we have $R_o\tau_oR^{-2}dR/dt = -4/3$ or $R_o(R_f^{-1}-R_i^{-1}) = 4/3\ (t_f-t_i)/\tau_o$, where now $R_i/R_o = (T_o/T)^2$ and $R_f/R_o = 1$. Therefore, this interval is $9/16\ \tau_o$, about 17 million years.
4. The entire contraction time, $5/8\ \tau_o$ is about 19 million years. It is dominated by the last contraction stage. The order of magnitude estimate, τ_o, is quite satisfactory.

Interpretation: The Sun starts shining as a protostar at $L \sim 10^2L_o$, $T \sim 3x10^3$ °K. At high luminosity the contraction is fast. More time is spent once the Sun has reached lower L.

Clearly, this computation oversimplifies the actual evolution of the Sun, which must be evaluated by detailed computer programs. For instance, the small wiggle in the Sun's track, just before reaching the main sequence, is caused by the nuclear fusion of deuterium, which "turns on" at a lower temperature than proton fusion. Since there is only little deuterium in the solar gases, this nuclear fuel is soon exhausted.

Stars are born all the time. But we have only a small chance to catch them in very brief stages of their lives, such as when they first contract at high luminosity. (If you could search your entire city, how many babies less than one week old would you find, relative to the whole population?) The later stages of contraction take longer. We have a better chance to observe stars in these stages. Indeed, stars in the later contraction stages are now observed. Hubble Space Telescope observed that many contracting stars are surrounded by a disk of gas and dust, a suitable place to form planets.

Didactics:

1. Check that students have written the formula for τ_o correctly before they proceed to construct the differential equations. For an order-of-magnitude estimate, factors like 3/5 are irrelevant.
2. Students find it difficult to eliminate σ and will keep it for several steps before simplifying the algebra. They need guidance to look for ways to make the algebra simpler and more understandable. The teacher's effort is worthwhile because it is a very useful habit for students to look for mathematical simplicity.
3. At any one time, the star has the gravitational energy $\Omega = -\int[GM(r)/r]dM(r)$. With $dM(r) = 4\pi\rho r^2 dr$ and uniform density ρ, the result is $\Omega = -3/5\ GM^2/R$. The coefficient 3/2 used here represents the central condensation of gas and the deeper potential well in real stars.
4. There is a very basic reason that stars without nuclear energy must gradually shrink and must radiate. These stars, being in hydrostatic equilibrium at any one time, must satisfy the Virial Theorem $\Omega = -2K$, where for stars K is the total thermal energy within the star (see the next paragraph). As gravitational energy is gradually released, only half can go into thermal energy. The formula $L = 1/2\ d\,|\Omega|/dt$ represents the other half of the energy, which must be lost as radiation. Conversely, given that a protostar radiates, it must shrink. All this changes when nuclear fusion provides energy and $d\,|\Omega|/dt = 0$.

Proof of the Virial Theorem for a star supported by gas pressure, that is, by thermal motions. Assume a spherically symmetric gas (such as the Sun) satisfying the equation of hydrostatic equilibrium. We chose to evaluate $\int r\,(dp/dr)\,4\pi r^2 dr = -\int g(r)\rho 4\pi r^3 dr$. Evaluate the integral on the right side using $g(r) = GM(r)/r^2$ and $dM(r) = 4\pi r^2\rho dr$. The integral becomes $\int GM(r)/r]dM(r) = \Omega$. Evaluate the integral on the left side using an integration by parts (with zero for the terms evaluated at r = o and r = R) and then use $p = \sum_n < p_x v_x > = \sum_n < 1/3\ mv^2 >$, where the sum extends over electrons and protons and the 1/3 comes from averaging over a nearly isotropic velocity distribution. The integral becomes $-\int\Sigma_{nm}$ $<v^2>$ $4\pi r^2 dr = -2K$. A general proof of the Virial Theorem is in Chapter 8.9.

5. Students may question why we can have the star in hydrostatic equilibrium and yet say that it is gradually contracting. If a star is not in hydrostatic equilibrium, pressures that are not balanced will create sound waves. The sound waves can cross the star in a matter of hours or days. Equilibrium is re-established when the sound waves have adjusted the pressures throughout the star and have been dissipated by viscosity. Hours or days are a short time compared to the contraction time τ_o. (An exception are stars that pulsate. Their periods of pulsation are of the order of the time needed for soundwaves to cross the star.)
6. Students' problem: A "brown dwarf" (a star which never reaches fusion temperature in the center) with $M = 0.08M_o$ contracts for $1.2x10^{10}$years with constant $T=0.5T_o$. What is its L at the end of that time? ($L \sim 1.1x10^{-4}L_o$) Finding such a star and measuring its L might tell us the age of the Milky Way Galaxy.
7. Students' problem: How long does it take a massive protostar, $M=20M_o$, to contract at constant $L = 10^5L_o$ from $T = 3x10^3$ °K to $T = 3.3x10^4$ °K? ($3x10^{-4}\tau_o \sim 10^4$yr)

3. The Sun in Middle Age

$E = mc^2$

The Problem: Given $L_o = 4x10^{26}$w and Einstein's relation $E = mc^2$, how much mass does the Sun convert to energy and lose per second ? At this rate, how many years would it take to lose all of the Sun's mass, $M_o=2x10^{30}$kg?

Given L_o, how much mass of hydrogen is converted into helium per second? (The resulting helium has 0.7% less mass than the original hydrogen.) At this rate, how many years would it take to convert one solar mass of hydrogen into helium?

Actual stars convert 10% of their hydrogen into helium during their lifetime. How many years can one expect for the Sun's lifetime?

The Setting: On the time scale of a human lifetime, stars appear to last "forever". Of course, stars must change with time because they exhaust their nuclear fuels. But, fortunately for the evolution of life on Earth, the Sun is very nearly constant for some 10^{10}years. This is the period during which its luminosity is derived from the conversion of hydrogen to helium. The most common reaction chain is: $p+p \Rightarrow d+e^++\nu$ (1.44 MeV) and $d+p \Rightarrow {}^3He+\gamma$(5.49 MeV); after both reactions have occurred twice, then ${}^3He+{}^3He \Rightarrow {}^4He+2p$ (12.9 MeV). Here p = proton, d = deuteron (proton plus one neutron), and the energy release is expressed in MeV = $1.6x10^{-13}$j. The most difficult step is the first step, the pp fusion, because the two protons repel each other by their electrical charges and yet they must be near each other long enough for a weak interaction to take place. The interaction requires quantum-mechanical tunneling. The Sun's structure is adjusted so that the temperature near the center is sufficient for the pp fusion to work. At the center of the Sun, approximately $T = 1.6\text{x}10^{7\circ}\text{K}$ and $\rho = 1.6\text{x}10^5\ \text{kg/m}^3$.

The Solution: The mass loss per second $dM/dt = L_0/c^2 = 5\text{x}10^9$ kg/s. The related time is $M_0c^2/L_0 = 1.5\text{x}10^{13}$years. If only 0.7% of mass is converted, the same amount of mass lasts only 10^{11}years. The expected life of the Sun is 10% of that, namely about10^{10}years.

Interpretation: The solar mass loss is enormous on a human scale, 5 million tons per second. Yet the life of the Sun is very long compared to a human lifetime.

Why do stars use only 10% of all hydrogen? Once the hydrogen is exhausted near the center and fusion occurs only slightly further out, the gas at the center becomes isothermal. The pressure gradient is reduced. If the mass of the isothermal core exceeds 0.1 of the star's mass, the core cannot be supported against gravity. This limit is named after Schoenberg and Chandrasekhar.

A correction is necessary here : The Sun is not exactly constant. After some hydrogen is used up at the center, the hydrogen fusion must proceed more vigorously to provide enough pressure, and the Sun becomes slightly more luminous with time . When the Sun was young $4\text{x}10^9$years ago, the Sun was less luminous than now by 30%. This creates a puzzle : It seems that the Earth $4\text{x}10^9$years ago should have been too cold for liquid water! Yet we know that primitive (one-celled) life occurred soon thereafter. Presumably this primitive life needed liquid water. What extra heat kept our atmosphere warm enough for liquid water and life at that time? We do not know.

Didactics: The pp reaction is an extremely improbable reaction. The electrostatic repulsion presents a 1-MeV barrier through which the typically 1-KeV protons must tunnel. Just how improbable the pp reaction is can be seen from the end result in the Sun: Since most protons near the center have fused after 10^{10} years, every proton near the center must move about and collide with other protons for $5\text{x}10^9$years before it has even a 50% chance of fusing with another proton.

A note on the "neutrino experiments": Solar neutrinos are created near the solar center. Therefore, if we can detect them, they provide an opportunity to "look" directly into the solar center. All current experiments seek to catch neutrinos from (here not listed) branches of the pp fusion chain. The observed number of neutrinos is between half and a third of the theoretically expected number. Is the discrepancy due to faulty solar theory? Sound waves observed to traverse the solar interior (helioseismology) confirm solar theory. Probably the neutrino experiments tell us a fundamental physical fact: neutrinos have mass. Indeed, the comparison of different experiments also suggests that "standard" electroweak theory cannot explain the data.

4. The Sun in middle age

One-step integration of hydrostatic equilibrium

The Problem:

Given: the equation for hydrostatic equilibrium of a plane atmosphere, $dp/dz = -g\rho$. At any place within a spherically symmetric star, the gravity g(r) is effectively due to all the mass M(r) inside a sphere of radius r, placed at the center. Write down the equation of hydrostatic equilibrium at some arbitrary r, dp/dr = function of G, M(r), ρ, and r.

Estimate the central pressure in a star of mass M, radius R, as follows. Assume p at the surface is negligible compared to p at the center. Then p(center) = $\int dp$. Evaluate this integral by a "one-step" integration of the equation of hydrostatic equilibrium, replacing M(r) by M, r and dr by R/2 (since most of the star's mass resides inside R/2), etc. Derive the equation for p(center) as function of M and R. Evaluate for the Sun. Compare to the actual solar value for the Sun, $2.5\times10^{16}\ N/m^2$.

The Setting: Until a few years ago, we observed only the surfaces of the stars. But theory could tell us much about the stellar interior even before we knew how nuclear fusion works. We needed to know merely that the Sun is made of gas (rather than, say, wood or coal). That fact we learned from stellar spectra well over a century ago. Within the steady gaseous Sun, the equation of hydrostatic equilibrium specifies that each layer of the star is supported by a pressure gradient to balance the force of gravity. We integrate the differential equation in a very crude way.

The solution: The description of g yields $g = GM(r)/r^2$ and the equation of hydrostatic equilibrium becomes $dp/dr = -GM(r)\rho/r^2$. With $M(r) \Rightarrow M$, $r \Rightarrow R/2$, $\rho \Rightarrow (3/4\pi)\, M/(R/2)^3$, we have p(center) = $-\int[GM(r)\rho/r^2]dr \sim (3/4\pi)GM^2/(R/2)^4$. For $M = M_o$, $R = R_o$, p(center) = $4\times10^{15} N/m^2$, about 1/6 of the actual value.

Interpretation: Our one-step integration implies that the Sun has a uniform density out to r = R/2 and has negligible density beyond r = R/2. It is a crude attempt to represent the actual concentration of the Sun's mass near the center. A detailed solar model shows that half the mass is inside 0.3Ro or 3% of the volume. Perhaps we should choose $r \Rightarrow R/3$, but this uncertainty merely shows how crude the one-step integration is. (In problem 8.6.2, we effectively used a radius 2/5 R in the formula for the gravitational energy).

The one-step integration does give correctly the proportionality of the central pressure to M^2/R^4. This proportionality holds true for all stars that have a similar degree of concentration. Thus, given p(center) for the Sun we can expect to scale p(center) for other main sequence stars in proportion to their known M^2/R^4.

Modern computers produce detailed models for stellar interiors. We can check them for the Sun because we observe pressure (sound) waves that travel through the Sun (helioseismology). The observations tell us the sound speed throughout the interior. The deduced sound speed agrees with the theoretical model within about 1% throughout the Sun.

Didactics: p(center) = 2.5×10^{11} times the pressure at Earth's surface! This huge value makes the "error" caused by the one-step integration seem unimportant. Despite this error, the central temperature estimated according to p(center) = $<\rho>kT/m_p$ comes out quite reasonable, 2×10^7 °K.

The approximation p(center) = $\int dp$ is very well justified. The pressure at the solar surface, $p\sim2\times10^4$ N/m^2 is twelve orders of magnitude smaller than at the solar center.

The gravity $g(r) = GM(r)/r^2$ is derived in Chapter 8.9 and was used earlier in problem 8.2.6.

An analytic limit for the central pressure: For those who dislike the one-step integration, here is a rigorous lower limit to p(center), namely p(center) > $GM^2/(12\pi R^4)$. This is again proportional to M^2/R^4. The mathematical steps are : p(center) > $<p>$ = $\int p\,dM(r)/M = -\int M(r)dp/M = \int M(r)g(r)\rho(r)dr/M = G\int [M^2(r)/4\pi r^4]dM(r)/M > (G/4\pi R^4)\int M(r)^2 dM(r)/M = GM^2/(12\,\pi R^4)$. Reasons for the steps in this proof: p(center) > any average of p must be true because $dp/dr < 0$ everywhere. Here, this average is chosen over mass. Next follows an integration by parts, neglecting values at the surface. Then the equation of hydrostatic equilibrium is used. Then $g(r) = GM(r)/r^2$ is used together with $dM(r)/dr = 4\,\pi r^2\rho$. Finally, any average over r^{-4} must exceed R^{-4}.

5. The Sun's Old Age

Quantum effects, pessure due to degenerate electrons

This problem is placed here to complete the story of the Sun's life. The physics here are probably the conceptually most complicated physics of the section. The next two problems return to the (radiation) physics of ordinary thermal gases. If the student has not done problem 8.6.4, then the one-step integration used here may need some elaboration.

Physics Introduction: Pressure is defined as momentum flux across a surface, $P = \int p_x v_x f(\underline{p})d^3p$, where $f(\underline{p})$ is the distribution function, $\int f(\underline{p})d^3p = n$. For an isotropic distribution, $p_x v_x f(\underline{p}) \Rightarrow 1/3\ pvf(p)$ and $d^3p \Rightarrow 4\pi p^2 dp$ on integration over angles. In most situations, the distribution is thermal, $P = 1/3\ m \langle v^2 \rangle = nkT$, where n is the total particle density. However, when electrons are very densely packed, their momentum is influenced by the Heisenberg uncertainty principle $\Delta x \Delta p_x \sim h/2\pi$, where $\Delta x \sim n^{-1/3}$, and now n = electron density. The higher n, the smaller the space available and the higher must be the momentum of most electrons. The distribution $f(p) = (3/4\pi)n/p_F^3$ for $0 < p < p_F = (3\pi^2 n)^{1/3}(h/2\pi)$ implies that all quantum states are filled by electrons up to momentum p_F, and higher momentum states are empty.

The problem:

1. Assume non-relativistic electrons, $v = p/m$. Evaluate the pressure as a function of n. Write n (the electron density) in terms of mass density ρ, assuming the star is made of carbon (one proton and one neutron) per electron. Use this pressure (including all its physical constants) in the equation for hydrostatic equilibrium and do a one-step integration, by replacing $r \Rightarrow R$, etc., still keeping all the physical constants. This should yield a relation between M and R. Evaluate R for $M = 0.5\ M_o$ and compare with the Earth's radius ($6x10^3$km).
2. Repeat the procedure for relativistic electrons, $v = c$. Show that the R cancel and you get only one possible value of M.

The Astrophysical setting: The Sun in middle age, as a main sequence star, maintains T(center) high enough for fusion of H ⇒He plus energy. When the H at the center becomes exhausted, the central gases gradually lose support and must shrink, heat up, and thus regenerate support. The gases heat up until they reach $T \sim 2x10^8\ {}^oK$. At that temperature, He nuclei (two positive electric charges) overcome their electrostatic repulsion so that a new version of fusion occurs, 3He ⇒C plus energy. This fusion provides a new source of energy to support the center. Meanwhile, the outer layers of the Sun have expanded so that the Sun is a red giant . Its huge luminosity implies the exhaustion of the He fuel in mere millions of years.

What happens when the He is exhausted at the center? Now the gases at the center are so dense that a new pressure is available to support the gas against gravity, the so-called electron degeneracy pressure. This pressure is strictly a quantum effect: the less physical space each electron occupies, the more momentum space it must occupy, according to Heisenberg's uncertainty principle. When electrons are squeezed into higher density, energy must be provided to give higher momenta to the electrons. This has nothing to do with temperature. Because of this new electron pressure, the solar center can adjust (actually rather violently) to support the rest of the star even when there is no more nuclear fusion at the center.

Once the Sun has been a red giant for a few million years, the Sun sheds its outer layers. The expelled gases become observable as a "planetary nebula". (Such nebulae appeared like planets in early telescopes, but they have nothing to do with planets physically.) While the nebula expands, the stellar core becomes visible as a very hot star. In a mere few thousand years, its surface temperature decreases until the star becomes what we call a "white dwarf". (The "white" is also an historical misnomer). The Sun at age $\sim 10^{10}$ years will become a white dwarf, cooling off ever more slowly, forever.

Probably the most famous white dwarf is Sirius B, the companion to the bright star Sirius. It was the first white dwarf to be detected visually in 1862, 18 years after Sirius was known to be a binary system. (Sirius B has $M = 1M_o$, surface temperature about $T = 2.7 \times 10^{4}$ °K.)

The Solution:

1. Plugging in $v = p/m_e$ and integrating over p up to p_F yields $P = n\, p_F^2/5m_e$, where $n = \rho/2m_p$, so that p is proportional to $\rho^{5/3}$. Used in $P/R \sim GM\rho/R^2$ with $M = (4\pi/3)\rho R^3$, one gets $R = M^{-1/3} G^{-1} (h/2\pi)^2 m_e^{-1} m_p^{-5/3} (9\pi/8)^{2/3} 10^{-1} = 1.5 \times 10^3$ km, about 1/4 R(Earth).
2. Plugging in $v = c$ and integrating over p up to p_F yields $P = 1/4\, n\, c\, p_F$, so that P is proportional to $\rho^{4/3}$. In $p/R \sim GM\rho/R^2$ the R cancel, and $M = (3/64)\pi^{1/2} (hc/2\pi G)^{3/2} m_p^{-2} = 0.15 M_o$.

Interpretation:

The radius of a nonrelativistic white dwarf decreases as the mass increases, quite unlike ordinary objects or even ordinary stars. Evidently, it is difficult to support an increasing mass. Moreover, as the density increases, the electrons with momenta near p_F become relativistic. Subsequently, further mass addition causes less pressure increase, and at some point no equilibrium is possible. The "Chandrasekhar limit" for the maximum mass of a white dwarf is $1.44 M_o$ (for negligible rotation and for 2 nucleons per electron, as for carbon). Our estimate was too small by a factor ten.

The appearance of h in the estimates for the radius and for the maximum mass of a white dwarf makes the white dwarf a macroscopic quantum object. The quantum nature of white dwarfs and their maximum mass due to relativity were first recognized and worked out by S. Chandrasekhar.

Isolated white dwarfs cannot contain hydrogen because any such hydrogen would have fused long ago (see didactic 6). If new hydrogen falls onto a white dwarf from a binary companion, the fusion is explosive. A supernova or a (recurrent) nova occurs, depending on the rate of accretion. Supernovae of this type are all very similar. They are visible far out in the Universe. Current research suggests that they provide the most accurate distance scale for distant clusters of galaxies. And that scale, in turn, is needed, together with other data, to determine the age of our Universe.

Didactics:

1. The highest energy of non-relativistic degenerate electrons, 1/2 p_F^2/m_e, is called the Fermi energy, hence the subscript F.
2. In a mixture of electrons and nuclei, such as in white dwarfs, the degenerate electrons provide the main pressure, while the non-degenerate nuclei maintain a Boltzmann velocity distribution. The ions are supported against gravity by an electric field (due to a very tiny charge separation) coupling the ions to the electrons, which are supported by their pressure.
3. For a non-relativistic white dwarf, the one-step integration is adequate. The factor 1/5 in P comes from $\int p^4 dp$. The actual radius for a $0.5M_o$ white dwarf is about 10^4km..
4. The factor $(hc/2\pi G)^{1/2} = 2.2\times10^{-8}$kg appearing in the maximum mass is called the Planck mass.
5. The cancellation of R for a relativistic white dwarfs means that there is no equilibrium radius. A slight expansion leads to permanent expansion, a slight contraction leads to collapse (until the physics change). A fundamental explanation why a white dwarf cannot be supported by relativistic degenerate electrons, based on the Virial Theorem, can be found in didactic 3) of problem 8.5.2.
6. Within a white dwarf, heat conduction is high, and the temperature of the nuclei is nearly uniform. The thermal energy of the nuclei gradually leaks out through a non-degenerate surface layer. The interior temperature can be estimated from $L = (4\pi r^2)(c/3n\sigma)adT^4/dr$ (see problem 8.6.6). A one-step integration with $dr \Rightarrow 10^{-2}R$ yields T $\sim10^7$ °K. At this interior temperature, there can be no hydrogen left, as it would have fused long ago.

6. The Short-lived Massive Stars

Radiation diffusion equation, scaling of parameters

Required: Students must do problem 8.6.4 first.

Physics Introduction: The diffusion of some quantity Q is given by $\delta Q/\delta t=\delta/\delta x[D\delta Q/\delta x]$, where the diffusion coefficient is D =1/3 vλ. Here, v = root-mean-square particle velocity and the mean free path $\lambda=1/(n\sigma)$, with σa collision cross section and n the density of particles that are targets for collisions. The flux of the quantity Q is $F= D\delta Q/\delta x$. F is independent of x when $\delta Q/\delta t = 0$.

The Problem: In a spherically symmetric star, radiation steadily diffuses outwards through each spherical layer of the star without accumulating anywhere (except near the center, where nuclear fusion adds heat which is turned into radiation). The diffusing quantity Q now represents the radiative energy density, $Q = aT^4$, and $F = L/(4\pi r^2) = D\delta Q/\delta r$. Write down an equation for L in terms of n, T, dT/dr, and the constants c, a, and σ.

Now do a one-step integration with r $\Rightarrow$R/2, kT $\Rightarrow$kT(center) = p(center)/<n> and use p(center) as derived in problem 8.6.4. Show that the dependence on R disappears and derive how L depends on M .

Given the solar lifetime of 10^{10} years and that stars use only 10% of their H during their lifetimes, estimate the lifetime of a star with $M = 20\ M_0$.

The Setting: Most stars, including the Sun, are found to be "main-sequence" stars. All main-sequence stars derive their energy from fusion of hydrogen into helium. The hotter and more luminous main sequence stars turn out to be more massive stars. We do not know why they obtained more mass during star formation, but there they are. Among stars whose masses can be measured (see problem 8.2.4), L is roughly proportional to M^3. Stars with the largest masses $M \sim 10^2\ M_0$ have luminosities $L \sim 10^6\ L_0$. If stars during their lifetimes all consume 10% of their hydrogen (see problem 8.6.3), the luminous stars must consume their nuclear fuel extremely rapidly compared to the Sun.

Nucleosynthesis: Massive stars are of special interest because they end their life by exploding as a supernova. Nuclear reactions during the explosion generate most of the elements. The elements that compose most of Earth and us humans, mainly carbon, nitrogen, oxygen, silicon, and iron, have all been made in such exploding stars. The oxygen we breathe was made in an exploding star. The oldest stars in our Galaxy, about $1.3\text{x}10^9$ years old, were formed before any stars had exploded, and they may be predicted to contain none of these elements. Indeed, they consist (almost) entirely of hydrogen and helium. The hydrogen and helium are left over from the Big Bang origin of the Universe.

The Solution: In the diffusion equation dealing with radiation, v $\Rightarrow$c. Therefore, $L = (4\pi r^2)(c/3n\sigma)a(dT^4/dr)$. Since the problem asks only for proportionality, we ignore not only all factors such as 2 or π, but also factors such as c, a, σ, or G. With r $\Rightarrow$R and (from problem 8.6.4) T $\Rightarrow$p(center)/<n> $\Rightarrow (M^2/R^4)(R^3/M) = M/R$, we obtain L $\Rightarrow R^2 (R^3/M)(M/R)^4/R = M^3$.

If we calibrate the relation using the Sun, then $L/L_0=(M/M_0)^3$. The lifetime of a star with structure similar to the Sun is proportional to the amount of nuclear fuel divided by the rate the fuel is used, thus to M/L and therefore to M^{-2}. A star with $M = 20M_0$ has a lifetime about $2.5\text{x}10^{-3}$ of the Sun's lifetime, or about 25 million years.

Interpretation: The derived relation is actually quite good. The observations fit reasonably to L proportional to $M^{3.3}$. The one-step integration works well because all the stars on the main sequence have a similar degree of concentration.

The main simplification in our solution is the assumption that σ is the same constant for all the stars. The atomic absorption of radiation is a strong function of both temperature and density. It differs significantly between stars of different mass. Within the Sun, the mean free path of an x-ray photon near the center is on the order of cm. As the radiation diffuses outward into less hot gases, the mean free path increases until it is of the order of 100 km for a visible photon near the surface .

Another structural factor ignored in the problem is convection. Where σ is too high, dT/dr becomes so high that convection starts and carries the heat upward. Massive stars are convective near the center; the Sun is convective for $r > 0.7R_o$; and low-mass stars are entirely convective. The transport of heat by conduction is negligible in main sequence stars.

Finally, the manner of hydrogen fusion changes for the more massive stars. Carbon, nitrogen and oxygen nuclei are used as catalysts to speed up the fusion of hydrogen into helium. Because the electrostatic repulsion of these nuclei is larger, this "CNO cycle" requires higher central temperatures than exist in the Sun. Indeed, with T(center) proportional to M/R and, observationally, R proportional to $M^{0.6}$, the more massive stars are hotter at the center. Because the tunneling probabilities increase very rapidly with temperature, even a small increase in central temperature, above that of the Sun, allows a much greater luminosity.

Didactics:

1. The physics introduction presented diffusion in a coordinate x. Strictly, we should write the diffusion equation appropriate to spherical symmetry, $\delta Q/\delta t = \mathrm{div}(\mathrm{Flux})$. In the problem, the spherical symmetry was introduced by defining $F = L/(4\pi r2)$.
2. Problem 8.6.2, didactic 6) deals with the time needed for protostars of various masses to reach the main sequence. This stage is much shorter for massive stars than for the Sun. In fact, every stage of a massive star's life is much shorter than the equivalent stage in the Sun's life. At the end, the massive stars explode violently and leave behind a neutron star or a black hole.
3. How long does radiation take to diffuse out of the Sun? We obtain this answer directly from problem 8.6.2: the protostars contract at a rate given by how fast radiation can leave the star. According to Problem 8.6.2, the time is roughly $\tau_o = GM_o^2/R_oL_o$. The radiation now leaving the Sun started out near the center some 30 million years ago.

7. The Most Luminous Stars

Thomson scattering cross section

Required: Students must do problems 8.6.4 and 8.6.6 first.

The problem: In problem 8.6.6 you derived the equation $L = (4\pi r^2)(c/3n\sigma)adT^4/dr$. The right side, derived in terms of radiation energy density aT^4, can also be expressed in terms of radiation pressure $P(rad) = 1/3\ aT^4$. Now suppose that gas pressure is much smaller than radiation pressure so that $dP(rad)/dr$ dominates the equation for hydrostatic equilibrium. Eliminate $dP(rad)/dr$ between the equation for L and the equation of hydrostatic equilibrium. Eliminate ρ in the equation of hydrostatic equilibrium by expressing it in terms of electron density n_e assuming pure ionized hydrogen, and assume that electrons scatter the radiation, so that $n = n_e$ in the equation for L. Set $M(r) = M$ for the outer parts of a star, and obtain an equation for L in terms of only M, σ, and constants. Evaluate this L using the Thomson scattering cross section $\sigma_T = 6.6\text{x}10^{-29}\text{m}^2$. Derive L/L_O in terms of M/M_O.

The Setting: If one makes models of stars on the main sequence, one finds that the hotter, more massive stars involve an increasing ratio of radiation pressure to gas pressure, throughout most of the star. The gradient in radiation pressure acts outward, as does the gas pressure. If the gradient in radiation pressure becomes sufficient, it overcomes gravity and the star can no longer be in hydrostatic equilibrium. This stage is reached when L equals the "Eddington luminosity", evaluated in this problem. Stars with higher luminosity (depending on M/M_O) cannot be stable.

The Solution: The equation of hydrostatic equilibrium is $dP(rad)/dr = -GM\rho/r^2$. If radiation scattered by electrons supports the gas, then $dP(rad)/dr = (L/4\pi r^2)(n_e\sigma/c)$. Also, $\rho = n_e m_p$. On canceling similar terms, $L/L_O = 4\pi GMm_pc/\sigma L_O = 4.5\text{x}10^4 M/M_O$.

Interpretation: The Eddington luminosity is the maximum luminosity any static star (of pure hydrogen) can have. For stars on the main sequence with $L/L_O = (M/M_O)^{3.3}$, this limit is reached for M/M_O about 100. Indeed, no steady stars have been measured to have larger M.

Many stars with L near the Eddington luminosity are observed to have strong winds. This confirms that gravity binds the gas only weakly. For example, the star Eta Carinae has $M = 150M_O$, $L = 6\text{x}10^6 L_O$ and has a huge mass loss rate. In the 1830's it brightened to be the second brightest star in the sky. Altogether it seems to be at the edge of stability. The "Pistol" star, recently observed by Hubble Space Telescope, has an unexplained $L = 10^7 L_O$. A surrounding nebula 4 light-years across contains about $10M_O$ of gas, which was probably ejected in two eruptions $4\text{x}10^3$ and $6\text{x}10^3$ years ago.

The Eddington luminosity applies to any radiating object. For quasars with a black hole of mass $M = 10^9\ M_o$, the Eddington luminosity is around $4x10^{13}\ L_o$. Indeed many quasars have such luminosities, far more than the luminosity of the galaxies whose centers they occupy. Quasars that exceed the Eddington luminosity cannot be static. Typically, gases from equatorial accretion disks are funneled into the powerful outflowing polar jets that help make quasars so spectacular.

Didactics:

1. The Eddington luminosity was derived for pure ionized hydrogen. For stars with 75% H and 25% He by mass (so that $n_e\ \rho/m_p$), $L/L_o = 3.8x10^4 M/M_o$.
2. The coefficient 1/3 in the diffusion coefficient D of problem 8.6.6 was chosen to match diffusion of nearly isotropic radiation (mean free path short compared to the pressure or temperature scale height). It stands for the hemispheric average of $\cos^2\theta$. The same factor 1/3 appears in the radiation pressure. Indeed throughout most of the star (all except the photosphere) the radiation is very nearly isotropic.
3. There is a good physical reason that stars cannot involve radiation pressure much larger than gas pressure. In equilibrium, according to the Virial Theorem, $\Omega = -2K$ (Problem 8.6.2, didactic 4). Imagine the whole star expanding or contracting. Ω changes in proportion to 1/R. If internally P changes in proportion to $\rho^a \sim R^{-3a}$, where $a = 5/3$ for an ideal monatomic gas, then K, an integral of pressure over the volume of the star, changes in proportion to R^{3-3a}. If $a = 5/3$, or more generally if $a > 4/3$, making R smaller makes K increase faster than Ω. Physically, that means the enhanced pressure will cause the star to expand again. As a result, the radius of the star will oscillate. The predicted period of oscillation is of the order of the time needed for a sound wave to cross the star. Indeed, many stars are observed to oscillate according to this prediction. However, if $a = 4/3$, then Ω and K change with R in the same manner. Therefore, $\Omega = -2K$ remains satisfied even as R changes. In principle, the star can be in equilibrium at any R. But if equilibrium is violated even slightly, the star will either continue to expand forever or continue to contract forever. Therefore, no stars with $a = 4/3$ can exist. Since radiation has $a = 4/3$, no star can exist supported purely by radiation. Similarly, in problem 8.6.5, a massive white dwarf with relativistic electrons has $a = 4/3$ and cannot exist.

Chapter 8.7

COSMIC MAGNETIC FIELDS

D.G. Wentzel

Introduction

In the laboratory, we are familiar with electrical currents, and with magnetic fields caused by the currents. Solenoids play an important role in technology. A well known cosmic analogy to solenoids are sunspots. There, we measure the magnetic field (by the Zeeman splitting of the spectrum of light from sunspots). We infer the electrical currents, and we find that a sunspot is much like a solenoid in the laboratory (see problem 8.7.1).

In the 1940's, new observations showed that magnetic fields must be truly cosmic: radio astronomers showed that much cosmic radio radiation is polarized, and optical astronomers showed that light from the stars in the Milky Way becomes polarized as it travels toward us. Only magnetic fields can explain the preferred directions inherent in the polarization of the radiation.

Magnetic fields can be detected everywhere, in and near planets, in and near stars, between stars, and throughout galaxies. The associated electrical currents and $\underline{\mathbf{I}} \times \underline{\mathbf{B}}$ forces must be inferred, with much uncertainty. Given that uncertainty, theory has emphasized rather simple concepts that are equivalent to mechanics and gas dynamics: magnetic energy can do work like any other reservoir of energy (problem 8.7.2), magnetic pressure acts much like a gas pressure (as 8.7.3, 8.7.4), and magnetic tension acts like the tension on a violin string. Moreover, the concept of magnetic induction can be used to visualize magnetic fields and to convert the magnetic fields from a mathematical construct to what is practically a physical quantity (problem 8.7.5). Thus the topic of

W. Wamsteker et al. (eds.), Developing Basic Space Science World-Wide, 477-488.

cosmic magnetic fields can be used in physics courses as example of energies, pressures, waves, and Faraday's law of induction.

1. Sunspots and their Tera-Amperes

Solenoid - magnetostatics - Zeeman effect

The Problem: Treat a sunspot as a solenoid much longer than its diameter: The spot magnetic field, vertical relative to the surface of the Sun, is B = 0.15 Tesla. Electrical current, horizontal relative to the surface of the Sun, encircles the magnetic field. Find the current per unit length of the solenoid (in Amp/m). Find the total current (in Amp) if the depth of the sunspot is $3x10^4$ km.

The Setting: Most of the solar surface is at a temperature of about 5800 °K. A photograph of the Sun shows "dark" spots. The fairly uniformly "dark" central portion, typically about 10^4 km in diameter, has temperatures as low as about 4000 °K. Thus it is less bright than most of the surface according to the Stefan-Boltzmann law, but the so-called darkness is only relative to the surroundings. A spectrum of the light from a sunspot shows spectrum lines split by the Zeeman effect. Typically, the magnetic field is vertical relative to the surface of the Sun, it has a relatively sharp boundary, roughly 10^3 km thick, and the value of the field strength in most spots is about 0.1 to 0.2 Tesla. Since all the Sun is gaseous, there cannot be any solid magnetic material. The magnetic field must be due to an electrical current running through the solar gas. The gas is sufficiently ionized so that free electrons can easily carry the current.

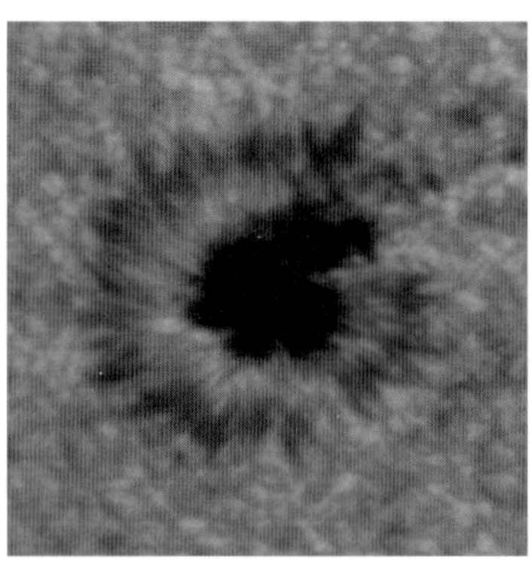

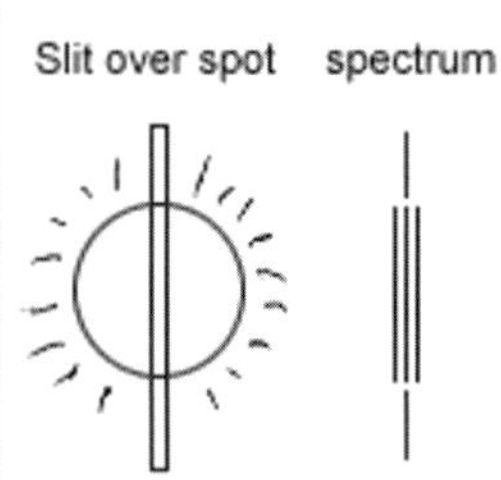

Figure 8.7-1. Credit: Lockheed Research Laboratories and NASA

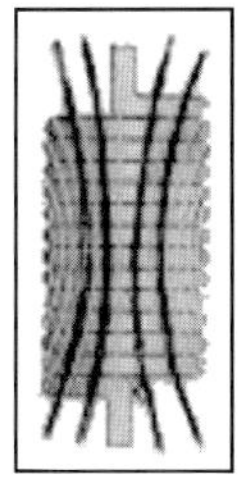

Figure 8.7-2

A model of the sunspot is that of a solenoid: wire wrapped tightly in the form of a cylinder. The "wire" is the gas in the boundary, ~10^3 km thick, and the cylinder is the sunspot, with diameter ~10^4 km. The length of the solenoid corresponds to the depth to which the sunspot reaches inside the Sun.

Strictly, the magnetic field inside a solenoid is uniform only if the solenoid is much longer than its diameter. For our estimate, we shall assume this is true. In reality, however, we know only that a typical spot is at least 10^4 km deep (based on the details observed at the surface), probably about 3×10^4 km deep (based on observations of sound waves interacting with the sunspot), and less than 10^5 km (=1/7 R_o) deep (based on the details of sunspots emerging from the interior).

The Solution: The formula for the magnetic field inside an "infinitely" long solenoid wound by N turns of wire per meter and carrying current I is $B = \mu NI$ where $\mu = 4\pi \times 10^{-7}$ if B is in Tesla and I in Amperes. Given B = 0.15Tesla, we find NI = 1.2×10^5 amperes/meter. This is the current encircling the solenoid for each meter of length. (The value of N is irrelevant for continuous gas.) For a depth of 3×10^4 km, the total current encircling the sunspot is 4×10^{12} Ampere!

Interpretation: Compared to electrical currents running in our cities, the currents running in the Sun are enormous. Quite generally, astronomical currents are well beyond our imagination. For instance, the magnetic field between the stars is roughly 3×10^{-10} Tesla; it is arranged in magnetic structures (not necessarily in the shape of a solenoid) with dimensions of the order of 10 light-years = 10^{17} m; therefore, currents of the order of 3×10^{13} Amperes must commonly run between the stars. Astrophysical gases must be thought of as giant electromagnets. Moreover, because the "wires" are so thick (10^3 km for sunspots, 10^{14} km between the stars), currents are not turned into heat (problem 8.7.5). In this sense, astrophysical gases are rather like superconducting electromagnets.

Didactics: Because there are no solid magnets in astrophysics and all the currents in gases are accounted for explicitly, there is no need to consider separately the magnetic field $\underline{\mathbf{H}}$ and the magnetic induction or flux density $\underline{\mathbf{B}}$. In astrophysics, $\underline{\mathbf{B}}$ is referred to as the magnetic field.

2. Solar Coronal Mass Ejections

Conservation of magnetic and kinetic energies

Physics Introduction: In the laboratory, a wire carrying current $\underline{\mathbf{I}}$ in a magnetic field $\underline{\mathbf{B}}$ has a force exerted on it which is $\underline{\mathbf{I}}\mathbf{x}\underline{\mathbf{B}}$ per unit length of wire. In astrophysics we deal with continuous gases. We replace $\underline{\mathbf{I}}$ by the current flowing through a unit cross-section, $\underline{\mathbf{j}}$. Then the force exerted on the gas is $\underline{\mathbf{j}}\mathbf{x}\underline{\mathbf{B}}$ per unit volume. In astrophysics, $\underline{\mathbf{j}}$ is not observed, but $\underline{\mathbf{B}}$ is observed, and $\underline{\mathbf{B}}$ can be easily thought of in terms of magnetic lines of force : Lines of force are familiar in terms of iron filings around a bar magnet and in terms of the Van Allen belts in the Earth's magnetosphere. Therefore, simple ways of thinking have been developed that allow us to focus on $\underline{\mathbf{B}}$ and on field lines rather than on $\underline{\mathbf{j}}$.

Simple ways of thinking always start with energy. We know from theory that magnetic fields involve energy, specifically an energy density $B^2/2\mu$ j/m^3. When $\underline{\mathbf{j}}\mathbf{x}\underline{\mathbf{B}}$ forces accelerate gas, they do work on the gas, and the kinetic energy of the gas comes from $B^2/2\mu$.

The Setting: If we cover up the bright solar disk and observe the edge of the Sun against the dark sky, we see many features that look like loops standing over the Sun. Their "feet" are on the solar surface, in regions with sunspots. The shape of the loops reminds us of magnetic field lines near magnets, consistent with sunspots. We recognize the loops because they contain somewhat denser gas than the surroundings.

Figure 8.7-3. Photo Credit: National Optical Astronomy Observatories

Many of these loops remain nearly constant for days. Why do the gases not fall to the surface? The shape of the loops suggests that the gases are held up against gravity by **j x B** forces. Electrical currents must be running there, but we visualize only the magnetic field.

When we image the Sun using x-rays, that is, when we look at hot coronal gas, then we also see many loops standing on the Sun. These, also, usually last many days. But almost daily, one or another of these loops is suddenly blown away from the solar corona, achieving speeds in the range 400 to 1000 km/s. The ejected gases involve a mass typically of the order of 10^{12}kg. How does the gas acquire so much kinetic energy? Apparently the magnetic fields in the solar corona, previously steadily holding up the embedded gas with **j** x **B** forces, suddenly lose their equilibrium. Then the **j** x **B** forces accelerate the gas.

These phenomena are called "coronal mass ejection", abbreviated as CME. When a CME hits the Earth's magnetosphere, the resulting electric currents may damage satellite electronics, disturb communications, cause electrical power failures, and more. This problem concerns the predicted speed of the CME.

The Problem: Assume that ionized hydrogen gas in the solar corona has a density n = 10^{15} protons m^{-3} (and the same number density of electrons) and a magnetic field of strength 10^{-3}Tesla. If all the magnetic energy in a cubic meter is converted into kinetic energy of the gas initially in that cubic meter, what speed does this gas attain?

The Solution:

Setting $1/2\rho v^2 = B^2/2\mu$ yields $v = B/(\mu\rho)^{1/2}$ = 700 km/s.

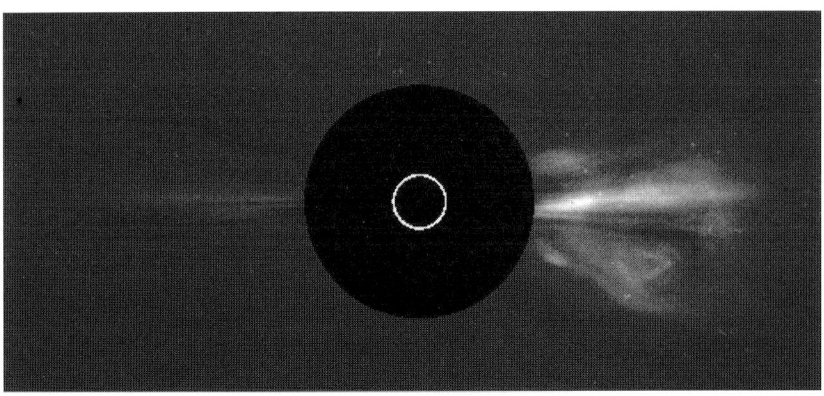

Figure 8.7-4. Credit: NASA

Interpretation: The estimate made in the problem is actually a fairly realistic estimate. The gas density is derived from the intensity of x-radiation. The magnetic field is measured at the surface (Zeeman splitting) and extrapolated. The predicted speed fits well into the range of observed speeds, 400 to 1000 km/s. The effect of these gases arriving at Earth appears in problem 8.7.3.

Didactics: The treatment of cosmic electrical currents and magnetic fields, when done in detail, involves nonlinear partial differential equations, which are very difficult to solve. Thus most of the theory of the last forty years has dealt primarily with a few simple soluble problems, such as energy estimates, as in this problem, or forces in very simple magnetic geometries, as in the next two problems. So far, no computer program can start with a steady loop in the corona, change it so as to lose its equilibrium, compute its acceleration, and follow the CME into space. But cameras on the spacecraft SOHO, which orbits permanently between Earth and Sun (at the Lagrangian point), can now warn us when CME's are approaching the Earth.

3. The Solar wind and the Earth's Magnetosphere

Physics Introduction: In Maxwell's equation, we neglect the term $c^{-2}\partial \underline{E}/\partial t$ term which was introduced by Maxwell. The physical approximation made by this neglect is outlined below. What remains of Maxwell's equation, $\text{curl}\underline{\mathbf{B}} = \mu \underline{j}$, $\mu = 4\pi \text{x} 10^{-7}$, is used to express the current density $\underline{j}$ in terms of $\underline{B}$. The force on a unit volume of gas can now be written $\underline{\mathbf{j}}\text{x}\underline{\mathbf{B}} = \text{curl}\underline{\mathbf{B}} \text{ x } \underline{\mathbf{B}}/\mu = -\text{grad}(B^2/2\mu) + (\underline{\mathbf{B}}\bullet\text{grad})\underline{\mathbf{B}}/\mu$. In the first term, $B^2/2\mu$ is identified as magnetic pressure, quite in analogy to the normal gas pressure. The second term corresponds to a tension (force per unit area) of B^2/μ along the lines of force. Since pressure and tension are familiar quantities from the laboratory, it is relatively easy to think of magnetic pressure and magnetic tension.

The tension acting on magnetic field lines is quite analogous to the tension on a violin string: The wave speed along the string is $(\text{tension/mass per unit length})^{1/2}$. For a continuous gas, we replace tension by force per unit area, B^2/μ, and we replace mass per unit length by mass per unit volume, ρ. Then the wave speed along the magnetic field line is $(B^2/\mu\rho)^{1/2}$. The wave is called an Alfvén wave, traveling at a speed called the Alfvén speed.

Here we deal with magnetic pressure.

The Problem: Estimate the distance from the Earth where the solar wind meets the Earth's magnetosphere. Specifically, how far from the Earth (r in units of Earth's radius r_E) at the magnetic equator is the magnetic pressure $B^2/2\mu$ of the Earth's dipole magnetic field equal to the dynamic pressure ρv^2 exerted by the solar wind? Use $B = 10^{-4}$T at the Earth's surface, and for wind parameters use pure ionized hydrogen, $n_p = 10^7 \text{m}^{-3}$, $v = 500$ km/s.

The Setting: The Earth's magnetic field shields the Earth's surface from the direct impact of the solar wind (or of a comet's poisonous tail). The volume controlled by the Earth's magnetic field, called the magnetosphere, has a tear-drop shape. It is round where it faces the solar wind. Its tail is very long and literally flaps like a flag in the solar wind. In first approximation, the magnetosphere is simply a blunt obstacle to the wind. At the sunward stagnation point, the pressure imparted by the wind must be balanced by magnetic pressure due to magnetospheric fields. This problem asks for an estimate on how far from the Earth the magnetic pressure can balance the wind pressure.

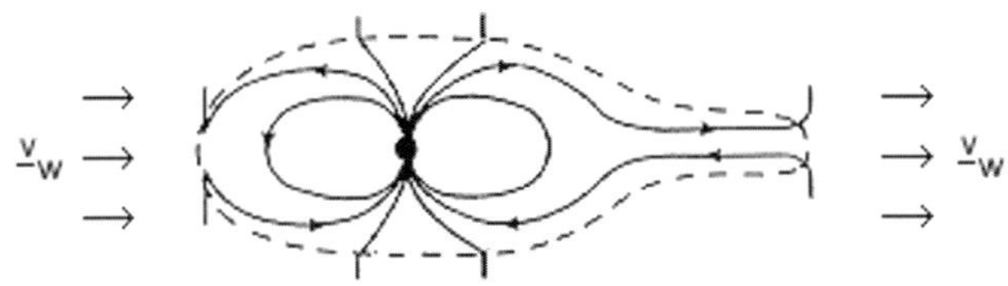

Figure 8.7-5.

The Solution: The strength of the Earth's magnetic field falls off as a dipole, proportional r^{-3}. Take $B = B_E\ (r_E/r)^{3}$. We need $B^2/2\mu = 4\text{x}10^{-3}\ (r_E/r)^{6} = \rho v^2 = 4\text{x}10^{-9}$. Thus $r = 10r_E$.

Interpretation: Obviously this is only an estimate. A dipole field carries zero electrical current and thus cannot produce a force. However, electrical currents are induced by the impact of the wind. The currents alter the Earth's magnetic field such that the new forces stop the wind. In first approximation, the currents run on the surface and the dipole field is terminated abruptly. Then magnetic pressure is the relevant force, balancing the external pressure of the wind. The estimate for the standoff distance is really quite good. However, the details of the boundary and its interactions with the wind are still not quite settled, even after three decades of satellite exploration, because the solar wind carries its own magnetic field. When a coronal mass ejection (CME, see problem 8.7.2) suddenly hits the magnetosphere, the front of the magnetosphere is compressed to a smaller radius. New electrical currents are induced throughout the magnetosphere, where they endanger satellite electronics, and in the ionosphere, where they interrupt radio communications, and on long conductors on Earth surface, where they can interrupt electrical power systems serving millions of people. CME's can now be watched by satellites all the way from their launch at the Sun to their arrival at Earth, but they can still provide surprises. One CME, in January 1997, seemed to pass quietly until an order-of-magnitude pressure pulse arrived at the end. That pulse compressed the magnetosphere so power-fully that the resulting currents short-circuited and destroyed a new communications satellite valued at 400 million dollars.

Didactics: Maxwell added the last term to what is now called Maxwell's equation, curl$\underline{\mathbf{B}} = \mu\underline{\mathbf{j}} + c^{-2}\partial\underline{\mathbf{E}}/\partial t$. Now we omit this term. Thus we cannot deal with vacuum electromagnetic radiation : In the vacuum, necessarily $\underline{j} = 0$, but curl$\underline{\mathbf{B}}$ is finite, so $\partial\underline{\mathbf{E}}/\partial t$ must be finite. The relation curl$\underline{\mathbf{B}} = \mu\underline{\mathbf{j}}$ implies div$\underline{j}$ = div curl$\underline{\mathbf{B}} = 0$ (because div$\underline{\mathbf{B}} = 0$). Thus we can deal only with problems where there is no accumulation of electric charge. The equation cannot describe most waves invoked in plasma physics. It can, however, describe most phenomena that happen on long time scales, that is, much longer than any plasma or cyclotron frequencies. This is the subject of magneto-hydrodynamics, usually abbreviated MHD, a subject begun about 50 years ago by H. Alfvén. It could have been investigated even before Maxwell.

4. Sunspots and the Earth's Climate

Magnetic pressure

The Problem: A cylindrical sunspot is in pressure equilibrium, such that the internal pressure $n_i kT_i + B^2/2\mu$ equals the external pressure $n_o kT_o$. The magnetic field strength outside the spot is zero, inside it is B = 0.15 Tesla. The temperatures are T_o =5800°K and T_i = 4800°K. The density outside is $n_o = 2\text{x}10^{23}$H atoms m^{-3}. Find the density inside, n_i. Compare n_i to n_o.

The Setting: Sunspots "live" some days or weeks. As the Sun turns, the spots move with the Sun. Seen from earth, the spots apparently move across the disk of the Sun. They become invisible before they are turned fully toward the edge of the Sun. Apparently, the surface of the sunspots is lower than the normal solar surface. Therefore, the gas at any one height must be more transparent within the sunspot than outside it. Why is that? Because of the magnetic pressure, the gas pressure inside the sunspot is much smaller than outside, as shown in the problem, and so is the gas density.

The Solution: Outside of the sunspot, $n_o kT_o = 1.7\text{x}10^4$, inside $B^2/2\mu = 0.9\text{x}10^4$, so that inside $n_i kT_i = 0.8\text{x}10^4$, and $n_i = 1.2\text{x}10^{23}m^{-3} = 0.6n_o$.

Interpretation: In the solar atmosphere, gas is stratified vertically according to the equation of hydrostatic equilibrium (problem 8.6.1). But the stratification can be different outside and inside a sunspot because of the spot magnetic field. At the height of the normal photosphere, i.e. from where most solar radiation can leave into space, the sunspot is relatively empty and transparent. The spot surface resides further down, where the density is more like that of the normal photosphere. Since the scale height of the pressure distribution in the solar surface is about 180 km (problem 8.6.1), the spot surface is about 200km below the solar surface. This roughly fits the observations.

Sunspots and Earth's climate: A sunspot is less bright than the surrounding photosphere because it is less hot. Why is it less hot? In the surroundings, convection brings up heat which must then be radiated away at the same rate. The energy flow determines the temperature at the solar surface (Stefan-Boltzmann law). The spot's magnetic field resists the convective motions, so that less energy is convected up from the interior within the spot. The arriving heat can be radiated away at a lower temperature.

The number of sunspots varies in a cycle of approximately 11 years. One might expect that Earth receives less sunlight in years when there are many spots. Indeed, when any one spot appears, we do receive less light than just before. Surprisingly, according to satellite measurements, we receive roughly 0.1% more energy on average during years with many spots, and less during years with very few spots. Indeed, historical periods with long absences of sunspots are times when the Earth's climate was relatively cool.

Why is the observation opposite to the prediction? In times when there are many spots, detailed observations of the solar surface also show very many very thin bundles of magnetic fields with B ~0.15 Tesla. These are solenoids similar to sunspots but merely some 100 km in diameter. Like sunspots, they have a depressed surface. However, they are so thin that most of their radiation is lost from the sides rather than from the depressed surface. We observe the walls of the solenoids. They are hotter than the solar surface, and so we receive more energy.

5. Magnetic fields of White Dwarfs and Neutron stars

Faraday's law of induction

Physics Introduction: In the laboratory, if a wire heats up because of electrical resistance, one uses a thicker wire. Cosmic magnetic fields imply electrical currents flowing in extremely thick gaseous "wires", so thick that dissipation of the currents into heat is negligible over time scales of interest. Effectively, there is no electrical resistance. Therefore, the electrical field in the frame of the ions vanishes. The electrical field seen in any other frame of reference is $\underline{\mathbf{E}} = -\underline{\mathbf{v}}\mathrm{x}\underline{\mathbf{B}}$, and Faraday's law of induction becomes $\delta\underline{\mathbf{B}}/\delta t = -\mathrm{curl}\underline{\mathbf{E}} = \mathrm{curl}(\underline{\mathbf{v}}\mathrm{x}\underline{\mathbf{B}})$. The integral form of Faraday's law is $\delta/\delta t\int\underline{\mathbf{B}}\bullet d\underline{\mathbf{s}} = \int\underline{\mathbf{v}}\mathrm{x}\underline{\mathbf{B}}\bullet d\underline{\mathbf{l}}$ where the integrals are over any fixed surface and around the edge of that surface. On interchanging • and x in the line integral, one can show geometrically that Faraday's law reduces to $d/dt\int\underline{\mathbf{B}}\bullet d\underline{\mathbf{s}} = 0$, where now the integral is over any surface moving with the gas. The magnetic flux through any surface moving with the gas is conserved.

Imagine two elements of gas connected by a field line at some time. Draw a surface, generally not flat, which includes this field line and which also includes neighboring field lines. This surface has zero magnetic flux through it. Let this surface move with the gas. By Faraday's law without current dissipation, the surface must continue to have zero magnetic flux through it. Therefore, the same field line still connects the two elements of gas. In effect, the field lines are attached to the gas. One says the magnetic field is "frozen into" the gas. Although magnetic field lines are mathematical constructs, in MHD they take on a real physical property. That is why, for instance, the vibration of field lines and gas participating in an Alfvén wave is completely analogous to the vibration of a string with tension as in a violin.

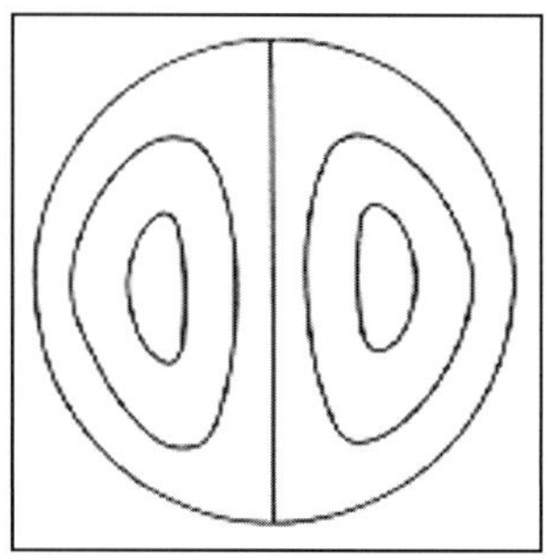

Figure 8.7-6.

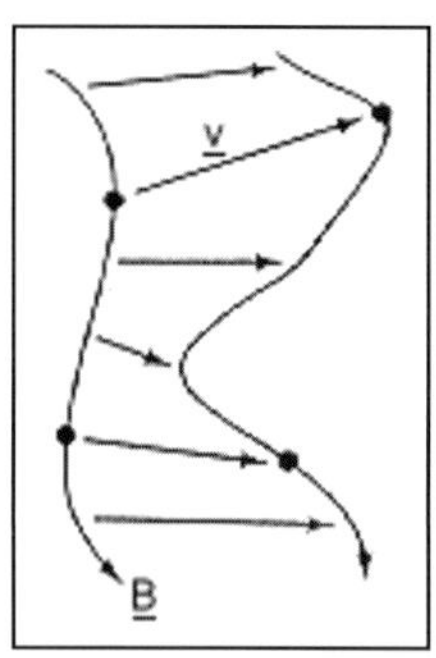

Figure 8.7-7

The problem: Imagine that the Sun, in its interior, contains a magnetic field encircling the axis of the Sun, parallel to the solar equator, much as if a giant solenoid were stretched around the Sun at its equator. Now imagine making the Sun smaller, keeping the same structure of density and frozen-in magnetic field. If the field strength B changes in proportion to R^n, what is n? If the field strength within the Sun is 10 Tesla, what is it after the Sun shrinks to the size of a white dwarf, and after the Sun shrinks to the size of a neutron star? For the stellar radii, use R = 10^6 km, 10^4 km, and 10 km, respectively.

The solution: It is simplest to consider any part of a plane through the axis of the star, that is, normal to the magnetic field, for instance a circle within that plane. Follow the motion of the gas on the boundary of that circle as the circle shrinks. The magnetic flux through the circle must remain constant. But the area of the circle, attached to the gas, shrinks as R^{-2}. Thus n = -2, and $B/B_o = (R_o/R)^2$. The white dwarf then is expected to have B = 10^5 Tesla, the neutron star B = 10^{11} Tesla.

Interpretation: The strongest magnetic field observed on a white dwarf is 5×10^4 Tesla. Magnetic fields of neutron stars observed as radio pulsars have values up to 4×10^9 Tesla. The highest field deduced in an x-ray pulsar is 1.6×10^{11} Tesla. So the estimates are quite satisfying. One can really expect only order-of-magnitude agreement because the solar B = 10 Tesla is estimated only rather indirectly, from the manner how sunspots emerge from the deep solar interior; because the white dwarf fields are left in the star after the star has shed its outer layers; and because the neutron star fields are left after the star exploded as a supernova. Moreover, it is possible that powerful convection during the supernova explosion amplified the field in the neutron star. Nevertheless, it seems reasonable that the observed fields on white dwarfs and neutron stars derive from the compression of magnetic fields that were within the stars when these stars were main sequence stars.

Exceptions to the frozen-in approximation occur at very powerful events, such as solar flares (whose explosive energy must derive from powerful localized electric currents suddenly dissipated in the solar corona), the turbulence during a supernova explosion, the interaction of the Crab Nebula with the pulsar at its center (since the magnetic fields in the nebula now are far too strong as to have expanded in frozen-in form from the original supernova and its progenitor star), and the turbulence of the gases between the stars (which must have created the observed interstellar galactic magnetic field), and at singular geometries such as the front and back of our magnetosphere (diagram in problem 8.7.3).

Didactics: When electrical currents are negligibly dissipated into heat, this is often interpreted as "zero resistance" or "infinite conductivity", inspiring the analogy with superconductors. However, the conductivity of many astrophysical gases is within an order of magnitude of the conductivity of copper in Earth's laboratories. Negligible dissipation in astrophysics is literally due to the extreme thickness of the "wires" (mathematically, the very large value of the length B/|curl**B**|).

Possible students' problem. (Students should discuss each part and agree on the answer before moving on). Suppose the magnetic field now in the Crab Nebula, B = $4x10^{-8}$ Tesla, were truly frozen into the expanding nebular gas, approximately a sphere of radius r = 3 light-years. Earlier, r was smaller and B was higher. With what exponent does B depend on r? (Answer: B is proportional to r^{-2}, as for the stars.) How do the magnetic energy density and the total magnetic energy in the nebula depend on r? (Answer: r^{-4} and r^{-1}, respectively). Estimate backwards in time: what was the total magnetic energy when r equaled the radius of an about-to-explode star, say r= $150R_o$= 10^{11}m? (Answer: Now the magnetic energy density = $2/\pi$ x 10^{-9}j m^{-3}, the volume = 3.6 $\pi x10^{49}m^3$, the total energy = $7.2x10^{40}$j; scaled by a factor $3x10^{16-11}$, the total magnetic energy then = $2x10^{46}$j.) Compare to the gravitational energy of the original star, assuming M = 10 M_o, and interpret. (Answer: GM^2/R ~$2.6x10^{41}$j <<$2x10^{46}$j. No stable star can exist with negligible gravitational energy). [The Virial theorem extended to MHD has the total kinetic energy K replaced by K + total magnetic energy.]

Chapter 8.8

HIGH-ENERGY ASTROPHYSICS, ELECTROMAGNETIC RADIATION

D.G. Wentzel

Introduction

Cosmic-ray physicists demonstrated that our Galaxy is filled with very energetic cosmic rays, mostly protons. Observations in the radio and x-ray range have made astronomers realize that much of the universe is filled with energetic electrons. Synchrotron radiation from relativistic electrons is a very common cosmic phenomenon (problems 8.8.1, 8.8.2). During the last decades, as observations were expanded from the x-rays toward ever higher energy gamma rays, the highest deduced energies of electrons have grown dramatically. Inverse-Compton radiation, which used to be mostly a topic of theoretical physics, has become very real (problem 8.8.3). Just how the electrons acquire their high energies is not clear, but magnetic energy is surely involved, and one part of the process is observed in the Crab Nebula (problem 8.8.4).

The radiation processes are derived here in a heuristic manner, using only rather simple concepts from radiation theory and relativity. It is much more important, in astrophysics, to keep in mind the relevant physics rather than worry about the factors of two arising from exact theory.

An overview of Radiation processes. The radiation from the hot gases in the solar corona, in accretion disks, and between the galaxies located in clusters of galaxies mimics the shape of the Planck spectrum, though not its intensity. This radiation is considered "thermal" radiation because the particles have a thermal Boltzmann velocity distribution. Collisions among these particles both create the Boltzmann velocity distribution and cause the emission of the "thermal" radiation.

In many astrophysical situations such as solar flares, supernova remnants, or quasars, electrons are accelerated by electromagnetic fields to energies orders of magnitude higher than their original thermal

W. Wamsteker et al. (eds.), Developing Basic Space Science World-Wide, 489-499.

energies. Typically, these electrons acquire a power-law distribution in energies. In solar flares, the electrons reach mildly relativistic energies. When these electrons migrate from the flare in the solar corona to the underlying denser gases, they collide with local protons and electrons and emit a power-law spectrum of x-rays. Such radiation is referred to as "non-thermal" x-rays. Both thermal and nonthermal x-rays are caused by electron collisions, and in both cases the emission process is often called Bremsstrahlung.

More frequently in astrophysics, the electrons reach highly relativistic energies. These electrons easily radiate on their own, without any collisions with other particles. They radiate via the cyclotron, synchrotron, and inverse-Compton radiation mechanisms. The cyclotron and synchrotron radiation mechanisms depend on the energy density in magnetic fields. These fields are present everywhere, but they tend to be stronger in those places where electrons are accelerated to high energies, thus helping the high efficiency with which these electrons emit cyclotron or synchrotron radiation. The inverse-Compton mechanism depends on the energy density of electromagnetic radiation, which is very high in powerful objects like quasars.

1. The Crab Nebula

Synchrotron radiation - relativistic beaming

Physics Needed: Lorentz transformations in special relativity; relativistic gyrofrequency.

Physics Introduction. A non-relativistic electron in a magnetic field of strength B has an angular frequency qB/m rad/s ($q=1.6\text{x}10^{-9}$C) or a frequency $f_o = 2.8\text{x}10^4$B MHz where B is in Teslas. The electron emits electromagnetic radiation at frequency f_o. When an electron is highly relativistic with energy $E = \gamma mc^2$, $\gamma >> 1$, the gyrofrequency is f_o/γ. [$mc^2 = 0.8\text{x}10^{-13}$j = 0.5 MeV, where eV = electron Volt.] The most intense radiation we observe from this electron is at a frequency different from the gyrofrequency. First, a Lorentz transformation alters the apparent angle at which a photon propagates. If photons are emitted roughly isotropically in the frame of an electron, they are observed by us as a beam of photons concentrated in the forward direction, with opening angle $\theta \sim 1/\gamma$ radians. In addition, there is the equivalent of a Doppler shift. Let the beam be emitted while the electron travels the distance $2R\theta$ between points a and b in time $\Delta t(el)=2R\theta/v$. This would be the duration of the beam we would observe if the speed of light were infinite. In reality, while the electrons travel from a to b, the first part of the beam, emitted when the electrons were at a, travels the distance $c\Delta t(el) > 2R\theta$. Thus the front of the beam is ahead of the back of the beam, emitted when the electrons are at b, by (c-v) $\Delta t(el)$ and the duration of the beam we observe when it arrives here is (1-v/c) $\Delta t(el)$. Therefore, the duration of the beam is shortened by a factor $1-v/c = (1-v^2/c^2)/(1+v/c) = 1/2\gamma^2$, where we have set $1+v/c = 2$ in the denominator. Accordingly we see a beam of a duration shortened by a factor of $2\gamma^2$relative to $\Delta t(el)$. Using $\theta \sim 1/\gamma$, we find that the pulse lasts only a fraction $\sim 1/\gamma^3$ of the gyroperiod $2\pi R/v = \gamma/f_o$, that is, it lasts only about $1/(f_o\gamma^2)$. Such a pulse, when Fourier analyzed, has its main signal at a frequency $\sim f_o\gamma^2$.

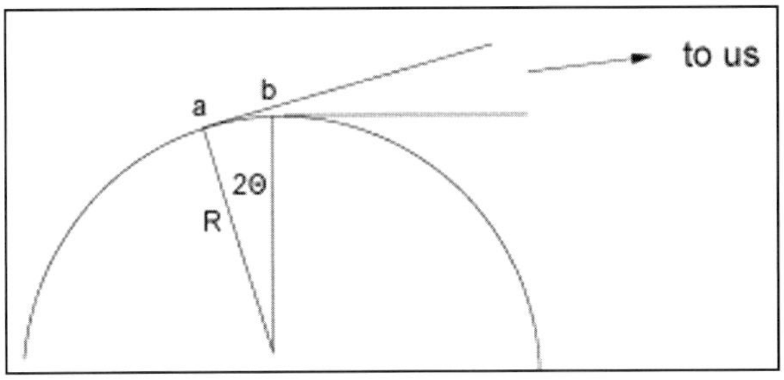

Figure 8.8-1.

Computed in more detail, the main emission is observed at about $f = 10^4\,\gamma^2 B$ MHz. Any one electron also radiates at somewhat higher and lower frequencies, say from 0.3 to 1.5 f. If we observe an object containing many electrons with a large range of energies, we actually detect at our selected observing frequency contributions from electrons with γ whose f is within 1/1.5 to 1/0.3 times the observing frequency. If we gradually change the observing frequency, we detect radiation from electrons with a gradually changing range of energies. The received radiation changes smoothly with observing frequency. We observe a continuous spectrum. This type of radiation is named synchrotron radiation.

The Problem: We observe the Crab Nebula ($B = 4 \times 10^{-8}$ Tesla) in the radio range at 300 MHz. What is the energy of the electrons whose radiation we observe? Compare to the energy 3/2 kT for $T = 10^8$ °K. We also observe the Crab Nebula in the x-ray range, say at wavelengths of 1 nm. What is the energy of the x-radiating electrons? Compare the energy to that of a flying mosquito.

The Astrophysical Setting: The Crab Nebula is the gaseous remnant of a supernova seen by humans in the year 1054 A.D. = 433 A.H. = 4815 A.M. (Picture in problem 8.4.4.) The Crab is named after a pattern of filaments, most obvious in red light due to atomic hydrogen emission at 656 nm. In addition, there is continuum radiation from all parts of the roughly circular nebula. The continuum is synchrotron radiation. This radiation was first detected and identified in the 1950's in the radio range, but now it is observed at frequencies up to the x-ray range. The pulsar which is left from the exploded star energizes the nebula (problems 8.8.4, 8.8.4) and indirectly accelerates electrons to the high energies evaluated here.

The Solution: f = $4x10^{-4}\gamma^2$ MHz. Radio range: $\gamma^2 = 0.8x10^6$, energy $0.5x10^9$eV = $0.8x10^{-10}$j »3/2 kT = $2x10^{-15}$j. X-ray range: f = c/λ = $3x10^{17}$Hz = $3x10^{11}$MHz, $\gamma^2 = 8x10^{14}$, energy $1.4x10^{13}$eV = $2.3x10^{-6}$j. This is for just one electron. For the mosquito, estimate v = 1 km/hr, mass 10^{-2}gm , kinetic energy = 1/2 10^{-5}kg $(0.3m/s)^2$= $0.5x10^{-6}$ j, comparable.

Interpretation: These electrons have energies far above any plausible thermal energies. These electrons radiate so powerfully that they lose their energy in a time shorter than the age of the Crab Nebula (see next problem). Therefore, such electrons must be accelerated in recent times. This was the first good evidence for what is now generally accepted : electrons are accelerated much more easily than skeptical theoreticians had expected.

Didactics: The derivation of f is only heuristic. Most advanced students in astrophysicists must suffer through a detailed, quite mathematical derivation and they (almost) all decide that the heuristic derivation is sufficient for their needs.

2. The Crab Nebula

Synchrotron radiation - total power

Physics Needed: Larmor formula for rate of radiation; relativistic time dilation and gyrofrequency. The student must first do problem 8.8.1.

Physics Introduction: Classically, the energy loss by radiation from an electron spiraling in a magnetic field of strength B is dE/dt = 2/3 $kq^2\underline{\mathbf{a}}\bullet\underline{\mathbf{a}}/c^3$ where non-relativistically $\underline{\mathbf{a}}$ = d$\underline{\mathbf{v}}$/dt with md$\underline{\mathbf{v}}$/dt = -q$\underline{\mathbf{v}}$x$\underline{\mathbf{B}}$, leading to a gyrofrequency qB/m rad/s. (Here k = $\mu c^2/4\pi$ = $9x10^9$, q = $1.6x10^{-19}$Coulomb.) For very energetic electrons, there are two relativistic effects. First, the gyrofrequency becomes qB/γm, introducing a factor 1/γ for each factor $\underline{\mathbf{a}}$. Second, since the time appears squared in $\underline{\mathbf{a}}$, time dilation introduces a factor γ^2 for each factor $\underline{\mathbf{a}}$.The above formula becomes dE/dt = $8\pi/3\ r_e^2 c\gamma^2\ (B^2/\mu)(v_{perp}/c)^2$ where $r_e = kq^2/mc^2 = 2.8x10^{-15}$ m is the classical electron radius. For an isotropic velocity distribution, averaging $(v_{perp}/c)^2$ over angles yields an additional factor 2/3, so that dE/dt = 4/3 σ_T c $\gamma^2 U_B$, where $\sigma_T = 8\pi/3\ r_e^2$ is the Thomson crossection, $U_B = B^2/2\mu$ is the energy density in the magnetic field, and v = c for highly relativistic electrons. This last form of dE/dt gives the total power radiated per electron in an isotropic distribution of electrons with energy γmc^2.

The Problem: The electrons lose kinetic energy by radiation. Derive the time scale $\tau = E/(dE/dt)$ using dE/dt averaged over isotropic electrons. Compute τ for both the electrons yielding the radio emissions and the electrons yielding the x-ray emissions in the Crab Nebula (see problem 8.8.1). Can the relevant electrons have existed for the entire age of the Crab Nebula ~950 years?

The Solution: $\tau = 3mc/(4\sigma_T U_B \gamma) = 0.48\text{x}10^{16}/\gamma$ seconds. For the radio emitting electrons, $\tau = 1.7\text{x}10^5$years »10^3years; yes; for the x-ray emitting electrons, $\tau = 5$ years «10^3years; no.

Interpretation: The radio emitting electrons have lost very little of their energy. They may have been dispersed throughout the nebula at the time of the supernova explosion. But the x-ray emitting electrons lose their energy very rapidly. Any such energetic electrons accelerated at the time of the explosion would have lost their energy long ago. They must have been accelerated during the last few years. Evidently the pulsar is still a very active object and the energy flowing from it (problems 8.4.4, 8.5.4) continuously accelerates the electrons.

Didactics:

1. Students may ask why we bother to express dE/dt in terms of r_e and the Thomson cross section, since we deal with the spiraling of an electron in a magnetic field, while the Thomson cross section measures an electron's response to the electric field of passing electromagnetic radiation. Physically, the magnetic field is just a low-frequency electromagnetic field. In fact, we shall use this knowledge in the next problem. Moreover, the use of a cross section also leads to the use of an energy density. It is always good in astrophysics to express quantities in terms of energies, since energies directly provide physical implications. Mathematically, the formula is in a form where one can easily check that the dimensions are correct, energy per second, without getting into the dimensions of electric charge.
2. The formula for dE/dt was derived assuming constant electron energy over the time of a few gyrations around the magnetic field, that is, during τ»R/v. This is very well satisfied. The result for dE/dt given here is rigorous, although the derivation is abbreviated. Radiation tends to make the electrons anisotropic, but they are scattered and kept nearly isotropic by Alfvén waves.
3. Most students will first evaluate dE/dt, then E, then divide. Let them do this. Afterwards suggest the simpler way of writing down the combined formula for τ before evaluating.

3. Gamma Rays from Quasars

Inverse Compton radiation

Physics Introduction: Highly relativistic electrons may collide with low-energy photons to create high-energy photons, the inverse-Compton (IC) effect. The resulting photon energy can be derived quite simply by two transformations of reference frames: a photon of frequency ν is "seen" by an electron with energy γmc^2 as having a frequency $\sim\gamma\nu$ in the frame of the electron, the electron scatters the photon at constant energy (true if the photon energy in this frame is less than mc^2 so that the electron suffers no recoil); finally, back in our frame of reference, we see the photon as having a frequency $\sim\gamma^2\nu$. This process can easily turn a radio photon into a gamma ray. The rate at which a fast electron turns low-energy photons into high-energy photons is best given by analogy to the rate of synchrotron loss, $dE/dt = 4/3\ \sigma_T\ c\ \gamma^2 U_{rad}$, where U_{rad} is now the energy density in the low-energy photons. The low-energy photons may belong to the cosmic background radiation (often called 3°K radiation but actually T = 2.7°K). This radiation has an energy density of $0.4\text{x}10^{-13}$j m^{-3}. Alternatively, the low-energy photons may be the synchrotron photons emitted by the energetic electrons. The latter process is called the synchrotron self-Compton process.

The Problem: The quasar 3C279 emits photons up to the high gamma-ray range, 4 GeV (GeV = 10^9 eV). Suppose that the 4 GeV photons have been IC-boosted, starting as microwave photons of 10^{11}Hz. What is the γ of the electrons? Suppose that the 4 GeV photons have been IC-boosted twice by the same electrons, starting from microwave photons of 10^{11}Hz. What is the required γ of the electrons?

The Setting: A number of quasars and "active galactic nuclei" emit several components of radiation. One component may be due to synchrotron radiation. Another, often the most energetic component, may be inverse-Compton radiation. Some quasars produce astoundingly intense high-energy radiation. The quasar 3C279 has been observed with the satellite called Compton Gamma-Ray Observatory up to photon energies of 4 GeV. Its spectrum between 10^{11}Hz and 4 GeV is close to a single power law f^{-a} that is flatter than a = 1. The object emits most of its energy in the gamma-ray range.

The Solution: $h\nu = 6.6\text{x}10^{-23}$j; $4\text{x}10^9$eV/$4\text{x}10^{-4}$eV = 10^{13}. For a single IC boost in energy, $\gamma^2 = 10^{13}$, $\gamma = 3\text{x}10^6$. For a double IC boost, $\gamma^4 = 10^{13}$, $\gamma = 2\text{x}10^3$.

Interpretation: The spectrum of 3C279 rises rapidly through the radio range and becomes much flatter at 10^{11}Hz (microwave range). Thus, in this problem, the microwave photons are taken to be the most abundant photons available for IC-boosting. But a single IC boost needs electrons at extremely high energies, for which there is no independent evidence. No theory yields sufficiently rapid acceleration to these very high energies. Therefore, double-IC-boosting has been investigated. It functions with electrons of more normal energies. So far, however, the details of the radio and gamma-ray spectrum of this particular quasar do not fit the theory of double IC boosting. (The theoretical details include the electron recoil at the second scattering.) Perhaps the compact jet emanating from this object provides a local intense source of higher-frequency photons that can be IC-boosted.

The Crab Nebula emits radiation up to 5×10^{13} eV! Perhaps this is IC radiation, but again the theory does not fit well to observations.

Didactics:

1. Ask your students to find the total power loss by integrating the spectrum f^{a} to ever higher f, using a < 1. They will, of course, find that the integral diverges. In some range of very large frequencies, the spectrum must decrease, a > 1. We do not know at what frequencies this occurs.
2. Quasars and related objects (with a host of names) are exciting because they still surprise us. Thirty years ago, their total power output seemed unbelievable, but astrophysicists gradually became used to the power released near a black hole, once some black holes (with either star or galaxy masses) were clearly identified. Today the excitement focuses on the extreme energy to which at least a fraction of all the fast electrons are accelerated and the powerful gas flows that make this acceleration possible. The work is challenging both observationally - to make and combine observations spanning fifteen orders of magnitude in frequency - and theoretically - to understand the electron-photon interactions that can focus so much energy into a few electrons, to understand the violent gas flows, especially jets, which probably provide the energy source for the acceleration and, ultimately, to identify how all this relates to the central power source surrounding the probable black hole.

Relativistic Motion (jets): When radio telescopes first resolved the angular structure of quasars on the sky, some of the radio sources were resolved into parts that appeared to separate on the sky faster than the speed of light. Indeed, a source moving toward us, nearly along our line of sight, at a speed v close to c can appear to move across the sky at a speed up to γv with $\gamma = 1/(1-v^2/c^2)^{1/2}$. When this explanation was first suggested, such fast objects seemed incredible. Now, however, relativistic motions with γ in the range 2 to 10 are also used to explain why electron energies in very powerful quasars and in apparently rapidly varying (thus apparently very compact) quasars are evidently not quickly drained by inverse - Compton radiation. (The deduced energy density of radiation is lower and τ proportional to $1/U_{rad}$ is longer if one invokes relativistic motion.) Relativistic jets now seem to be common to many quasars and centers of active galaxies. In our Galaxy, we find relativistic jets on a stellar scale. The large variety of powerful objects may perhaps all be reduced to one kind of object, with power emanating from a central "engine", and with jets that are merely seen from different directions.

4. Magnetic field of Pulsars

Low-frequency magnetic "dipole" radiation

Didactic Purpose: An example of approximations that retain the essential physics.

Physics needed: Energy density and propagation speed of an electromagnetic wave.

The Setting: The Crab nebula needs a prodigious source of energy, about $1.2x10^5$ L_0. Indeed, the central rotating neutron star is slowing down and provides the energy needed for the nebula (problem 8.4.4). But how is the energy carried from the pulsar to the nebula? The rotating pulsar causes the surrounding magnetic field to vary in time. Thus the pulsar emits electromagnetic radiation at the pulsar rotation period. The pulsar radiates as much energy as the kinetic energy lost by the pulsar's slow-down if the magnetic field at the pulsar's surface has the appropriate value. This problem leads the student through the appropriate (approximate) steps to evaluate the necessary pulsar magnetic field.

The Problem: A neutron star (of mass M, radius R, rotation period P) carries a dipole-like magnetic field with strength B_p at the surface, tilted with respect to the rotation axis. Radio radiation is emitted in a cone centered on the magnetic axis. As the cone sweeps past Earth, we observe a pulse of radiation. The low-frequency radiation at period P removes rotational energy from the pulsar and causes a slow-down in the rotation rate, $dP/dt > 0$. Evaluate the low-frequency radiation loss by the pulsar, as follows: Near the pulsar, write $B(r)=B_p(r/R)^n$. What is n for a dipole-like field? Assume that this relation is valid out to the "light cylinder", that value of r where an object circling the neutron star with period P moves at the speed of light, c. Evaluate B at the light cylinder, roughly at the equator, in terms of B_p, R, P, and c.

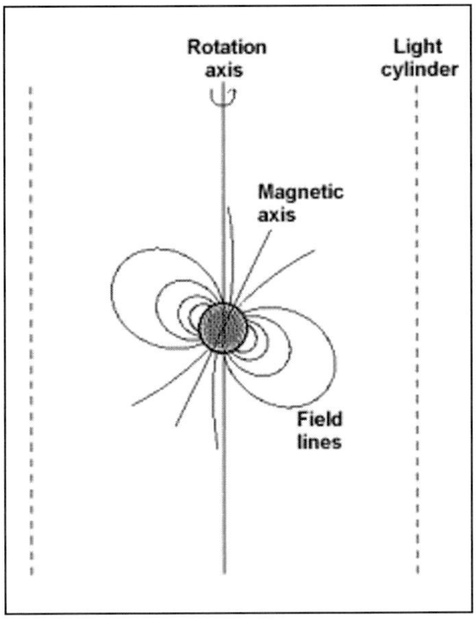

Figure 8.8-2.

Now assume that, exactly at the light cylinder, B makes a transition from a static dipole inside the light cylinder to a propagating electromagnetic field outside. The electromagnetic wave consists mostly of magnetic field and very little electric field. The wave moves from the light cylinder into space at speed c. What is the energy density of the electromagnetic field at the light cylinder, expressed in terms of B_p, R, P, and c? What is the energy flux (i.e. energy crossing unit area per second)?.

Suppose this radiation is emitted from the light cylinder all around the equator and for a distance R above and below the equator. What is the total energy, dE/dt, emitted over this surface, expressed in terms of B_p, R, c, and P?.

Now equate your formula for dE/dt to the kinetic energy lost by the neutron star as its rotation slows down (derived in problem 8.4.4), dE/dt = $-8/5\ \pi^2 MR^2\ (dP/dt)\ /\ P^3$. Derive an equation for B_p involving M, R, c, P and dP/dt. Evaluate Bp using the same numerical values appropriate for the Crab Nebula pulsar: P = 0.0333 sec , dP/dt = 4.21×10^{-13} sec/sec, R = 10 km and, as used in problem 8.4.4, $M = 1.4\ M_o$.

The Solution: n=3. The radius of the light cylinder is $Pc/2\pi$. B at the light cylinder is $B_p(2\pi R/Pc)^3$. The energy density there is $[B_p(2\pi R/Pc)^3]^2/2\mu$, the flux is energy density x c, the area is $4\pi\ (Pc/2\pi)^2$, and so $dE/dt = -(2\pi c/\mu)B_p^2\ (2\pi/Pc)^4\ R^6$.

On equating the two expressions for dE/dt, $B_p^2 = (\mu/20\pi^3)(Mc^3P\ dP/dt\ /R^4)$. The numbers for the Crab pulsar yield $B_p = 4.6x10^8$Tesla.

Interpretation: The space surrounding the neutron star is clearly not a vacuum but is permeated by a plasma, possibly electrons and protons. After a few theoretical improvements for the radiation loss and plasma flow, the most widely accepted value for the Crab pulsar is Bp = $4x10^8$ Tesla, remarkably close to our estimate. Magnetic fields estimated in this way for various radio pulsars range from $3x10^4$ to $4x10^9$ Tesla. The lower values are for the very fast (millisecond) pulsars that have acquired gases from a companion star.

One obvious simplification here is the assumed dipole magnetic field inside the light cylinder. A dipole field involves no electrical current, it cannot do work, and thus it cannot emit radiation. It is more accurate, but much more complicated, to describe the field as nearly dipole near the neutron star and becoming gradually more distorted, current-carrying, time-dependent and electromagnetic at larger distances from the pulsar. If one attempts this, one must promptly ask: "What is "near", relative to what distance?". The electromagnetic nature of the desired waves then implicates the distance of the light cylinder, and that distance is then chosen for the assumed sudden transition.

Even the surface of the neutron star is by no means free of electrical current. Because of the rapid rotation, the electric field there is so strong that it pulls either ions or electrons from the surface. The details of the resulting magnetosphere, including the electrical currents, are not yet solved self-consistently. (MHD is not adequate.) Certainly, particle energies are high enough for emission of gamma rays, which then can create electron-positron pairs. Probably an electron-positron wind, aided by the low-frequency magnetic waves, permeates the entire Crab Nebula (except the filaments for which the Crab is named). Perhaps shocks in this wind accelerate a fraction of the electrons and positrons to sufficiently high energies so that they emit the observed synchrotron radiation. There is not yet any observational evidence whether the synchrotron radiation is emitted by only electrons or by both electrons and positrons. The low-frequency waves, being mostly magnetic, somehow also maintain the magnetic field of the Crab Nebula.

Didactics: A great art in frontier physics, and especially in astrophysics, is to make those simplifications that retain the important physics. The example here concerns the assumed sudden transition from the dipole to the radiation field. There have been many other examples earlier. These simplifications allow some understanding of new phenomena and allows asking questions for new observations. Of course, afterwards it is important to verify the simplifications in detail.

Sometimes, the simplifications turn out to be wrong. But not in this example of the Crab pulsar's energy loss.

Chapter 8.9

MATHEMATICAL PROOFS

D.G. Wentzel

1. *The consistency of Newton's laws with an elliptical orbit:* In cylindrical coordinates, assume some orbit $r(\theta)$ with time t a free parameter. For a central force, $L = mr^2 d\theta/dt$ = constant. Hence $Ldt = mr^2 d\theta$, $d/dt = (L/mr^2)d/d\theta$, and specifically $dr/dt = (L/mr^2)dr/d\theta = -(L/m)du/d\theta$ where $u = 1/r$. Then the equation of motion $m\, d^2r/dt^2 - mr(d\theta/dt)^2 = -k/r^2$ is transformed to $d^2u/d\theta^2 + u = km/L^2$. With $y = u - km/L^2$, y = constant x $\cos(\theta - \theta_o)$, and $u = 1/r = mk/L^2$ [1+constant x $(L^2/mk)\cos(\theta - \theta_o)$], the equation for an ellipse.
2. *Kepler's Third Law (for the two-body problem):* We start with the total energy of the system, $E = 1/2\, \mu v^2 - GM\mu/r$, where $M = m_1+m_2$, $\mu=m_1 m_2/M$, and $L = \mu rv$ with $v = v_1+v_2$, the velocities evaluated in the center-of-mass system. From the geometry of the ellipse, with a and p standing for aphelion and perihelion, $r_p=a(1-e)$, $r_a=a(1+e)$, $r_p/r_a = (1-e)/(1+e) = v_a/v_p$. If E is now expressed once in terms of parameters at aphelion, once perihelion, then one obtains $v_p^2 =(GM/a)(1+e)/(1-e)$, and $L = \mu v_p r_p = \mu\, [GMa(1-e^2)]^{1/2}$, and finally $E = -Gm_1m_2/2a$. Since the area swept out is $dA/dt = 1/2\, r^2 d\theta/dt$, we have also $L = 2A\mu/P = 2\pi a^2 (1-e^2)^{1/2}\mu/P$. The two forms of L yield $P^2 =4\pi^2a^3/GM$. The energy equation becomes $1/2\, \mu v^2 = Gm_1m_2\,(1/r - 1/2a)$.
3. *Gravity within a Spherical system:*
   11. The local gravity can be derived quite elegantly in terms of the differential equation **div grad**$\varphi = -4\pi\rho$ where φ is the gravitational potential and ρ the mass density. Assuming spherical symmetry makes gradφ into a radial vector. Integration of both sides over a spherical volume out to radius r yields $\int$**div grad** $\varphi dV = \int$**grad**$\varphi \bullet d\underline{\mathbf{A}} = 4\pi r^2$**grad**$\varphi = -4\pi G\int\rho dV$, or grad$\varphi = - GmM(r)/r^2$.

W. Wamsteker et al. (eds.), Developing Basic Space Science World-Wide, 501-502.

12. The local gravity can be derived on a lower mathematical level by adding up all the gravitational forces (one needs only the components toward the center) acting on a particle of mass m at distance r from a thin ring of some mass dM that resides at an arbitrary distance r'<r from the center. (All parts of the ring are equidistant from the particle.) When one adds the effect of all the rings contributing to a shell at r', of mass dM(r'), the force becomes $GmdM(r')/r^2$ The gravity from that shell is the same as if its mass were at the center. Then it is easy to add the effect of all shells at r'<r, $GmM(r)/r^2$. If the limits of integration are handled carefully for shells at r'>r, indeed they turn out not to contribute.

4. *The Virial Theorem.* The proof for any system of self-gravitating particles is based on evaluating dQ/dt, with the definition $Q = \Sigma \mathbf{\underline{p}}_i \bullet \mathbf{\underline{r}}_i = 1/2\ dI/dt$, where the sum is over all particles and I is the moment of inertia. The differentiation of $\mathbf{\underline{r}}_i$ leads to 2K. The differentiation of $\mathbf{\underline{p}}_i$ leads to $\Sigma d\mathbf{\underline{p}}_i/dt \bullet \mathbf{\underline{r}}_i = \Sigma \mathbf{\underline{F}}_i \bullet \mathbf{\underline{r}}_i = \Sigma\Sigma F_{ij} \bullet \mathbf{\underline{r}}_i = 1/2\ \Sigma\Sigma\ (\mathbf{\underline{F}}_{ij} - F_{ji}) \bullet \mathbf{\underline{r}}_i = 1/2\ \Sigma\Sigma F_{ij}$ $(\mathbf{\underline{r}}_i - \mathbf{\underline{r}}_j)$ where ΣΣ is a sum over all pairs of stars *i* and *j* except $i = j$, $F_{ij} = -F_{ji}$ has been used where the factor 1/2 appears, and the indices *i* and *j* have been interchanged in the last step. With the gravitational force between particles $\mathbf{\underline{F}}_{ij}$ proportional to $(\mathbf{\underline{r}}_i - \mathbf{\underline{r}}_j)/|\mathbf{\underline{r}}_i - \mathbf{\underline{r}}_j|^3$, the quantity to be summed turns into a scalar quantity which is the total gravitational energy. The end result is $d^2I/dt^2 = \Omega + 2K$. In equilibrium, $\Omega = -2K$.

Author Index